T0305508

Fixed Point Theorems with Applications

As a very important part of nonlinear analysis, fixed point theory plays a key role in the solvability of many complex systems from mathematics applied to chemical reactors, neutron transport, population biology, infectious diseases, economics, applied mechanics, and more.

The main aim of *Fixed Point Theorems with Applications* is to explain new techniques for investigation of different classes of ordinary and partial differential equations. The development of the fixed point theory parallels the advances in topology and functional analysis. Recent research investigated not only the existence but also the positivity of solutions for various types of nonlinear equations. This book will be of interest to those working in functional analysis and its applications.

Combined with other nonlinear methods such as variational methods and the approximation methods, the fixed point theory is powerful in dealing with many nonlinear problems from the real world.

The book can be used as a textbook to develop an elective course on nonlinear functional analysis with applications in undergraduate and graduate programs in mathematics or engineering programs.

Karima Mebarki is a professor in the Department of Mathematics, Bejaia University, Algeria. Her research interests are: fixed point theory, fixed point index theory, nonlinear ordinary differential equations, and boundary value problems.

Svetlin Georgiev is a mathematician who has worked in various aspects of mathematics. Currently, he focuses on harmonic analysis, ordinary differential equations, partial differential equations, time scale calculus, integral equations, and nonlinear analysis. He has published several books with Taylor & Francis/CRC Press.

Smail Djebali works on fixed point theory and applications in differential equations. He is currently a professor at Imam Mohammad Ibn Saud Islamic University, Riyadh, Saudi Arabia.

Khaled Zennir earned his PhD in mathematics in 2013 from Sidi Bel Abbès University, Algeria. In 2015, he received his highest diploma in Habilitation in mathematics from Constantine University, Algeria. He is currently an assistant professor at Qassim University in the Kingdom of Saudi Arabia. His research interests lie in the subjects of nonlinear hyperbolic partial differential equations: global existence, blowup, and long time behavior.

Fixed Point Theorems
with Applications

Karima Mebarki
Svetlin Georgiev
Smail Djebali
Khaled Zennir

CRC Press
Taylor & Francis Group
Boca Raton London New York

CRC Press is an imprint of the
Taylor & Francis Group, an **informa** business

A CHAPMAN & HALL BOOK

First edition published 2023
by CRC Press
6000 Broken Sound Parkway NW, Suite 300, Boca Raton, FL 33487-2742

and by CRC Press
4 Park Square, Milton Park, Abingdon, Oxon, OX14 4RN

CRC Press is an imprint of Taylor & Francis Group, LLC

© 2023 Karima Mebarki, Svetlin Georgiev, Smail Djebali and Khaled Zennir

Reasonable efforts have been made to publish reliable data and information, but the author and publisher cannot assume responsibility for the validity of all materials or the consequences of their use. The authors and publishers have attempted to trace the copyright holders of all material reproduced in this publication and apologize to copyright holders if permission to publish in this form has not been obtained. If any copyright material has not been acknowledged please write and let us know so we may rectify in any future reprint.

Except as permitted under U.S. Copyright Law, no part of this book may be reprinted, reproduced, transmitted, or utilized in any form by any electronic, mechanical, or other means, now known or hereafter invented, including photocopying, microfilming, and recording, or in any information storage or retrieval system, without written permission from the publishers.

For permission to photocopy or use material electronically from this work, access www.copyright. com or contact the Copyright Clearance Center, Inc. (CCC), 222 Rosewood Drive, Danvers, MA 01923, 978-750-8400. For works that are not available on CCC please contact mpkbookspermissions@tandf.co.uk

Trademark notice: Product or corporate names may be trademarks or registered trademarks and are used only for identification and explanation without intent to infringe.

ISBN: 978-1-032-46496-1 (hbk)
ISBN: 978-1-032-46499-2 (pbk)
ISBN: 978-1-003-38196-9 (ebk)

DOI: 10.1201/9781003381969

Typeset in Nimbus font
by KnowledgeWorks Global Ltd.

Contents

Preface

As a very important part of nonlinear analysis, fixed point theory plays a key role with regards to the solvability of many complex systems from applied mathematics (chemicals reactors, neutron transport, population biology, infection diseases, economics, applied mechanics, and more). The development of the fixed point theory parallels the advances in topology and functional analysis. During the last couple of years, the theory developed quickly in many directions starting from Brouwer's fixed point theorem (1910), Banach's contraction principle (1922), and Schauder's fixed point theorem for compact mappings (1930). An extension and a combination of the last two results to the sum of a contraction and a compact mapping were discovered by Krasnosel'skii in 1955. Indeed, many problems in science can be mathematically recast as nonlinear equations of the form $Tx + Fx = x$ and posed in some closed convex subset of a Banach space. Notice further that the positivity of solutions of nonlinear equations, especially ordinary, fractional, partial differential equations, and integral equations, is a very important issue in applications where a positive solution may represent a density, temperature, velocity, density, gravity, and more. That is why many recent research investigated not only the existence but also the positivity of solutions for various types of nonlinear equations. Combined with other nonlinear methods such that the variational methods, the approximation methods, and more, the fixed point theory turns out to be powerful in dealing with many nonlinear problems from the real world.

This book is devoted to the study of the existence, multiplicity, positivity, and localization of fixed points for some operators that are of the form $T + F$ where $(I - T)$ is a Lipschitz invertible mapping and F is a k-set contraction. The text offers the reader an overview of recent developments of positive fixed point theorems and their applications.

The book consists of six chapters. In Chapter 1, some preliminaries and basic concepts used throughout this book are collected. The chapter opens with the topology of the normed and Banach spaces and the linear operators defined therein. However, the measure of noncompactness (MNC for short) of a set in a metric space occupies the major part of this chapter. The abstract definition is followed by the classical Kuratowski and Hausdorff measures where the fundamental properties are investigated in detail. MNCs are involved in many convergence results in ordinary differential equations (ODEs) and partial differential equations (PDEs), particularly to derive compactness of some nonlinear operators. MNCs are also connected with strict-set contractions, 1-set contraction and condensing maps for which numerous fixed point theorems have been established so far. Chapter 1 ends with some useful compactness criteria that can be regarded as extensions of the classical Ascoli-Arzelà Theorem.

The fixed point index is a generalization of the Leray-Schauder degree. Chapter 2 starts with a reminder of the main properties of the fixed point index for strict set contractions set on bounded convex or on translates of cones of some Banach spaces. As a consequence, the celebrated Krasnosel'skii's compression–expansion fixed point theorem is derived. However, the core of Chapter 2 is the fixed point index for some classes of sums of two mappings, one being h-expansive and the other k-set contraction. The definition of a generalized fixed point index as well as some of its properties are presented. Then several conditions allowing computation of this index are shown.

In Chapter 3, there are formulated and proved various theorems of existence of positive fixed points for the sum of two operators. In particular, several new versions of Krasnosel'skii's expansion-compression fixed point theorems type are presented (fixed point in conical annulus, extensions to more general region of cones and more general class of mappings, extensions to translates of cones, and more) as well as a vector version useful in regard to the solvability of some nonlinear systems. The chapter ends with the proof of a new version of the original Leggett-Williams fixed point theorem and some fixed point theorems on open sets of cones for special mappings that include 1-set contractions.

The last three chapters are exclusively concerned with some applications of the theory developed in Chapters 2–3 to some ordinary and partial differential equations.

Mathematical problems formulated as ODEs are investigated in Chapter 4. We consider the existence of the classical solutions for some classes of systems ODEs (and even nth-order ODEs) and some boundary value problems for ODEs. The questions of periodicity, positivity, and multiplicity of solutions are investigated.

Chapter 5 mainly discusses some classes of parabolic equations. Some criteria for the existence of classical solutions for some classes' initial boundary value problems for nonlinear parabolic equations are presented. Subsequently, an application for existence of classical solutions for the Burgers-Fisher equation is provided. Existence of positive solutions is also proven.

As for the final Chapter 6, it deals with applications with some applications of the index fixed point theory to the solvability of some classes of initial value problems (IVPs) associated with some hyperbolic equations represented by the wave equation in different dimensions.

August 2022

Karima Mebarki
Svetlin Georgiev
Smail Djebali
Khaled Zennir

1

Preliminaries

1.1 Normed Linear Spaces

Definition 1.1.1 *A set X with a collection $\mathscr{F}$ of its subsets is called a topological space if $\mathscr{F}$ possesses the following properties.*

  1. $\emptyset, X \in \mathscr{F}$.

  2. An intersection of a finite number of sets of $\mathscr{F}$ belongs to $\mathscr{F}$.

  3. A union of any subcollection of $\mathscr{F}$ belongs to $\mathscr{F}$.

The elements of $\mathscr{F}$ are called open sets. A subset $U \subset X$ is called a neighborhood of a point $x \in X$ if there is an open set $G \subset X$ such that $x \in G \subset U$. The collection $\mathscr{F}$ is called a topology of X.

Example 1.1.2 *Let X be a given set. Take $\mathscr{F} = \{\emptyset, X\}$. Then X is a topological space. The topology $\mathscr{F}$ is the trivial topology.*

Example 1.1.3 *The discrete topology on X is defined by letting every subset of X be open.*

Definition 1.1.4 *A set X with a real-valued function $d : X \times X \rightarrow [0, \infty)$ for which*

  1. (identity of indiscernibles) $d(x,y) = 0$ if and only if $x = y \in X$,

  2. (symmetry property) $d(x,y) = d(y,x)$ for any $x, y \in X$,

  3. (triangle inequality) $d(x,y) \leq d(x,z) + d(z,y)$ for any $x, y, z \in X$,

is called a metric space. The function d is called a metric of X.

DOI: 10.1201/9781003381969-1

Example 1.1.5 *The set of real numbers is a metric space with a metric*

$$d(x,y) = |x - y|, \quad x,y \in \mathbb{R}.$$

Example 1.1.6 *The set of positive real numbers with a metric*

$$d(x,y) = \left| \log \frac{y}{x} \right|, \quad x,y \in (0,\infty),$$

is a metric space.

Example 1.1.7 *The set $\mathscr{C}([a,b])$ of continuous functions on $[a,b]$ with a metric*

$$d(f,g) = \max_{x \in [a,b]} |f(x) - g(x)|, \quad f,g \in \mathscr{C}([a,b]).$$

Definition 1.1.8 *If d is a metric on X, then*

$$B(x,r) = \{y \in X : d(x,y) < r\}$$

is called an open ball centered at x with radius $r > 0$.

Definition 1.1.9 *Open sets in a metric space X are defined as subsets $G \subset X$ which have the following property: for every $x \in X$, there is a $\delta > 0$ such that $B(x,\delta) \subset G$.*

Definition 1.1.10 *A subset G of a topological space X is called a closed set if $X \backslash G$ is open.*

Definition 1.1.11 *If $A \subset X$, then the intersection of all closed sets containing A is called the closure of A and it is denoted by $\overline{A}$, i.e.,*

$$\overline{A} = \bigcap \{B : \quad A \subset B, \quad B \quad is \quad closed\}.$$

A dual notion is the interior of A, i.e.,

$$\overset{\circ}{A} = \bigcup \{B : \quad B \subset A, \quad B \quad is \quad open\}.$$

The boundary ∂A of A is defined by

$$\partial A = \overline{A} \bigcap \overline{(X \backslash A)}.$$

Definition 1.1.12 *A subset A of X is said to be dense if $\overline{A} = X$.*

In a metric space X, we have the following equivalences.

1. $x \in \overline{A}$ if and only if there exists $\{x_n\}_{n=1}^{\infty} \subset A$ so that $\lim_{n \to \infty} d(x_n, x) = 0$.

2. $x \in \overset{\circ}{A}$ if and only if there exists a $\delta > 0$ such that $B(x, \delta) \subset A$.

Definition 1.1.13 *A metric space X is said to be separable if there is a countable dense subset of X.*

Now, we are ready with the main object in this section. It is well known that the notion of norm is of fundamental importance in discussing linear topological spaces. We shall begin with the definition of the semi-norm.

Definition 1.1.14 *A real-valued function v defined on a vector space E is called a semi-norm on E, if the following conditions are satisfied:*

1. $v(x+y) \leq v(x) + v(y)$ for any $x, y \in E$ (sub-additivity),

2. $v(\alpha x) = |\alpha| v(x)$ for any $x \in E$ and $\alpha \in \mathbb{R}$ (homogeneity),

3. $v(x) \geq 0$ for any $x \in E$ (non-negativity).

Theorem 1.1.15 *If E is a real vector space and $v : E \mapsto \mathbb{R}$ is a seminorm. Then*
$$v(x-y) \geq |v(x) - v(y)|,$$
for any $x, y \in E$.

Proof 1.1.16 *We have*
$$v(x) \leq v(x-y) + v(y)$$
for any $x, y \in E$, so
$$v(x) - v(y) \leq v(x-y) \tag{1.1}$$

for any $x,y \in E$. Since

$$v(x-y) \quad = \quad |-1|v(y-x)$$

$$\geq \quad v(y) - v(x)$$

for any $x,y \in E$, we have

$$-(v(x) - v(y)) \leq v(x) - v(y) \tag{1.2}$$

for any $x,y \in E$. Inequalities (1.1) and (1.2) give the desired inequality.

Definition 1.1.17 *A normed space is an ordered pair $(E, \|\cdot\|)$, where E is a vector space (also called a linear space) over F and $\|\cdot\|$ is a norm on E, i.e., a function $\|\cdot\| : E \longmapsto \mathbb{R}$ such that for any $x,y,z \in E$ the following holds.*

1. *$\|x\| \geq 0$ (non-negativity),*
2. *$\|x\| = 0$ iff $x = 0$ (separate points),*
3. *$\|\lambda x\| = |\lambda|\|x\|$ for any $\lambda \in F$ (homogeneity of the norm),*
4. *$\|x+y\| \leq \|x\| + \|y\|$ (triangle inequality).*

Note that the first condition follows from the other three. To see this, take $x \in E$ arbitrarily. Then

$$\|0\| \quad = \quad \|x + (-x)\| \leq \|x\| + \|-x\| = \|x\| + \|(-1)x\|$$

$$= \quad \|x\| + |-1|\|x\| = 2\|x\|.$$

Hence, using the second condition, we get that $\|x\| \geq 0$.

Example 1.1.18 *In E_n (n-dimensional Euclidean space) define a norm as follows.*

$$\|x\| = \left(\sum_{l=1}^{n} |x_l|^2\right)^{\frac{1}{2}}, \quad x_l \in F, \quad l \in \{1,\dots,n\}, \quad x = (x_1,\dots,x_n).$$

$$\tag{1.3}$$

We will check that (1.3) satisfies all axioms for a norm. Take $x,y \in E_n$, $x = (x_1,\dots,x_n)$, $y = (y_1,\dots,y_n)$, arbitrarily. Then

1. $\|x\| \geq 0$.

2. $\|x\| = 0$ *iff* $\left(\sum\limits_{l=1}^{n} |x_l|^2 \right)^{\frac{1}{2}} = 0$ *iff* $x_l = 0$ *for any* $l \in \{1, \ldots, n\}$.

3.

$$
\begin{aligned}
\|\lambda x\| &= \left(\sum_{l=1}^{n} |\lambda x_l|^2 \right)^{\frac{1}{2}} \\
&= \left(\sum_{l=1}^{n} |\lambda|^2 |x_l|^2 \right)^{\frac{1}{2}} \\
&= |\lambda| \left(\sum_{l=1}^{n} |x_l|^2 \right)^{\frac{1}{2}} \\
&= |\lambda| \|x\|,
\end{aligned}
$$

for any $\lambda \in F$.

4. *Applying Minkowski's inequality, we get*

$$
\begin{aligned}
\|x+y\| &= \left(\sum_{l=1}^{n} |x_l + y_l|^2 \right)^{\frac{1}{2}} \\
&\leq \left(\sum_{l=1}^{n} |x_l|^2 \right)^{\frac{1}{2}} + \left(\sum_{l=1}^{n} |y_l|^2 \right)^{\frac{1}{2}} \\
&= \|x\| + \|y\|.
\end{aligned}
$$

Example 1.1.19 *In the space* $\mathscr{C}^k([a,b])$ *define a norm*

$$
\|f\| = \sum_{l=0}^{k} \max_{t \in [a,b]} \left| f^{(l)}(t) \right|, \quad f \in \mathscr{C}^k([a,b]). \tag{1.4}
$$

We will check that (1.4) satisfies all axioms for a norm. Let $f, g \in \mathscr{C}^k([a,b])$ *and* $\lambda \in F$ *be chosen arbitrarily. Then*

1. $\|f\| \geq 0$.

2. $\quad 0 \;=\; \|f\| \;\; iff \;\; \sum_{l=0}^{k} \max_{t\in[a,b]} \left| f^{(l)}(t) \right| = 0 \;\; iff$

$\max_{t\in[a,b]} \left| f^{(l)}(t) \right| = 0 \; for\; any\; l \in \{0,\dots,k\} \; iff\; f \equiv 0 \; on\; [a,b]$.

3.
$$\|\lambda f\| \;=\; \sum_{l=0}^{k} \max_{t\in[a,b]} \left| (\lambda f)^{(l)}(t) \right|$$
$$=\; \sum_{l=0}^{k} \max_{t\in[a,b]} \left| \lambda f^{(l)}(t) \right|$$
$$=\; |\lambda| \sum_{l=0}^{k} \max_{t\in[a,b]} \left| f^{(l)}(t) \right|$$
$$=\; |\lambda|\,\|f\|.$$

4.
$$\|f+g\| \;=\; \sum_{l=0}^{k} \max_{t\in[a,b]} \left| (f+g)^{(l)}(t) \right|$$
$$=\; \sum_{l=0}^{k} \max_{t\in[a,b]} \left| f^{(l)}(t) + g^{(l)}(t) \right|$$
$$\leq\; \sum_{l=0}^{k} \max_{t\in[a,b]} \left| f^{(l)}(t) \right| + \sum_{l=0}^{k} \max_{t\in[a,b]} \left| g^{(l)}(t) \right|$$
$$=\; \|f\| + \|g\|.$$

Example 1.1.20 With l_p, $p \geq 1$, we denote the set of all sequences $x = \{x_l\}_{l\in\mathbb{N}}$ for which $\sum_{l=1}^{\infty} |x_l|^p < \infty$. Note that l_p is a vector space. In l_p, $1 \leq p < \infty$, we define

$$\|x\| = \left(\sum_{l=1}^{\infty} |x_l|^p \right)^{\frac{1}{p}}, \quad x = \{x_l\}_{l\in\mathbb{N}} \in l_p. \tag{1.5}$$

We will check that (1.5) satisfies all axioms for a norm. Let $x, y \in l_p$,
$x = \{x_l\}_{l \in \mathbb{N}}$, $y = \{y_l\}_{l \in \mathbb{N}}$, $\lambda \in F$ *be arbitrarily chosen. Then*

1. $\|x\| \geq 0$.

2.

$$0 \;=\; \|x\| \quad iff \quad \left(\sum_{l=1}^{\infty} |x_l|^p \right)^{\frac{1}{p}} = 0 \quad iff$$

$$\sum_{l=1}^{\infty} |x_l|^p = 0 \quad iff \quad x_l = 0 \quad for \quad any \quad l \in \mathbb{N} \quad iff \quad x = 0.$$

3.

$$\|\lambda x\| \;=\; \left(\sum_{l=1}^{\infty} |\lambda x_l|^p \right)^{\frac{1}{p}}$$

$$=\; \left(\sum_{l=1}^{\infty} |\lambda|^p |x_l|^p \right)^{\frac{1}{p}}$$

$$=\; |\lambda| \left(\sum_{l=1}^{\infty} |x_l|^p \right)^{\frac{1}{p}}$$

$$=\; |\lambda| \|x\|.$$

4. *Since for any $m \in \mathbb{N}$, using Minkowski's inequality, we
have*

$$\left(\sum_{l=1}^{m} |x_l + y_l|^p \right)^{\frac{1}{p}} \leq \left(\sum_{l=1}^{m} |x_l|^p \right)^{\frac{1}{p}} + \left(\sum_{l=1}^{m} |y_l|^p \right)^{\frac{1}{p}},$$

we conclude that

$$\left(\sum_{l=1}^{\infty} |x_l + y_l|^p \right)^{\frac{1}{p}} \leq \left(\sum_{l=1}^{\infty} |x_l|^p \right)^{\frac{1}{p}} + \left(\sum_{l=1}^{\infty} |y_l|^p \right)^{\frac{1}{p}}.$$

Therefore,

$$\|x+y\| = \left(\sum_{l=1}^{\infty} |x_l + y_l|^p \right)^{\frac{1}{p}}$$

$$\leq \left(\sum_{l=1}^{\infty} |x_l|^p \right)^{\frac{1}{p}} + \left(\sum_{l=1}^{\infty} |y_l|^p \right)^{\frac{1}{p}}$$

$$= \|x\| + \|y\|.$$

Exercise 1.1.21 *Check that*

$$\|f\| = |f(a)| + |f'(a)| + \max_{t \in [a,b]} |f''(t)|, \quad f \in \mathscr{C}^2([a,b]),$$

satisfies all axioms for a norm.

Example 1.1.22 *In $\mathscr{C}^1([a,b])$ we define*

$$\|f\| = \max_{t \in [a,b]} |f'(t)|. \tag{1.6}$$

Since

$$\|f\| = 0 \quad iff \quad \max_{t \in [a,b]} |f'(t)| = 0 \quad iff$$

$$f'(t) = 0 \quad for \quad any \quad t \in [a,b] \quad iff \quad f \equiv const \quad on \quad [a,b],$$

(1.6) does not satisfy the axioms for a norm.

Exercise 1.1.23 *Check if*

$$\|f\| = \max_{t \in [a,b]} |f'(t)| + |f(b) - f(a)|$$

satisfies all axioms for a norm in $\mathscr{C}^1([a,b])$.

Answer. No.

Note that in a normed vector space over F, a metric can be defined by

$$d(x,y) = \|x - y\|.$$

It is evident that the defined metric satisfies all axioms for a metric.
Below we will suppose that E is a normed space with a norm $\|\cdot\|$.

Lemma 1.1.24 *For every $x, y \in E$ the following inequality*

$$|\|x\| - \|y\|| \leq \|x - y\|$$

holds.

Proof 1.1.25 *We have*

$$\|x\| = \|x - y + y\|$$

$$\leq \|x - y\| + \|y\|.$$

Therefore,

$$\|x\| - \|y\| \leq \|x - y\|.$$

If we interchange the positions of x and y in the last inequality, we get

$$\|y\| - \|x\| \leq \|y - x\|$$

$$= \|x - y\|.$$

This completes the proof.

Definition 1.1.26

1. *An element $x_0 \in E$ will be called a limit of a sequence $\{x_n\}_{n \in \mathbb{N}} \subset E$, if*

$$\|x_n - x_0\| \to 0, \quad as \quad n \to \infty.$$

We will write $x_n \to x_0$, as $n \to \infty$ or $\lim\limits_{n \to \infty} x_n = x_0$.

2. *For $r > 0$ the set*

$$S_r(x_0) = \{x \in E : \|x - x_0\| < r\} \, (S_r[x_0] = \{x \in E : \|x - x_0\| \leq r\})$$

will be called an open (closed) ball with a center x_0 and radius r. Sometimes, we will say that $S_r(x_0)$ is a neighborhood of x_0.

3. *A set $M \subset E$ is said to be bounded, if there exists a positive constant c such that $\|x\| \leq c$ for any $x \in M$.*

Theorem 1.1.27 *Every convergent sequence in E is a bounded sequence.*

Proof 1.1.28 *Let $\{x_n\}_{n\in\mathbb{N}}$ be a convergent sequence in E to the element $x_0 \in E$. Let $\varepsilon > 0$ be arbitrarily chosen and fixed. Then there exists an $N = N(\varepsilon) \in \mathbb{N}$ such that*

$$\|x_n - x_0\| < \varepsilon$$

for any $n > N$, $n \in \mathbb{N}$. This with Lemma 1.1.24 yield

$$\|x_n\| - \|x_0\| < \varepsilon \quad or \quad \|x_n\| < \varepsilon + \|x_0\|$$

for any $n > N$, $n \in \mathbb{N}$. Let

$$c = \max\{\|x_1\|, \ldots, \|x_N\|, \varepsilon + \|x_0\|\}.$$

Then

$$\|x_n\| \le c$$

for any $n \in \mathbb{N}$. This completes the proof.

Theorem 1.1.29 *Let $\{x_n\}_{n\in\mathbb{N}}$ be a convergent sequence in E to the element $x_0 \in E$.*

> *1. For any $r > 0$ there is an $N = N(r) \in \mathbb{N}$ such that $x_n \in S_r(x_0)$ for any $n > N$.*
>
> *2. Every subsequence $\{x_{n_k}\}_{k\in\mathbb{N}}$ of the sequence $\{x_n\}_{n\in\mathbb{N}}$ is convergent to x_0.*
>
> *3. If $\{\lambda_n\}_{n\in\mathbb{N}} \subset F$ and $\lambda_n \to \lambda_0$ as $n \to \infty$, $\lambda_0 \in F$, then $\lambda_n x_n \to \lambda_0 x_0$ as $n \to \infty$.*
>
> *4. If $\{y_n\}_{n\in\mathbb{N}} \subset E$ and $y_n \to y_0$ as $n \to \infty$, $y_0 \in E$, then $x_n + y_n \to x_0 + y_0$, as $n \to \infty$.*
>
> *5. $\|x_n\| \to \|x_0\|$, as $n \to \infty$.*
>
> *6. x_0 is unique.*

Proof 1.1.30

1. *Let $r > 0$ be arbitrarily chosen and fixed. Then there is an $N = N(r) \in \mathbb{N}$ such that*

$$\|x_n - x_0\| \leq r$$

for any $n > N$, $n \in \mathbb{N}$, i.e., $x_n \in S_r(x_0)$ for any $n > N$, $n \in \mathbb{N}$.

2. *Let $\varepsilon > 0$ be arbitrarily chosen and fixed. Then there exists an $N = N(\varepsilon) \in \mathbb{N}$ such that*

$$\|x_n - x_0\| < \varepsilon \qquad (1.7)$$

for any $n > N$, $n \in \mathbb{N}$. Also, there is a $K = K(\varepsilon) \in \mathbb{N}$ such that $n_k > N$ for any $k > K$, $k \in \mathbb{N}$. Hence, and (1.7), we get

$$\|x_{n_k} - x_0\| < \varepsilon$$

for any $k > K$, $k \in \mathbb{N}$.

3. *Since $x_n \to x_0$ and $\lambda_n \to \lambda_0$ as $n \to \infty$, there exist positive constants c_1 and c such that*

$$|\lambda_n| \leq c_1 \quad and \quad \|x_n\| \leq c$$

for any $n \in \mathbb{N}$.

(a) Let $\lambda_0 = 0$. Then

$$
\begin{aligned}
\|\lambda_n x_n\| &= |\lambda_n| \|x_n\| \\[2mm]
&\leq c|\lambda_n| \\[2mm]
&\to 0, \quad as \quad n \to \infty.
\end{aligned}
$$

(b) Let $\lambda_0 \neq 0$. Then

$$
\begin{aligned}
\|\lambda_n x_n - \lambda_0 x_0\| &= \|\lambda_n x_n - \lambda_n x_0 + \lambda_n x_0 - \lambda_0 x_0\| \\[2mm]
&\leq \|\lambda_n x_n - \lambda_n x_0\| + \|\lambda_n x_0 - \lambda_0 x_0\| \\[2mm]
&= |\lambda_n| \|x_n - x_0\| + |\lambda_n - \lambda_0| \|x_0\| \\[2mm]
&\leq c_1 \|x_n - x_0\| + |\lambda_n - \lambda_0| \|x_0\| \\[2mm]
&\to 0, \quad as \quad n \to \infty.
\end{aligned}
$$

4. We have

$$\|x_n + y_n - x_0 - y_0\| = \| (x_n - x_0) + (y_n - y_0) \|$$

$$\leq \|x_n - x_0\| + \|y_n - y_0\|$$

$$\to 0, \quad as \quad n \to \infty.$$

5. By Lemma 1.1.24, we get

$$|\|x_n\| - \|x_0\|| \leq \|x_n - x_0\| \to 0, \quad as \quad n \to \infty.$$

*6. Assume that there exists $y_0 \in E$ so that $x_n \to y_0$, as $n \to \infty$.
Then*

$$\|y_0 - x_0\| = \|y_0 - x_n + x_n - x_0\|$$

$$\leq \|x_n - y_0\| + \|x_n - x_0\|$$

$$\to 0, \quad as \quad n \to \infty.$$

Therefore, $x_0 = y_0$.

This completes the proof.

Definition 1.1.31 *A set $M \subset E$ will be called open, if for every $x_0 \in M$ there exists $r_0 > 0$ such that $S_{r_0}(x_0) \subset M$.*

Theorem 1.1.32 *Let $A_1, \ldots, A_l \subset E$ be open sets. Then $\bigcap_{k=1}^{l} A_k$ is an open set in E.*

Proof 1.1.33 *Let $x \in \bigcap_{k=1}^{l} A_k$ be arbitrarily chosen. Then $x \in A_k$ for any $k \in \{1, \ldots, l\}$. Since A_k, $k \in \{1, \ldots, l\}$, are open sets in E, there are $r_k > 0$ so that $S_{r_k}(x) \subset A_k$. Let $r = \min_{1 \leq k \leq l} r_k$. Then $S_r(x) \subset A_k$ for any $k \in \{1, \ldots, l\}$. Therefore, $S_r(x) \subset \bigcap_{k=1}^{l} A_k$. This completes the proof.*

Theorem 1.1.34 *Let $\{A_k\}_{k\in\mathbb{N}}$ be open sets in E. Then $\bigcup_{k\in\mathbb{N}} A_k$ is an open set in E.*

Proof 1.1.35 *Let $x \in \bigcup_{k\in\mathbb{N}} A_k$ be arbitrarily chosen and fixed. Then there is a $k_0 \in \mathbb{N}$ such that $x \in A_{k_0}$. Since A_{k_0} is an open set in E, there is an $r_0 > 0$ such that $S_{r_0}(x) \subset A_{k_0}$. From here, $S_{r_0}(x) \subset \bigcup_{k\in\mathbb{N}} A_k$. This completes the proof.*

Definition 1.1.36 *A point $a \in E$ will be called a limit point for a set $M \subset E$ if for any $r > 0$ there is $x \in S_r(a)\bigcap M$, $x \neq a$.*

Theorem 1.1.37 *A point $a \in E$ is a limit point for the set $M \subset E$ if and only if there is a sequence $\{x_n\}_{n\in\mathbb{N}} \subset M$ that converges to a and $x_n \neq a$ for any $n \in \mathbb{N}$.*

Proof 1.1.38

1. Let $a \in E$ be a limit point for the set M. Then for any $n \in \mathbb{N}$ there are $x_n \in S_{\frac{1}{n}}(a)\bigcap M$, $x_n \neq a$. In this way, we obtain a sequence $\{x_n\}_{n\in\mathbb{N}}$ such that

$$\|x_n - a\| \to 0, \quad as \quad n \to \infty, \quad x_n \neq a,$$

i.e., $x_n \to a$ as $n \to \infty$ and $x_n \neq a$.

2. Let there is a sequence $\{x_n\}_{n\in\mathbb{N}} \subset M$ such that $x_n \neq a$ for any $n \in \mathbb{N}$ and $x_n \to a$ as $n \to \infty$. Hence, for any $r > 0$ there is an $N = N(r) \in \mathbb{N}$ such that $x_n \in S_r(a)$ for any $n > N$ and $x_n \neq a$.

This completes the proof.

Definition 1.1.39 *A set $M \subset E$ is said to be closed if it contains all its limit points.*

Theorem 1.1.40 *Let $A_1,\ldots,A_l$ be closed sets in E. Then $\bigcup_{k=1}^{l} A_k$ is a closed set in E.*

Proof 1.1.41 *Let $a \in E$ be a limit point for $\bigcup_{k=1}^{l} A_k$. Then there exists a sequence $\{x_n\}_{n \in \mathbb{N}} \subset \bigcup_{k=1}^{l} A_k$ such that $x_n \to a$, as $n \to \infty$. Hence, there is an $m \in \{1, \ldots, l\}$ and a subsequence $\{x_{n_s}\}_{s \in \mathbb{N}}$ of the sequence $\{x_n\}_{n \in \mathbb{N}}$ such that $\{x_{n_s}\}_{s \in \mathbb{N}} \subset A_m$. We have that $x_{n_s} \to a$, as $s \to \infty$ and $x_n \neq a$. Hence by Theorem 1.1.37, it follows that a is a limit point for A_m. Because A_m is a closed set in E, we conclude that $a \in A_m$. Therefore, $a \in \bigcup_{k=1}^{l} A_k$ and $\bigcup_{k=1}^{l} A_k$ is a closed set in E. This completes the proof.*

Theorem 1.1.42 *Let $\{A_k\}_{k \in \mathbb{N}}$ be closed sets in E. Then $\bigcap_{k \in \mathbb{N}} A_k$ is a closed set in E.*

Proof 1.1.43 *Let $a \in E$ be a limit point for $\bigcap_{k \in \mathbb{N}} A_k$. Then there exists a sequence $\{x_n\}_{n \in \mathbb{N}} \subset \bigcap_{k \in \mathbb{N}} A_k$ such that $x_n \to a$, as $n \to \infty$. Hence, $\{x_n\}_{n \in \mathbb{N}} \subset A_k$, $x_n \to a$, as $n \to \infty$ for any $k \in \mathbb{N}$. Therefore, a is a limit point of A_k for any $k \in \mathbb{N}$. Because A_k, $k \in \mathbb{N}$ are closed sets in E, we have that $a \in A_k$ for any $k \in \mathbb{N}$. Therefore, $a \in \bigcap_{k \in \mathbb{N}} A_k$ and $\bigcap_{k \in \mathbb{N}} A_k$ is a closed set in E. This completes the proof.*

Definition 1.1.44 *Let $M \subset E$.*

1. The set M together with all of its limit points is called the closure of M. It will be denoted by $\overline{M}$.

2. The set $E \backslash M$ will be called the completion of the set M to E.

3. A point $x_0 \in E$ will be called an interior point for the set M, if there is an $r > 0$ such that $S_r(x_0) \subset M$.

4. A point $x_0 \in E$ will be called an exterior point of the set M, if there is an $r > 0$ such that $S_r(x_0) \bigcap M = \emptyset$.

5. A point $x_0 \in E$ will be called a boundary point of the set
M, if for every $r > 0$ we have

$$S_r(x_0) \bigcap M \neq \emptyset \quad and \quad S_r(x_0) \bigcap (E \backslash M) \neq \emptyset.$$

6. The set of all boundary points of the set M will be called
the boundary of the set M and it will be denoted by ∂M.

Remark 1.1.45 *Note that we have the following possibilities.*

$$\partial M \subset M \quad or \quad \partial M \bigcap M = \emptyset \quad or \quad \partial M \bigcap M \neq \partial M.$$

Definition 1.1.46 *Two norms $\| \cdot \|_1$ and $\| \cdot \|_2$ in E will be called equivalent, if there are positive constants c_1 and c_2 such that*

$$c_1 \|x\|_2 \leq \|x\|_1 \leq c_2 \|x\|_2$$

for any $x \in E$. We will write $\| \cdot \|_1 \sim \| \cdot \|_2$.

Theorem 1.1.47 *In every finite-dimensional vector space every norms are equivalent.*

Proof 1.1.48 *Let U be a finite-dimensional vector space over F. With $\{\phi_l\}_{l=1}^m$ we will denote a basis in U. Then every $x \in U$ has the following representation*

$$x = \sum_{k=1}^{m} \xi_k \phi_k, \quad \xi_k \in F, \quad k \in \{1, \ldots, m\}.$$

In U we define a norm

$$\|x\| = \left(\sum_{l=1}^{m} |\xi_l|^2 \right)^{\frac{1}{2}} \quad for \quad x \in U. \tag{1.8}$$

We take an arbitrary norm $\| \cdot \|_1$ in U. Let

$$c_2 = \left(\sum_{l=1}^{m} \|\phi_l\|_1^2 \right)^{\frac{1}{2}}.$$

Then for $x = \sum\limits_{l=1}^{m} \xi_l \phi_l,\ \xi_l \in F,\ l \in \{1,\ldots,m\},\ \text{we have}$

$$
\begin{aligned}
\|x\|_1 &= \left\|\sum_{l=1}^{m} \xi_l \phi_l\right\|_1 \\
&\leq \sum_{l=1}^{m} |\xi_l| \|\phi_l\|_1 \\
&\leq \left(\sum_{l=1}^{m} |\xi_l|^2\right)^{\frac{1}{2}} \left(\sum_{l=1}^{m} \|\phi_l\|_1^2\right)^{\frac{1}{2}} \\
&= c_2 \|x\|,
\end{aligned}
$$

i.e.,

$$\|x\|_1 \leq c_2 \|x\|. \tag{1.9}$$

On the other hand, by Lemma 1.1.24 and (1.9), we get

$$
\begin{aligned}
|\|x\|_1 - \|y\|_1| &\leq \|x - y\|_1 \\
&\leq c_2 \|x - y\|
\end{aligned}
$$

for any $x, y \in U$. *Therefore, the function* $\|\cdot\|_1$ *is a continuous function in* U. *Hence, there exists*

$$c_1 = \inf_{\|x\|=1} \|x\|_1.$$

Consequently

$$\left\|\frac{x}{\|x\|}\right\|_1 \geq c_1 \quad \text{or} \quad \|x\|_1 \geq c_1 \|x\|.$$

This completes the proof.

Exercise 1.1.49 *Prove that (1.8) satisfies all axioms for a norm.*

Theorem 1.1.50 *Let L be a linear subspace of E which is a closed set in E. Then*

$$\|l\|_{E/L} = \inf_{x \in l} \|x\|, \quad l \in E/L, \tag{1.10}$$

is a norm in the quotient space E/L.

Proof 1.1.51 *Firstly, we will prove that every $l \in E/L$ is a closed set. Let $\{x_n\}_{n \in \mathbb{N}}$ be a sequence of elements of l such that $x_n \to x_0$, as $n \to \infty$. We fix $m \in \mathbb{N}$ and consider $x_m - x_n$ for $n \in \mathbb{N}$. We have that $x_m - x_n \in L$ for any $n \in \mathbb{N}$ and $x_m - x_n \to x_m - x_0$, as $n \to \infty$. Hence, $x_m - x_0 \in L$. Because $x_m \in l$, we get that $x_0 \in l$.*
Let $l, m \in E/L$ be arbitrarily chosen.

1. *$\|l\|_{E/L} \geq 0$.*

2. *We will prove that $\|l\|_{E/L} = 0$ if and only if $l = L$.*

(a) *Let $\|l\|_{E/L} = 0$. Then there exists a sequence $\{x_n\}_{n \in \mathbb{N}}$ of elements of l such that $x_n \to 0$ as $n \to \infty$. Because l is a closed set, in L, we obtain that $0 \in l$ and hence $l = L$.*

(b) *Let $l = L$. Then $0 \in l$ and $\|l\| = 0$.*

3. *Let $\lambda \in F$ be arbitrarily chosen. Then*
$$\lambda l = \{\lambda x : x \in l\}$$
and

$$
\begin{aligned}
\|\lambda l\|_{E/L} &= \inf_{x \in l} \|\lambda x\| \\
&= |\lambda| \inf_{x \in l} \|x\| \\
&= |\lambda| \|l\|_{E/L}.
\end{aligned}
$$

4. *We have*

$$
\begin{aligned}
\|l + m\|_{E/L} &= \inf_{x \in l+m} \|x\| \\
&\leq \inf_{\substack{x = x_1 + x_2 \\ x_1 \in l, x_2 \in m}} (\|x_1 + x_2\|) \\
&\leq \inf_{\substack{x = x_1 + x_2 \\ x_1 \in l, x_2 \in m}} (\|x_1\| + \|x_2\|) \\
&\leq \inf_{x_1 \in l} \|x_1\| + \inf_{x_2 \in m} \|x_2\| \\
&= \|l\|_{E/L} + \|m\|_{E/L}.
\end{aligned}
$$

This completes the proof.

Theorem 1.1.52 *Let L be a closed linear subspace of E. Then the sequence $\{l_n\}_{n\in\mathbb{N}}$ of elements of E/L is convergent to l if and only if there exists a sequence $\{x_n\}_{n\in\mathbb{N}}$ of elements $x_n \in l_n$ such that $x_n \to x$, as $n \to \infty$, $x \in l$.*

Proof 1.1.53

1. Let $\{l_n\}_{n\in\mathbb{N}}$ be a sequence of elements of E/L that converges to l. Then, we get

$$\|l_n - l\|_{E/L} \to 0, \quad as \quad n \to \infty,$$

that is

$$\|l_n - l\|_{E/L} = \varepsilon_n, \quad \varepsilon_n \to 0, \quad as \quad n \to \infty.$$

Hence, there exist $y_n \in l_n$ and $x \in l$ such that

$$\|y_n - x\| < 2\varepsilon_n.$$

Let $x_0 \in l$ be arbitrarily chosen. Then

$$\|y_n - x\| = \|(y_n - x + x_0) - x_0\|$$

$$< 2\varepsilon_n.$$

Since $x_0, x \in l$, we have that $x - x_0 \in L$. Therefore,

$$x_n = y_n - x + x_0 \in l_n.$$

Consequently for every $x_0 \in l$ there exists a sequence $\{x_n\}_{n\in\mathbb{N}}$, $x_n \in l_n$, such that $x_n \to x_0$, as $n \to \infty$.

2. Let there exists a sequence $\{x_n\}_{n\in\mathbb{N}}$, $x_n \in l_n$, such that $x_n \to x_0$, as $n \to \infty$, $x_0 \in l$. Then, using (1.10),

$$\|l_n - l\| \le \|x_n - x_0\| \to 0, \quad as \quad n \to \infty.$$

This completes the proof.

Definition 1.1.54 *Let L be a linear subspace of E. We define a distance from $x \in E$ to L as follows*

$$dist(x,L) = \inf_{y \in L} \|x - y\|.$$

By Definition 1.1.54, we get

1. $dist(x,L) \geq 0$,

2. for any $y \in L$, we have

$$dist(x,L) \leq \|x - y\|,$$

3. for any $\varepsilon > 0$ there exists $y_\varepsilon \in L$ such that

$$\|x - y_\varepsilon\| < \varepsilon + dist(x,L).$$

Theorem 1.1.55 *Let L be a closed linear subspace of E. If $x \notin L$, then $dist(x,L) > 0$.*

Proof 1.1.56 *Assume that $dist(x,L) = 0$. Then there exists a sequence $\{y_n\}_{n \in \mathbb{N}}$ of elements of L such that*

$$\|y_n - x\| < \frac{1}{n}$$

for any $n \in \mathbb{N}$. Since L is closed, we get that $x \in L$, which is a contradiction. This completes the proof.

Theorem 1.1.57 *Let L be a finite-dimensional linear subspace of E. Then for any $x \in E$ there exists $x^* \in L$ such that*

$$dist(x,L) = \|x - x^*\|.$$

Proof 1.1.58 *Suppose that L is m-dimensional.*

1. *If $x \in L$, then $dist(x,L) = 0$ and $x = x^*$.*

2. Let $x \notin L$. Then $d = dist(x, L) > 0$. We take $\{\phi_l\}_{l=1}^m$ to be a basis in L. Then any $x \in L$ can be represented in the following way

$$y = \sum_{l=1}^m y_l \phi_l, \quad y_l \in F, \quad l \in \{1, \ldots, m\}.$$

Define a norm in L in the following way

$$\|y\|_c = \left(\sum_{l=1}^m |y_l|^2 \right)^{\frac{1}{2}} \quad for \quad y = \sum_{l=1}^m y_l \phi_l \in L.$$

Because L is finite-dimensional, all norms in L are equivalent. For a norm $\|\cdot\|$ in L there exist positive constants α and β such that

$$\alpha \|z\|_c \leq \|z\| \leq \beta \|z\|_c$$

for any $z \in L$. Take

$$r = \frac{d + 1 + \|x\|}{\alpha}$$

and let $y \in L$ be arbitrarily chosen. If $\|y\|_c > r$, then

$$
\begin{aligned}
\|x - y\| &\geq & \|y\| - \|x\| \\
&\geq & \alpha \|y\|_c - \|x\| \\
&> & \alpha r - \|x\| \\
&= & d + 1.
\end{aligned}
$$

Therefore, d is achieved for $\|y\| \leq r$. Since $\{y \in L : \|y\|_c \leq r\}$ is a closed and bounded set in L, and $x \mapsto \|x\|$ is a continuous function on it, there exists $x^* \in L$ such that

$$\inf_{\|y\|_c \leq r} \|x - y\| = \|x - x^*\|.$$

This completes the proof.

Definition 1.1.59 *The normed space E will be called strongly normed space if the equality*

$$\|x+y\| = \|x\| + \|y\|$$

holds if and only if $y = \lambda x$, $\lambda > 0$, $y, x \in E$.

Theorem 1.1.60 *Let E be a strongly normed space and L be a finite-dimensional linear subspace of E. If for $x \in E$ there exists $x^* \in L$ such that*

$$\|x - x^*\| = \inf_{y \in L} \|x - y\|,$$

then x^* is unique.

Proof 1.1.61 *If $dist(x, L) = 0$, then $x = x^*$. Suppose that $d = dist(x, L) > 0$. Assume that there are $x_1^*, x_2^* \in L$ such that*

$$d = \|x - x_1^*\|$$

$$= \|x - x_2^*\|.$$

Then

$$\left\| x - \frac{x_1^* + x_2^*}{2} \right\| = \left\| \frac{x - x_1^*}{2} + \frac{x - x_2^*}{2} \right\|$$

$$\leq \frac{1}{2}\|x - x_1^*\| + \frac{1}{2}\|x - x_2^*\|$$

$$= \frac{1}{2}d + \frac{1}{2}d$$

$$= d.$$

Hence,

$$\|2x - (x_1^* + x_2^*)\| = 2d$$

$$= \|(x - x_1^*) + (x - x_2^*)\|$$

$$= \|x - x_1^*\| + \|x - x_2^*\|.$$

Since E is a strongly normed space, there exists $\lambda > 0$ such that

$$x - x_1^* = \lambda \left(x - x_2^* \right).$$

If $\lambda \neq 1$, then

$$x = \frac{1}{1 - \lambda} \left(x_1^* - \lambda x_2^* \right) \in L,$$

which is a contradiction. Therefore, $\lambda = 1$ and $x_1^* = x_2^*$. This completes the proof.

Lemma 1.1.62 (Riesz's Lemma) *Let L be a closed linear subspace of E and $L \neq E$. Then for any $\varepsilon \in (0,1)$ there exists $z_\varepsilon \notin L$, $\|z_\varepsilon\| = 1$, such that*

$$\text{dist} \left(z_\varepsilon, L \right) > 1 - \varepsilon.$$

Proof 1.1.63 *Since $L \neq E$, there exists $x \in E$ and $x \notin L$. Let $d = \inf_{y \in L} \|x - y\|$. We have $d > 0$. Then for any $\varepsilon \in (0,1)$ there exists $y_\varepsilon \in L$ such that*

$$d \;\leq\; \|y_\varepsilon - x\|$$

$$< \; \frac{d}{1 - \varepsilon}.$$

Let

$$z_\varepsilon = \frac{y_\varepsilon - x}{\|y_\varepsilon - x\|}.$$

We have that $\|z_\varepsilon\| = 1$. If we suppose that $z_\varepsilon \in L$, then $y_\varepsilon - x \in L$. Hence, $x \in L$, which is a contradiction. Therefore, $z_\varepsilon \notin L$. For $y \in L$, we have

$$\begin{aligned}
\|z_\varepsilon - y\| &= \left\| \frac{y_\varepsilon - x}{\|y_\varepsilon - x\|} - y \right\| \\
&= \frac{\|x - (y_\varepsilon - y\|y_\varepsilon - x\|)\|}{\|y_\varepsilon - x\|} \\
&\geq \frac{d}{\|y_\varepsilon - x\|} \\
&> 1 - \varepsilon.
\end{aligned}$$

Therefore, $\text{dist} \left(z_\varepsilon, L \right) > 1 - \varepsilon$. This completes the proof.

1.2 Banach Spaces

Definition 1.2.1 *A normed vector space E that is complete in the sense of convergence in norm is called a Banach space.*

Example 1.2.2 *The space E_n is a Banach space with a norm*

$$\|x\| = \left(\sum_{l=1}^{n} \xi_l^2 \right)^{\frac{1}{2}}, \quad x = (\xi_1, \dots, \xi_n) \in E_n.$$

Example 1.2.3 *The space $\mathscr{C}([a,b])$ is a Banach space with a norm*

$$\|f\| = \max_{a \leq t \leq b} |f(t)|.$$

Example 1.2.4 *The vector space l_p, $p \in (1, \infty)$ is a Banach space with a norm*

$$\|x\| = \left(\sum_{l=1}^{\infty} |x_l|^p \right)^{\frac{1}{p}}.$$

Theorem 1.2.5 *Let E be a Banach space and L be a closed linear subspace in it. Then E/L is a Banach space.*

Proof 1.2.6 *Let $\{l_n\}_{n \in \mathbb{N}}$ be a Cauchy sequence in E/L. We take $x_n \in l_n$ so that*

$$\|x_n - x_m\| \leq 2\|l_n - l_m\|_{E/L}.$$

In this way we get a Cauchy sequence $\{x_n\}_{n \in \mathbb{N}}$ of elements of E. Because E is a Banach space, the sequence $\{x_n\}_{n \in \mathbb{N}}$ is convergent to an element $x \in E$. Let l be the class containing x. Hence, and Theorem 1.1.52, we conclude that the sequence $\{l_n\}_{n \in \mathbb{N}}$ is convergent to l. Therefore, E/L is a Banach space. This completes the proof.

Definition 1.2.7 *Let $x_1, x_2, \dots, x_n, \dots$ be elements of a Banach space E. An expression of the form $\sum_{l=1}^{\infty} x_l$ is called a series, made up of the elements of the space E. Let $s_n = \sum_{l=1}^{n} x_l$. If the sequence $\{s_n\}_{n \in \mathbb{N}}$ converges, then $\sum_{l=1}^{\infty} x_l$ is said to be a convergent series.*

Theorem 1.2.8 *Let $a_n \in F$, $n \in \mathbb{N}$, and $\sum\limits_{l=1}^{\infty} a_l$ be a convergent series. Let also, E be a Banach space and $x_n \in E$, $\|x_n\| \leq |a_n|$, $n \in \mathbb{N}$. Then $\sum\limits_{n=1}^{\infty} x_n$ is a convergent series.*

Proof 1.2.9 *For any $n, p \in \mathbb{N}$, we have*

$$
\|s_{n+p} - s_n\| = \left\| \sum_{l=n+1}^{n+p} x_l \right\|
$$

$$
\leq \sum_{l=n+1}^{n+p} \|x_l\|
$$

$$
\leq \sum_{l=n+1}^{n+p} |a_l|.
$$

Therefore, $\{s_n\}_{n \in \mathbb{N}}$ is a Cauchy sequence in E. Because E is a Banach space, we conclude that the sequence $\{s_n\}_{n \in \mathbb{N}}$ is convergent. This completes the proof.

An operator is generally a mapping that acts on the elements of a vector space to produce other elements of the same or other vector space. The most common operators are linear operators, which act on vector spaces.

Suppose that X and Y are vector spaces over F.

Definition 1.2.10 *The operator $\mathbb{A} : X \longmapsto Y$ will be called a linear operator, if*

1. *it is additive, i.e.,*

$$
\mathbb{A}(x_1 + x_2) = \mathbb{A}x_1 + \mathbb{A}x_2, \quad x_1, x_2 \in X,
$$

2. *it is homogeneous, i.e.,*

$$
\mathbb{A}(\lambda x) = \lambda \mathbb{A}x, \quad \lambda \in F, \quad x \in X.
$$

Example 1.2.11 *Let $K(t,s)$ be a continuous function on the square $0 \leq t, s \leq 1$. For $x \in \mathscr{C}([0,1])$, define the operator*

$$y(t) = \int_0^1 K(t,s)x(s)ds, \quad t \in [0,1], \quad y = \mathbb{A}x.$$

Let $X = Y = \mathscr{C}([0,1])$. It is evident that $\mathbb{A} : X \longmapsto Y$. We will prove that it is a linear operator.

1. Let $x_1, x_2 \in X$ be arbitrarily chosen. Then

$$\mathbb{A}x_1(t) = \int_0^1 K(t,s)x_1(s)ds,$$

$$\mathbb{A}x_2(t) = \int_0^1 K(t,s)x_2(s)ds,$$

$$\mathbb{A}(x_1 + x_2)(t) = \int_0^1 K(t,s)(x_1(s) + x_2(s))ds$$

$$= \int_0^1 K(t,s)x_1(s)ds + \int_0^1 K(t,s)x_2(s)ds$$

$$= \mathbb{A}x_1(t) + \mathbb{A}x_2(t), \quad t \in [0,1].$$

2. Let $\lambda \in F$ and $x \in X$ be arbitrarily chosen and fixed. Then

$$\mathbb{A}(\lambda x)(t) = \int_0^1 K(t,s)\lambda x(s)ds$$

$$= \lambda \int_0^1 K(t,s)x(s)ds$$

$$= \lambda \mathbb{A}x(t), \quad t \in [0,1].$$

Therefore, $\mathbb{A} : X \longmapsto Y$ is a linear operator.

Example 1.2.12 *For $x \in \mathscr{C}^1([0,1])$, define the operator*

$$y(t) = \frac{d}{dt}x(t), \quad t \in [0,1], \quad y = \mathbb{A}x.$$

Let $X = \mathscr{C}^1([0,1])$, $Y = \mathscr{C}([0,1])$. It is evident that $\mathbb{A} : X \longmapsto Y$. We will prove that it is a linear operator.

1. Let $x_1, x_2 \in X$ be arbitrarily chosen and fixed. Then

$$\mathbb{A}x_1(t) = \frac{d}{dt}x_1(t),$$

$$\mathbb{A}x_2(t) = \frac{d}{dt}x_2(t),$$

$$\mathbb{A}(x_1 + x_2)(t) = \frac{d}{dt}(x_1 + x_2)(t)$$

$$= \frac{d}{dt}x_1(t) + \frac{d}{dt}x_2(t)$$

$$= \mathbb{A}x_1(t) + \mathbb{A}x_2(t), \quad t \in [0, 1].$$

2. Let $\lambda \in F$ and $x \in X$ be arbitrarily chosen. Then

$$\mathbb{A}(\lambda x)(t) = \frac{d}{dt}(\lambda x)(t)$$

$$= \lambda \frac{d}{dt}x(t)$$

$$= \lambda \mathbb{A}x(t), \quad t \in [0, 1].$$

Therefore, $\mathbb{A} : X \longmapsto Y$ is a linear operator.

Example 1.2.13 *For $x \in \mathscr{C}([0, 1])$, define the operator*

$$y = \int_0^1 (x(t))^2 \, dt, \quad y = \mathbb{A}x.$$

Let $X = \mathscr{C}([0, 1])$, $Y = F$. It is evident that $\mathbb{A} : X \longmapsto Y$. Let

$$x_1(t) = 1, \quad x_2(t) = t, \quad t \in [0, 1].$$

Then

$$\mathbb{A}x_1 = \int_0^1 dt$$

$$= \left. t \right|_{t=0}^{t=1}$$

$$= 1,$$

$$\mathbb{A}x_2 = \int_0^1 t^2 dt$$

$$= \left. \frac{1}{3}t^3 \right|_{t=0}^{t=1}$$

$$= \frac{1}{3},$$

$$\mathbb{A}x_1 + \mathbb{A}x_2 = 1 + \frac{1}{3}$$

$$= \frac{4}{3},$$

$$\mathbb{A}(x_1 + x_2) = \int_0^1 (t+1)^2 dt$$

$$= \int_0^1 \left(t^2 + 2t + 1 \right) dt$$

$$= \int_0^1 t^2 dt + 2\int_0^1 t \, dt + \int_0^1 dt$$

$$= \frac{7}{3}.$$

Therefore,

$$\mathbb{A}(x_1 + x_2) \neq \mathbb{A}x_1 + \mathbb{A}x_2.$$

Consequently $\mathbb{A} : X \longmapsto Y$ is not a linear operator.

Exercise 1.2.14 *For $x \in \mathscr{C}([0,1])$, define the operator*

$$y(t) = x\left(t^2 \right), \quad t \in [0,1], \quad y = \mathbb{A}x.$$

Prove that $\mathbb{A} : \mathscr{C}([0,1]) \to \mathscr{C}([0,1])$ is a linear operator.

1.3   Linear Operators in Normed Vector Spaces

In this section, suppose that X and Y are normed vector spaces over F. The convergence in X and Y is a norm convergence.

Definition 1.3.1 *We say that the linear operator $A : X \longmapsto Y$ is continuous at $x \in X$, if for any $\varepsilon > 0$ there is a $\delta = \delta(\varepsilon)$ such that*

$$\|Ax_1 - Ax\| < \varepsilon$$

whenever $\|x_1 - x\| < \delta$, $x_1 \in X$. In other words, the linear operator $A : X \longmapsto Y$ is said to be continuous at $x \in X$ if $Ax_n \to Ax$ in Y, as $n \to \infty$ whenever $x_n \to x$ in X, as $n \to \infty$, where $\{x_n\}_{n\in\mathbb{N}}$ is a sequence of elements of X. We say that the linear operator $A : X \longmapsto Y$ is continuous in X, if it is continuous at every point of X.

Example 1.3.2 *Let $X = Y = \mathscr{C}([0,1])$. Consider the operator*

$$Ax(t) = t^2 \int_0^1 x(s)ds, \quad t \in [0,1], \quad x \in X.$$

Let $x \in X$ be arbitrarily chosen and fixed. We take $\varepsilon > 0$ arbitrarily and $x_n \in X$ such that

$$\|x_n - x\| = \max_{t \in [0,1]} |x_n(t) - x(t)| < \varepsilon.$$

Hence,

$$
\begin{aligned}
|Ax_n(t) - Ax(t)| &= \left| t^2 \int_0^1 x_n(s)ds - t^2 \int_0^1 x(s)ds \right| \\
&= \left| t^2 \int_0^1 (x_n(s) - x(s))\,ds \right| \\
&\leq t^2 \int_0^1 |x_n(s) - x(s)|\,ds \\
&\leq \int_0^1 \|x_n - x\|ds
\end{aligned}
$$

$$< \quad \varepsilon, \quad t \in [0,1].$$

Because $\varepsilon > 0$ was arbitrarily chosen, we conclude that A is continuous at x. Since $x \in X$ was arbitrarily chosen, we get that the operator A is continuous in X.

Now we suppose that $A : X \longmapsto Y$ is a linear continuous operator. We take $x = y + z$, $y, z \in X$. Then

$$
\begin{aligned}
Ax &= A(y+z) \\[2ex]
&= Ay + Az \\[2ex]
&= Ay + A(x-y).
\end{aligned}
$$

Therefore,

$$A(x-y) = Ax - Ay. \tag{1.11}$$

We set $x = y$ in (1.11) and we get

$$
\begin{aligned}
A0 &= Ax - Ax \\[2ex]
&= 0.
\end{aligned}
$$

We set $x = 0$ in (1.11) and we obtain

$$
\begin{aligned}
A(-y) &= A0 - Ay \\[2ex]
&= -Ay.
\end{aligned}
\tag{1.12}
$$

Theorem 1.3.3 *Let $A : X \longmapsto Y$ be linear operator, which is continuous at a single point $x_0 \in X$. Then it is continuous on the entire space X.*

Proof 1.3.4 *Let $\{x_n\}_{n \in \mathbb{N}}$ be a sequence of elements of X such that $x_n \to x$, as $n \to \infty$, in X, $x \in X$. Hence,*

$$x_n - x + x_0 \to x_0, \quad as \quad n \to \infty.$$

Therefore,

$$A(x_n - x + x_0) \to Ax_0, \quad as \quad n \to \infty. \tag{1.13}$$

Since $A : X \longmapsto Y$ is a linear operator, we get

$$A(x_n - x + x_0) = Ax_n - Ax + Ax_0, \quad n \in \mathbb{N}.$$

From here and (1.13), *we obtain*

$$Ax_n - Ax + Ax_0 \to Ax_0, \quad as \quad n \to \infty.$$

Therefore,

$$Ax_n \to Ax, \quad as \quad n \to \infty.$$

This completes the proof.

Definition 1.3.5 *Let $A, B : X \longmapsto Y$ be linear operators. We define the addition of the operators A and B by*

$$(A + B)x = Ax + Bx, \quad x \in X,$$

and the scalar multiplication by

$$(\lambda A)x = \lambda Ax, \quad x \in X, \quad \lambda \in F.$$

The zero operator O is defined by

$$Ox = 0$$

for any $x \in X$. The identity operator I is defined by

$$Ix = x$$

for any $x \in X$.
     Let $A, B : X \longmapsto X$. Define

$$(AB)x = A(Bx),$$

$$A^2 x = A(Ax),$$

$$A^n x = A(A^{n-1}x), \quad n \geq 3, \quad x \in X.$$

Remark 1.3.6 *If $A, B, C : X \longmapsto Y$, then*

$$(AB)C = A(BC),$$
$$(A+B)C = AC + BC,$$
$$C(A+B) = CA + CB.$$

In the general case, we have

$$AB \neq BA.$$

Definition 1.3.7 *Let $A : X \longmapsto Y$ be a linear operator. We say that the linear operator $B : X \longmapsto Y$ is a left inverse of the operator A, if*

$$BA = I.$$

We say that the linear operator $C : X \longmapsto Y$ is a right inverse of the operator A, if

$$AC = I.$$

Let $B, C : X \longmapsto Y$ be left and right inverse, respectively, of the linear operator $A : X \longmapsto Y$. Then

$$B = BI$$

$$= B(AC)$$

$$= (BA)C$$

$$= IC$$

$$= C.$$

In this case it is said that the operator A has an inverse denoted by A^{-1}. Thus, if A^{-1} exists, we have

$$A^{-1}A = AA^{-1}$$

$$= I.$$

A linear operator $A : X \longmapsto Y$ can have at most one inverse.

Theorem 1.3.8 *Let $A : X \longmapsto Y$ be a linear operator that is continuous at 0. Then A is continuous in X.*

Proof 1.3.9 *Let $x \in X$ be arbitrarily chosen. Since A is continuous at 0, we get*

$$\|A(x_n - x) - A0\| \to 0, \quad as \quad \|x_n - x\| \to 0,$$

or

$$\|Ax_n - Ax\| \to 0, \quad as \quad \|x_n - x\| \to 0.$$

Therefore, A is continuous at x. Because $x \in X$ was arbitrarily chosen, we conclude that A is continuous in X. This completes the proof.

Definition 1.3.10 *A linear operator $A : X \longmapsto Y$ will be called bounded if there is a constant $M \geq 0$ such that*

$$\|Ax\| \leq M\|x\|$$

for any $x \in X$.

Example 1.3.11 *Let $X = Y = \mathscr{C}([0,1])$, $a \in X$. Consider the operator*

$$Ax(t) = \int_0^t a(s)x(s)ds, \quad t \in [0,1], \quad x \in \mathscr{C}([0,1]).$$

We have that $A : X \longmapsto Y$. In X we define a norm as follows,

$$\|x\| = \max_{t \in [0,1]} |x(t)|, \quad x \in X.$$

Because $a \in \mathscr{C}([0,1])$, there is a positive constant M such that

$$|a(t)| \leq M$$

for any $t \in [0,1]$. Let $x \in X$ be arbitrarily chosen. Hence, for any $t \in [0,1]$, we have

$$|Ax(t)| = \left| \int_0^t a(s)x(s)ds \right|$$

$$\leq \int_0^t |a(s)||x(s)|ds$$

$$\leq M\|x\| \int_0^t ds$$

$$\leq M\|x\|.$$

From here,

$$\max_{t\in[0,1]} |Ax(t)| \leq M\|x\|,$$

or

$$\|Ax\| \leq M\|x\|.$$

Because $x \in X$ was arbitrarily chosen, we conclude that $A : X \longmapsto Y$ is a bounded operator.

Theorem 1.3.12 *A linear operator $A : X \longmapsto Y$ is bounded if and only if it is continuous.*

Proof 1.3.13

1. Let $A : X \longmapsto Y$ be a continuous operator. Assume that it is not bounded. Then, there is a sequence $\{x_n\}_{n\in\mathbb{N}}$ of elements of X such that

$$\|Ax_n\| > n\|x_n\|, \quad x_n \neq 0,$$

for any $n \in \mathbb{N}$. We set

$$\xi_n = \frac{x_n}{n\|x_n\|}.$$

Then

$$\|A\xi_n\| > 1, \quad n \in \mathbb{N}. \tag{1.14}$$

On the other hand,

$$\|\xi_n\| = \left\| \frac{x_n}{n\|x_n\|} \right\|$$

$$= \frac{\|x_n\|}{n\|x_n\|}$$

$$= \frac{1}{n}$$

$$\to 0, \quad as \quad n \to \infty.$$

Because $A : X \longmapsto Y$ is continuous, we get

$$\|A\xi_n\| \to 0, \quad as \quad n \to \infty.$$

This contradicts with (1.14). Therefore, $A : X \longmapsto Y$ is bounded.

2. Let $A : X \longmapsto Y$ be a bounded operator. Then there exists a positive constant M such that

$$\|Ax\| \leq M\|x\|$$

for any $x \in X$. Let $x_n \to x$ as $n \to \infty$, i.e., $\|x_n - x\| \to 0$ as $n \to \infty$. Then

$$\|Ax_n - Ax\| \quad = \quad \|A(x_n - x)\|$$

$$\leq \quad M\|x_n - x\|$$

$$\to \quad 0, \quad as \quad n \to \infty.$$

Therefore, $A : X \longmapsto Y$ is continuous. This completes the proof.

The space of all linear bounded operators $A : X \longmapsto Y$ will be denoted with $\mathscr{L}(X,Y)$. Note that $\mathscr{L}(X,Y)$ is a vector space.

Theorem 1.3.14 *Let X be a Banach space and $A : X \longmapsto Y$ be a linear operator. By X_n we denote the set of those $x \in X$ for which $\|Ax\| \leq n\|x\|$. Then $X = \bigcup_{n=1}^{\infty} X_n$ and at least one of the sets X_n is everywhere dense in X.*

Proof 1.3.15 *Note that the sets X_n, $n \in \mathbb{N}$ are not empty because $0 \in X_n$ for any $n \in \mathbb{N}$. Also, any $x \in X$, $x \neq 0$, occurs in one of the sets X_n because it is sufficient to take n as the least integer, greater than $\frac{\|Ax\|}{\|x\|}$.*

Therefore, $X \subset \bigcup_{n=1}^{\infty} X_n$. By the definition of the sets X_n, we have that $\bigcup_{n=1}^{\infty} X_n \subset X$. Consequently $X = \bigcup_{n=1}^{\infty} X_n$. Since a complete space cannot be a countable sum of nowhere dense sets, there is an $n_0 \in \mathbb{N}$ and there is an open ball $S_r(x_0)$ containing $S_r(x_0) \bigcap X_{n_0}$ everywhere dense. We take $x_1 \in S_r(x_0) \bigcap X_{n_0}$ and let $S_{r_1}[x_1]$ be a closed ball such that

$$S_{r_1}[x_1] \subset S_r(x_0).$$

Let $x \in X$ be an element for which $\|x\| = r_1$. Since

$$\|(x_1 + x) - x_1\| = \|x\|$$
$$= r_1,$$

we conclude that $x + x_1 \in S_{r_1}[x_1]$. There is a sequence $\{z_k\}_{k \in \mathbb{N}}$ of elements of $S_{r_1}[x_1] \bigcap X_{n_0}$, (this sequence can be stationary if $x_1 + x \in X_{n_0}$) such that $z_k \to x + x_1$ as $k \to \infty$. Consequently

$$x_k = z_k - x_1 \to x, \quad as \quad k \to \infty.$$

Because $x_k \to x$ as $k \to \infty$ and $\|x\| = r_1$, we can assume that $\frac{r_1}{2} \leq \|x_k\| \leq r_1$ for any enough large $k \in \mathbb{N}$. Now, using that $z_k, x_1 \in X_{n_0}$, we get

$$\|Ax_k\| = \|Az_k - Ax_1\|$$
$$\leq \|Az_k\| + \|Ax_1\|$$
$$\leq n_0(\|z_k\| + \|x_1\|).$$

Also,

$$\|z_k\| = \|x_k + x_1\|$$

$$\leq \quad \|x_k\| + \|x_1\|$$

$$\leq \quad r_1 + \|x_1\|.$$

Therefore, for any enough large $k \in \mathbb{N}$,

$$\|Ax_k\| \quad \leq \quad n_0 \left(r_1 + 2\|x_1\| \right)$$

$$\leq \quad \frac{2n_0 \left(r_1 + 2\|x_1\| \right)}{r_1} \|x_k\|.$$

Let n be the least integer greater than $\dfrac{2n_0 \left(r_1 + 2\|x_1\| \right)}{r_1}$. Then

$$\|Ax_k\| \leq n\|x_k\|$$

for any enough large $k \in \mathbb{N}$. Consequently $x_k \in X_n$ for any enough large $k \in \mathbb{N}$. Thus every element $x \in X$ with norm equal to r_1 can be approximated by elements of X_n. Let now $x \in X$, $x \neq 0$ be arbitrarily chosen. We set $\xi = r_1 \dfrac{x}{\|x\|}$. Then $\|\xi\| = r_1$. As in above, there is a sequence $\{\xi_k\}_{k \in \mathbb{N}} \subset X_n$ convergent to ξ. Then

$$x_k \quad = \quad \xi_k \frac{\|x\|}{r_1}$$

$$\rightarrow \quad x,$$

$$\|Ax_k\| \quad = \quad \frac{\|x\|}{r_1} \|A\xi_k\|$$

$$\leq \quad \frac{\|x\|}{r_1} n \|\xi_k\|$$

$$= \quad n\|x_k\|, \quad k \in \mathbb{N}.$$

Thus, $x_k \in X_n$, $k \in \mathbb{N}$. Consequently, X_n is everywhere dense in X. This completes the proof.

Definition 1.3.16 *Let $A : X \longmapsto Y$ be a bounded linear operator. The smallest number M for which $\|Ax\| \leq M\|x\|$ for any $x \in X$ will be called the norm of the operator A. It is denoted by $\|A\|$.*

By Definition 1.3.16, it follows

1. $\|Ax\| \leq \|A\|\|x\|$ for any $x \in X$.

2. For any $\varepsilon > 0$, there is an element $x_\varepsilon \in X$, $x_\varepsilon \neq 0$, such that

$$\|Ax_\varepsilon\| > (\|A\| - \varepsilon)\|x_\varepsilon\|.$$

Theorem 1.3.17 *Let $A : X \longmapsto Y$ be a bounded linear operator. Then*

$$\|A\| = \sup_{\|x\| \leq 1} \|Ax\|. \qquad (1.15)$$

Proof 1.3.18 *For any $x \in X$, $\|x\| \leq 1$, we have*

$$
\begin{aligned}
\|Ax\| &\leq \|A\|\|x\| \\
&\leq \|A\|.
\end{aligned}
\qquad (1.16)
$$

Let $\varepsilon > 0$ be arbitrarily chosen. Then there exists $x_\varepsilon \in X$, $x_\varepsilon \neq 0$, such that

$$\|Ax_\varepsilon\| > (\|A\| - \varepsilon)\|x_\varepsilon\|.$$

We take $\xi_\varepsilon = \dfrac{x_\varepsilon}{\|x_\varepsilon\|}$. Then $\|\xi_\varepsilon\| = 1$ and

$$
\begin{aligned}
\|A\xi_\varepsilon\| &= \left\| A\left(\frac{x_\varepsilon}{\|x_\varepsilon\|}\right) \right\| \\
&= \left\| \frac{1}{\|x_\varepsilon\|} Ax_\varepsilon \right\| \\
&= \frac{1}{\|x_\varepsilon\|} \|Ax_\varepsilon\| \\
&> \|A\| - \varepsilon.
\end{aligned}
$$

Because $\|\xi_\varepsilon\| = 1$, *from the last inequality, we get*

$$\sup_{\|x\|\leq 1} \|Ax\| \geq \|A\| - \varepsilon.$$

Hence, using that $\varepsilon > 0$ *was arbitrarily chosen, we get*

$$\sup_{\|x\|\leq 1} \|Ax\| \geq \|A\|.$$

From the last inequality and from (1.16), *we obtain the inequality* (1.15). *This completes the proof.*

Remark 1.3.19 *By Theorem 1.3.17, it follows that*

$$\|A\| = \sup_{x\in X, x\neq 0} \frac{\|Ax\|}{\|x\|}.$$

Example 1.3.20 *Let* $X = Y = \mathscr{C}([0,1])$. *Consider the operator*

$$Ax(t) = \int_0^1 x(s)ds, \quad x \in X, \quad t \in [0,1].$$

In X we take the norm $\|x\| = \max_{t\in[0,1]} |x(t)|$. *We have* $A : X \longmapsto X$. *We will find* $\|A\|$. *For any* $x \in X$, $\|x\| \leq 1$, *we have*

$$\begin{aligned}
|Ax(t)| &= \left|\int_0^1 x(s)ds\right| \\
&\leq \int_0^1 |x(s)|\,ds \\
&\leq \int_0^1 \max_{s\in[0,1]} |x(s)|\,ds \\
&= \|x\| \\
&\leq 1, \quad t \in [0,1].
\end{aligned}$$

Hence,

$$\max_{t\in[0,1]} |Ax(t)| \leq 1$$

or

$$\|Ax\| \leq 1.$$

Therefore,

$$\|A\| \leq 1. \tag{1.17}$$

Now we take $y(t) = 1$, $t \in [0,1]$. *Then* $\|y\| = 1$ *and*

$$Ay(t) = \int_0^1 ds$$

$$= 1, \quad t \in [0,1].$$

Hence, using that $\|y\| = 1$, *we get*

$$\sup_{\|x\| \leq 1} \|Ax\| \geq 1,$$

i.e., $\|A\| \geq 1$. *From here and from* (1.17), *we obtain that* $\|A\| = 1$.

Example 1.3.21 *Let* $K \in \mathscr{C}\left([0,1] \times [0,1]\right)$, $X = Y = \mathscr{C}\left([0,1]\right)$. *Consider the operator*

$$Ax(t) = \int_0^1 K(t,s)x(s)ds, \quad x \in X, \quad t \in [0,1].$$

In X we take the norm $\|x\| = \max_{t \in [0,1]} |x(t)|$. *We have* $A : X \longmapsto Y$. *We will find* $\|A\|$. *Let* $x \in X$ *be arbitrarily chosen. Then*

$$|Ax(t)| = \left| \int_0^1 K(t,s)x(s)ds \right|$$

$$\leq \int_0^1 |K(t,s)| \, |x(s)| ds$$

$$\leq \|x\| \max_{t \in [0,1]} \int_0^1 |K(t,s)| ds, \quad t \in [0,1].$$

Hence,

$$\max_{t \in [0,1]} |Ax(t)| \leq \|x\| \max_{t \in [0,1]} \int_0^1 |K(t,s)| ds$$

or

$$\|Ax\| \le \|x\| \max_{t \in [0,1]} \int_0^1 |K(t,s)| ds.$$

Then

$$
\begin{aligned}
\sup_{\|x\| \le 1} \|Ax\| &= \sup_{\|x\| \le 1} \left(\max_{t \in [0,1]} |Ax(t)| \right) \\
&\le \sup_{\|x\| \le 1} \left(\|x\| \max_{t \in [0,1]} \int_0^1 |K(t,s)| ds \right) \\
&\le \max_{t \in [0,1]} \int_0^1 |K(t,s)| ds.
\end{aligned}
$$

Therefore,

$$\|A\| \le \max_{t \in [0,1]} \int_0^1 |K(t,s)| ds. \qquad (1.18)$$

For each $n \in \mathbb{N}$, we take a set $X_n \subset [0,1]$ such that

$$meas(X_n) \le \frac{1}{2Mn},$$

where $M = \max_{(t,s) \in [0,1] \times [0,1]} |K(t,s)|$. Since $t \longmapsto \int_0^1 |K(t,s)| ds$ is a continuous function on $[0,1]$, there exists $t_0 \in [0,1]$ such that

$$\int_0^1 |K(t_0,s)| ds = \max_{t \in [0,1]} \int_0^1 |K(t,s)| ds.$$

Let

$$z_0(s) = signK(t_0,s), \quad s \in [0,1].$$

For each $n \in \mathbb{N}$, we take a continuous function x_n on $[0,1]$ such that $|x_n(s)| \le 1$ for any $s \in [0,1]$ and $x_n(s) = z_0(s)$ for $s \in [0,1] \backslash X_n$. We have that

$$
\begin{aligned}
|x_n(s) - z_0(s)| &\le |x_n(s)| + |z_0(s)| \\
&\le 1 + 1 \\
&= 2
\end{aligned}
$$

for any $s \in X_n$. Then

$$\left| \int_0^1 K(t,s) z_0(s) ds - \int_0^1 K(t,s) x_n(s) ds \right| = \left| \int_0^1 K(t,s) \left(z_0(s) - x_n(s) \right) ds \right|$$

$$\leq \int_0^1 |K(t,s)| \, |z_0(s) - x_n(s)| \, ds$$

$$= \int_{X_n} |K(t,s)| \, |z_0(s) - x_n(s)| \, ds$$

$$\leq 2Mm(X_n)$$

$$\leq \frac{1}{n}$$

for any $t \in [0,1]$ and for any $n \in \mathbb{N}$. Therefore,

$$\int_0^1 K(t,s) z_0(s) ds \ \leq \ \int_0^1 K(t,s) x_n(s) ds + \frac{1}{n}$$

$$= \ A x_n(t) + \frac{1}{n}$$

$$\leq \ \|A\| \|x_n\| + \frac{1}{n}$$

$$\leq \ \|A\| + \frac{1}{n}$$

for any $t \in [0,1]$ and for any $n \in \mathbb{N}$. In particular,

$$\int_0^1 K(t_0,s) z_0(s) ds \leq \|A\| + \frac{1}{n}$$

for any $n \in \mathbb{N}$. Therefore,

$$\int_0^1 K(t_0,s) z_0(s) ds \leq \|A\|,$$

i.e.,

$$\max_{t\in[0,1]}\int_0^1 |K(t,s)|ds \le \|A\|.$$

From the last inequality and from (1.18), we get

$$\|A\| = \max_{t\in[0,1]}\int_0^1 |K(t,s)|ds.$$

Example 1.3.22 *Let $X = Y = \mathscr{C}([0,1])$. Consider the operator*

$$Ax(t) = \frac{t^2}{2}\int_0^t x(s)ds, \quad t \in [0,1], \quad x \in X.$$

In X we take the norm $\|x\| = \max_{t\in[0,1]}|x(t)|$. We have $A : X \longmapsto Y$. We will find $\|A\|$. Let $x \in X$ be arbitrarily chosen. Then

$$
\begin{aligned}
|Ax(t)| &= \left|\frac{t^2}{2}\int_0^t x(s)ds\right| \\
&\le \frac{t^2}{2}\int_0^t |x(s)|ds \\
&\le \frac{1}{2}\|x\|
\end{aligned}
$$

for any $t \in [0,1]$. Hence,

$$
\begin{aligned}
\|Ax\| &= \max_{t\in[0,1]}|Ax(t)| \\
&\le \frac{1}{2}\|x\|.
\end{aligned}
$$

Then

$$
\begin{aligned}
\|A\| &= \sup_{|x|\le 1}\|Ax\| \\
&\le \frac{1}{2}.
\end{aligned}
$$

(1.19)

For each $n \in \mathbb{N}$ we take a set $X_n \subset [0,1]$ such that $\operatorname{meas}(X_n) \leq \dfrac{1}{n}$. Also, for each $n \in \mathbb{N}$ we take a continuous function x_n on $[0,1]$ such that $|x_n(s)| \leq 1$ for any $s \in [0,1]$ and $x_n(s) = 1$ for $s \in [0,1] \backslash X_n$. Then

$$|x_n(s) - 1| \leq 2$$

for any $s \in X_n$. Hence,

$$
\begin{aligned}
\left| \frac{t^3}{2} - \frac{t^2}{2} \int_0^t x_n(s)ds \right| &= \left| \frac{t^2}{2} \int_0^t (1 - x_n(s))ds \right| \\
&\leq \frac{t^2}{2} \int_0^t |1 - x_n(s)| ds \\
&\leq \frac{t^2}{2} \int_0^1 |1 - x_n(s)| ds \\
&= \frac{t^2}{2} \int_{X_n} |1 - x_n(s)| ds \\
&\leq m(X_n) \\
&\leq \frac{1}{n}
\end{aligned}
$$

for any $t \in [0,1]$ and for any $n \in \mathbb{N}$. Hence,

$$
\begin{aligned}
\frac{t^3}{2} &\leq \frac{1}{n} + \frac{t^2}{2} \int_0^t x_n(s)ds \\
&= \frac{1}{n} + Ax_n(t) \\
&\leq \frac{1}{n} + \|A\| \|x_n\| \\
&\leq \frac{1}{n} + \|A\|
\end{aligned}
$$

for any $t \in [0,1]$ and for any $n \in \mathbb{N}$. In particular, for $t = 1$, we get

$$\frac{1}{2} \leq \frac{1}{n} + \|A\|$$

for any $n \in \mathbb{N}$. Consequently

$$\frac{1}{2} \leq \|A\|.$$

From the last inequality and from (1.19), we get $\|A\| = \frac{1}{2}$.

Exercise 1.3.23 *Let $X = Y = \mathscr{C}([0,1])$, $\|x\| = \max\limits_{t \in [0,1]} |x(t)|$, $x \in X$. Find $\|A\|$, where*

$$Ax(t) = \frac{t^2}{3} \int_0^t s^2 x(s)ds, \quad x \in X, \quad t \in [0,1].$$

Answer. $\frac{1}{9}$.

Theorem 1.3.24 *Let $A : X \longmapsto Y$ be a linear operator. Then A is a bounded operator if and only if there is a constant $M \geq 0$ such that $\|A\| \leq M$.*

Proof 1.3.25

1. Let A be a bounded operator. Then there is a constant $M \geq 0$ such that

$$\|Ax\| \leq M\|x\|$$

for any $x \in X$. Hence,

$$\begin{aligned} \|A\| &= \sup_{\|x\| \leq 1} \|Ax\| \\ &\leq \sup_{\|x\| \leq 1} (M\|x\|) \\ &= M. \end{aligned}$$

2. Let there is a constant $M \geq 0$ such that $\|A\| \leq M$. Then

$$\|Ax\| \leq \|A\|\|x\| \leq M\|x\|$$

for any $x \in X$. Therefore, A is a bounded operator. This completes the proof.

Theorem 1.3.26 *Let $A, B : X \longmapsto Y$ be linear operators. Then*

$$\|A + B\| \leq \|A\| + \|B\| \quad and \quad \|\lambda A\| = |\lambda| \|A\|$$

for any $\lambda \in F$.

Proof 1.3.27 *We have*

$$
\begin{aligned}
\|A + B\| &= \sup_{x \in X, \|x\| \leq 1} \| (A + B) x \| \\
&= \sup_{x \in X, \|x\| \leq 1} \|Ax + Bx\| \\
&\leq \sup_{x \in X, \|x\| \leq 1} (\|Ax\| + \|Bx\|) \\
&\leq \sup_{x \in X, \|x\| \leq 1} \|Ax\| + \sup_{x \in X, \|x\| \leq 1} \|Bx\| \\
&= \|A\| + \|B\|,
\end{aligned}
$$

and

$$
\begin{aligned}
\|\lambda A\| &= \sup_{x \in X, \|x\| \leq 1} \|\lambda Ax\| \\
&= \sup_{x \in X, \|x\| \leq 1} (|\lambda| \|Ax\|) \\
&= |\lambda| \sup_{x \in X, \|x\| \leq 1} \|Ax\| \\
&= \lambda \|A\|
\end{aligned}
$$

for any $\lambda \in F$. This completes the proof.

Definition 1.3.28 *We say that a sequence $\{A_n\}_{n \in \mathbb{N}}$ of elements of $\mathscr{L}(X, Y)$ is uniformly convergent to $A \in \mathscr{L}(X, Y)$ if*

$$\|A_n - A\| \to 0, \quad as \quad n \to \infty.$$

We will write $A_n \to A$, as $n \to \infty$, or $\lim_{n \to \infty} A_n = A$.

Theorem 1.3.29 *Let* $\{A_n\}_{n\in\mathbb{N}} \subset \mathscr{L}(X,Y)$ *and* $A \in \mathscr{L}(X,Y)$. *Then* $A_n \to A$, *as* $n \to \infty$, *uniformly if and only if*

$$\|A_n x - Ax\| \to 0, \quad as \quad n \to \infty,$$

for any $x \in X$ *such that* $\|x\| \leq 1$.

Proof 1.3.30   *1. Let* $A_n \to A$, *as* $n \to \infty$, *uniformly. Then*

$$\|A_n - A\| \to 0, \quad as \quad n \to \infty.$$

Hence, for $x \in X$, $\|x\| \leq 1$, *we get*

$$\|A_n x - Ax\| \quad \leq \quad \|A_n - A\|\|x\|$$

$$\leq \quad \|A_n - A\|$$

$$\to \quad 0, \quad as \quad n \to \infty,$$

for any $x \in X$ *such that* $\|x\| \leq 1$.

2. Let $\|A_n x - Ax\| \to 0$, *as* $n \to \infty$, *for any* $x \in X$ *such that* $\|x\| \leq 1$. *Hence,*

$$\sup_{x\in X, \|x\|\leq 1} \|A_n x - Ax\| \leq \frac{\varepsilon}{2}$$

for any $n \geq N$. *Therefore,*

$$\|A_n - A\| \to 0, \quad as \quad n \to \infty.$$

This completes the proof.

Corollary 1.3.31 *Let* $\{A_n\}_{n\in\mathbb{N}} \subset \mathscr{L}(X,Y)$ *be such that* $A_n \to A$, *as* $n \to \infty$, *for* $A \in \mathscr{L}(X,Y)$. *Let also, U be an arbitrary bounded set in* X. *Then*

$$\|A_n x - Ax\| \to 0, \quad as \quad n \to \infty$$

for any $x \in U$.

Proof 1.3.32 *Because U is a bounded set in X, there is an $R > 0$ such that $\|x\| \leq R$ for any $x \in U$. Let $\varepsilon > 0$ be arbitrarily chosen. Since $A_n \to A$, as $n \to \infty$, we have*

$$\|A_n - A\| \to 0, \quad as \quad n \to \infty.$$

Hence, for $x \in U$, we have

$$\|A_n x - Ax\| \quad \leq \quad \|A_n - A\| \|x\|$$

$$\leq \quad R \|A_n - A\|$$

$$\to \quad 0, \quad as \quad n \to \infty.$$

This completes the proof.

Definition 1.3.33 *We say that a sequence $\{A_n\}_{n \in \mathbb{N}}$ of elements of $\mathscr{L}(X,Y)$ is a Cauchy sequence, if for any $\varepsilon > 0$ there is an $N \in \mathbb{N}$ such that*

$$\|A_{n+p} - A_n\| < \varepsilon$$

for any $n, p > N$.

Theorem 1.3.34 *Let Y be a Banach space. Then $\mathscr{L}(X,Y)$ is a Banach space.*

Proof 1.3.35 *Let $\{A_n\}_{n \in \mathbb{N}}$ be a Cauchy sequence of elements of $\mathscr{L}(X,Y)$. We fix $\varepsilon > 0$. Then there is an $N \in \mathbb{N}$ such that*

$$\|A_{n+p} - A_n\| < \varepsilon$$

for any $n, p \geq N$. Let $x \in X$ be arbitrarily chosen. We have

$$\|A_{n+p} x - Ax\| \quad = \quad \|(A_n - A)x\|$$

$$\leq \quad \|A_n - A\| \|x\|$$

for any $n, p \geq N$. Therefore, $\{A_n x\}_{n \in \mathbb{N}}$ is a Cauchy sequence in Y. Since Y is a Banach space, we have that there exists $\lim\limits_{n \to \infty} A_n x$. We define the operator $A : X \to Y$ as follows

$$Ax = \lim_{n \to \infty} A_n x, \quad x \in X.$$

For any $\alpha, \beta \in F$ and $x_1, x_2 \in X$, we have

$$
\begin{aligned}
A(\alpha x_1 + \beta x_2) &= \lim_{n \to \infty} A_n(\alpha x_1 + \beta x_2) \\[2mm]
&= \lim_{n \to \infty}(\alpha A_n x_1 + \beta A_n x_2) \\[2mm]
&= \alpha \lim_{n \to \infty} A_n x_1 + \beta \lim_{n \to \infty} A_n x_2 \\[2mm]
&= \alpha A x_1 + \beta A x_2.
\end{aligned}
$$

Therefore, $A : X \longmapsto Y$ is a linear operator. Note that

$$\left| \|A_{n+p}\| - \|A_n\| \right| \leq \|A_{n+p} - A_n\|$$

for any $n, p \in \mathbb{N}$. Therefore, $\{\|A_n\|\}_{n \in \mathbb{N}}$ is Cauchy sequence in $\mathbb{R}$. Hence, it is a bounded sequence in $\mathbb{R}$ and there exists a constant $c > 0$ such that

$$\|A_n\| \leq c$$

for any $n \in \mathbb{N}$. Hence,

$$\|A_n x\| \leq c \|x\|$$

for any $x \in X$ and for any $n \in \mathbb{N}$. Consequently

$$\|Ax\| \leq c \|x\|$$

for any $x \in X$. From here, $A \in \mathscr{L}(X, Y)$. This completes the proof.

Definition 1.3.36 *Let $A_n \in \mathscr{L}(X, Y)$, $n \in \mathbb{N}$.*

1. The series $\sum\limits_{n=1}^{\infty} A_n$ is said to be uniformly convergent, if the sequence $\left\{ S_n = \sum\limits_{k=1}^{n} A_k \right\}_{n \in \mathbb{N}}$ is uniformly convergent.

2. The series $\sum\limits_{n=1}^{\infty} A_n$ is said to be absolutely convergent if the series $\sum\limits_{n=1}^{\infty} \|A_n\|$ is convergent.

Theorem 1.3.37 *Let $\mathscr{L}(X,Y)$ be a Banach space. If the series $\sum\limits_{n=1}^{\infty} A_n$ is an absolutely convergent series, then it is uniformly convergent.*

Proof 1.3.38 *Since the series $\sum\limits_{n=1}^{\infty} A_n$ is absolutely convergent, we have that the series $\sum\limits_{n=1}^{\infty} \|A_n\|$ is convergent. We fix $\varepsilon > 0$. Then there is an $N \in \mathbb{N}$ such that*

$$\sum_{n=N+1}^{N+p} \|A_n\| < \varepsilon$$

for any $p \in \mathbb{N}$. Hence,

$$\begin{aligned} \|S_{N+p} - S_N\| &= \left\| \sum_{n=N+1}^{N+p} A_n \right\| \\ &\leq \sum_{n=N+1}^{N+p} \|A_n\| \\ &< \varepsilon \end{aligned}$$

for any $p \in \mathbb{N}$. Therefore, $\{S_n\}_{n \in \mathbb{N}}$ is a Cauchy sequence in $\mathscr{L}(X,Y)$. Because $\mathscr{L}(X,Y)$ is a Banach space, we have that $\{S_n\}_{n \in \mathbb{N}}$ is uniformly convergent. This completes the proof.

Below with $\mathscr{L}(X)$ we will denote the space $\mathscr{L}(X,X)$.

Exercise 1.3.39 *Let $A, B \in \mathscr{L}(X)$. Prove that $AB \in \mathscr{L}(X)$ and $A^k \in \mathscr{L}(X)$ for any $k \in \mathbb{N}$.*

Exercise 1.3.40 *For $A \in \mathscr{L}(X)$, define*

$$e^A = \sum_{k=1}^{\infty} \frac{A^k}{k!}.$$

Prove that $e^A \in \mathscr{L}(X)$ and $\left\|e^A\right\| \leq e^{\|A\|}$.

Theorem 1.3.41 *Let $\{A_n\}_{n\in\mathbb{N}}, \{B_n\}_{n\in\mathbb{N}} \subset \mathscr{L}(X)$, $A,B \in \mathscr{L}(X)$. If $A_n \to A$, $B_n \to B$, as $n \to \infty$, then $A_nB_n \to AB$, as $n \to \infty$.*

Proof 1.3.42 *Because $B_n \in \mathscr{L}(X)$, $n \in \mathbb{N}$, $B_n \to B$, as $n \to \infty$, the sequence $\{\|B_n\|\}_{n\in\mathbb{N}}$ is a bounded sequence. Therefore, there is a constant $M_1 > 0$ such that $\|B_n\| \leq M_1$ for any $n \in \mathbb{N}$. Since $A \in \mathscr{L}(X)$, then there is a constant $M_2 > 0$ such that $\|A\| \leq M_2$. We take $\varepsilon > 0$ arbitrarily. Since $A_n \to A$, $B_n \to B$, as $n \to \infty$, there is an $N \in \mathbb{N}$ such that*

$$\|A_n - A\| < \frac{\varepsilon}{2M_1} \quad and \quad \|B_n - B\| < \frac{\varepsilon}{2M_2}$$

for any $n \geq N$. Hence,

$$\begin{aligned}
\|A_nB_n - AB\| &= \|A_nB_n - AB_n + AB_n - AB\| \\
&\leq \|(A_n - A)B_n\| + \|A(B_n - B)\| \\
&\leq \|A_n - A\|\,\|B_n\| + \|B_n - B\|\,\|A\| \\
&< \frac{\varepsilon}{2M_1}M_1 + \frac{\varepsilon}{2M_2}M_2 \\
&= \frac{\varepsilon}{2} + \frac{\varepsilon}{2} \\
&= \varepsilon
\end{aligned}$$

for any $n \geq \mathbb{N}$. This completes the proof.

Definition 1.3.43 *We will say that the sequence* $\{A_n\}_{n\in\mathbb{N}} \subset \mathscr{L}(X,Y)$ *is strongly convergent to the operator* $A \in \mathscr{L}(X,Y)$, *if for any* $x \in X$ *we have*

$$\|A_n x - Ax\| \to 0$$

as $n \to \infty$.

Theorem 1.3.44 *If* $\{A_n\}_{n\in\mathbb{N}} \subset \mathscr{L}(X,Y)$ *is uniformly convergent to* $A \in \mathscr{L}(X,Y)$, *then it is strongly convergent to A.*

Proof 1.3.45 *Take* $\varepsilon > 0$ *arbitrarily. Then there is an* $N \in \mathbb{N}$ *such that*

$$\|A_n - A\| < \varepsilon$$

for any $n \geq N$. *Hence,*

$$\|A_n x - Ax\| \leq \|A_n - A\|\,\|x\|$$

$$< \varepsilon\|x\|$$

for any $n \geq N$ *and for any* $x \in X$. *This completes the proof.*

Theorem 1.3.46 *Let* $\{A_n\}_{n\in\mathbb{N}} \subset \mathscr{L}(X,Y)$ *and there exists* $c > 0$ *and a closed ball* $S_r[x_0]$ *such that* $\|A_n x\| \leq c$ *for any* $x \in S_r[x_0]$. *Then the sequence* $\{\|A_n\|\}_{n\in\mathbb{N}}$ *is bounded.*

Proof 1.3.47 *Let* $x \in X$, $x \neq 0$. *Then* $x_0 + \frac{x}{\|x\|}r \in S_r[x_0]$. *Hence,*

$$c \geq \left\| A_n\left(r\frac{x}{\|x\|} + x_0\right)\right\|$$

$$= \left\| \frac{r}{\|x\|}A_n x + A_n x_0\right\|$$

$$\geq \left\| \frac{r}{\|x\|}A_n x\right\| - \|A_n x_0\|$$

$$\geq \frac{r}{\|x\|}\|A_n x\| - c,$$

whereupon

$$\frac{r}{\|x\|}\|A_n x\| \le 2c \quad or \quad \frac{\|A_n x\|}{\|x\|} \le \frac{2c}{r}.$$

Therefore,

$$\|A_n\| \le \frac{2c}{r}.$$

This completes the proof.

Theorem 1.3.48 (Uniform Boundedness Principle) *Let X be a Banach space and $\{A_n\}_{n\in\mathbb{N}} \subset \mathscr{L}(X,Y)$. If $\{A_n x\}_{n\in\mathbb{N}}$ is bounded for any fixed $x \in X$, then the sequence $\{\|A_n\|\}_{n\in\mathbb{N}}$ is bounded.*

Proof 1.3.49 *Assume the contrary. Suppose that there is a closed ball $S_r[x_0]$ such that the sequences $\{\|A_n x\|\}_{n\in\mathbb{N}}$ is unbounded for some $x \in S_r[x_0]$. Then there is $x_1 \in S_r[x_0]$ and $n_1 \in \mathbb{N}$ such that $\|A_{n_1} x_1\| > 1$. Since A_{n_1} is continuous, there is a closed ball $S_{r_1}[x_1] \subset S_r[x_0]$ such that $\|A_{n_1}x\| > 1$ for any $x \in S_{r_1}[x_1]$. By Theorem 1.3.46, it follows that the sequences $\{\|A_n x\|\}_{n\in\mathbb{N}}$ is unbounded for any $x \in S_{r_1}[x_1]$. Then there is $x_2 \in S_{r_1}[x_1]$ and $n_2 > n_1$, $n_2 \in \mathbb{N}$ such that $\|A_{n_2} x_2\| > 2$. Since A_{n_2} is continuous, there is a closed ball $S_{r_2}[x_2] \subset S_{r_1}[x_1]$ such that $\|A_{n_2}x\| > 2$ for any $x \in S_{r_2}[x_2]$ and so on. In this way get the sequences $\{x_k\}_{k\in\mathbb{N}}$ and $\{S_{r_k}[x_k]\}_{k\in\mathbb{N}}$ such that*

$$\ldots \subset S_{r_k}[x_k] \subset S_{r_{k-1}}[x_{k-1}] \subset \ldots \subset S_{r_1}[x_1]$$

and $\|A_{n_k} x\| \ge k$ for any $k \in \mathbb{N}$ and for any $x \in S_{r_k}[x_k]$. Hence, it follows that there is an $\bar{x} \in X$ such that $\bar{x} \in S_{r_k}[x_k]$ for any $k \in \mathbb{N}$. Then $\|A_n \bar{x}\| \ge k$ for any $k \in \mathbb{N}$, i.e., the sequence $\{\|A_n \bar{x}\|\}_{k\in\mathbb{N}}$ is unbounded. This is a contradiction. Consequently $\{\|A_n\|\}_{n\in\mathbb{N}}$ is bounded. This completes the proof.

Theorem 1.3.50 (Banach-Steinhaus Theorem) *Let $\{A_n\}_{n\in\mathbb{N}} \subset \mathscr{L}(X,Y)$ and $A \in \mathscr{L}(X,Y)$. Then $A_n \to A$, as $n \to \infty$, strongly if and only if*

(H1) *the sequence $\{\|A_n\|\}_{n\in\mathbb{N}}$ is bounded and*

$$\|A_n x - Ax\| \to 0, \quad as \quad n \to \infty$$

for any $x \in X'$, where X' is dense in X.

Proof 1.3.51

1. Let $A_n \to A$, as $n \to \infty$, strongly. Then $A_n x \to Ax$, as $n \to \infty$, for any $x \in X$. Therefore, $\{\|A_n x\|\}_{n \in \mathbb{N}}$ is a bounded sequence for any $x \in X$. Hence, and by Theorem 1.3.48, it follows that the sequence $\{\|A_n\|\}_{n \in \mathbb{N}}$ is a bounded sequence. As X' we can take X.

2. Suppose (**H1**) *holds. We set*

$$c = \sup_{n \in \mathbb{N}} \|A_n\|,$$

where $A_0 = A$. We take $x \in X$, $x \notin X'$. Let $\varepsilon > 0$ be arbitrarily chosen. Because X' is dense in X, there is an element $x_1 \in X'$ such that $\|x - x_1\| < \dfrac{\varepsilon}{3c}$. Also, there is an $N \in \mathbb{N}$ such that

$$\|A_n x_1 - Ax_1\| < \frac{\varepsilon}{3}$$

for any $n \geq N$. From here,

$$
\begin{aligned}
\|A_n x - Ax\| &= \|A_n (x - x_1) + (A_n x_1 - Ax_1) + A(x_1 - x)\| \\[2mm]
&\leq \|A_n (x - x_1)\| + \|A_n x_1 - Ax_1\| + \|A(x - x_1)\| \\[2mm]
&\leq \|A_n\| \, \|x - x_1\| + \|A_n x_1 - Ax_1\| + \|A\| \, \|x - x_1\| \\[2mm]
&\leq 2c \, \|x - x_1\| + \|A_n x_1 - Ax_1\| \\[2mm]
&< 2c \frac{\varepsilon}{3c} + \frac{\varepsilon}{3} \\[2mm]
&= \varepsilon
\end{aligned}
$$

for any $n \geq N$. This completes the proof.

Now we suppose that $A : D(A) \longmapsto Y$ is a linear operator, where $D(A) \subset X$. For A, we can define its norm in the following way

$$\|A\| = \sup_{\substack{x \in D(A) \\ \|x\| \leq 1}} \|Ax\|$$

and we will say that A is bounded if $\|A\| < \infty$.

Theorem 1.3.52 *Let Y be a Banach space and $A : D(A) \longmapsto Y$ be a linear bounded operator, where $D(A) \subset X$ and $D(A)$ is dense in X. Then there exists a linear bounded operator $\widetilde{A} : X \longmapsto Y$ such that $\widetilde{A}x = Ax$ for any $x \in D(A)$ and $\|A\| = \left\|\widetilde{A}\right\|$.*

Proof 1.3.53 *For $x \in D(A)$ we define $\widetilde{A}x = Ax$. Let $x \notin D(A)$ and $x \in X$. Since $D(A)$ is dense in X, there exists a sequence $\{x_n\}_{n \in \mathbb{N}}$ of elements of $D(A)$ such that $x_n \to x$, as $n \to \infty$. Then we set*

$$\widetilde{A}x = \lim_{n \to \infty} Ax_n. \tag{1.20}$$

We will prove that $\widetilde{A}$ is well-defined, i.e., we will prove that the limit (1.20) exists and it does not depend on the choice of the sequence $\{x_n\}_{n \in \mathbb{N}}$. Note that

$$\|Ax_n - Ax_m\| \leq \|A\| \|x_n - x_m\|$$

$$\to 0, \quad as \quad m, n \to \infty.$$

Therefore, $\{Ax_n\}_{n \in \mathbb{N}}$ is a Cauchy sequence in Y. Since Y is a Banach space, it is convergent. Consequently (1.20) exists. Suppose that $\{x_n\}_{n \in \mathbb{N}}$ and $\{x'_n\}_{n \in \mathbb{N}}$ are two sequences of elements of X such that $x_n \to x$, $x'_n \to x$, as $n \to \infty$. Let

$$\alpha = \lim_{n \to \infty} Ax_n, \quad \beta = \lim_{n \to \infty} Ax'_n.$$

Then

$$\|\alpha - \beta\| = \|\alpha - Ax_n + Ax_n - Ax'_n + Ax'_n - \beta\|$$

$$\leq \quad \|\alpha - Ax_n\| + \|Ax_n - Ax'_n\| + \|\beta - Ax'_n\|$$

$$\leq \quad \|Ax_n - \alpha\| + \|A\| \|x_n - x'_n\| + \|Ax'_n - \beta\|$$

$$\rightarrow \quad 0, \quad as \quad n \rightarrow \infty,$$

i.e., $\alpha = \beta$. *Next*,

$$\|Ax_n\| \leq \|A\| \|x_n\|.$$

Hence,

$$\lim_{n \to \infty} \|Ax_n\| \leq \lim_{n \to \infty} \|A\| \|x_n\|,$$

whereupon

$$\left\|\widetilde{A}x\right\| \leq \|A\| \|x\|.$$

Consequently

$$\left\|\widetilde{A}\right\| \leq \|A\|. \tag{1.21}$$

On the other hand,

$$\left\|\widetilde{A}\right\| = \sup_{\substack{\|x\| \leq 1, \\ x \in X}} \left\|\widetilde{A}x\right\|$$

$$\geq \sup_{\substack{\|x\| \leq 1, \\ x \in D(A)}} \left\|\widetilde{A}x\right\|$$

$$= \sup_{\substack{\|x\| \leq 1, \\ x \in D(A)}} \|Ax\|$$

$$= \|A\|.$$

From the last inequality and from (1.21), we get that $\|A\| = \left\|\widetilde{A}\right\|$. *This completes the proof.*

1.4　Inverse Operators

Suppose that X and Y are normed vector spaces on F.

Theorem 1.4.1 *Let* $A : X \longmapsto Y$ *be a linear operator.*

$$R(A) = \{A(x) : x \in X\}.$$

Suppose that A^{-1} *exists. Then* $A^{-1} : R(A) \longmapsto X$ *is a linear operator.*

Proof 1.4.2 *Let* $y_1, y_2 \in R(A)$. *Then there exist* $x_1, x_2 \in X$ *such that*

$$Ax_1 = y_1, \quad Ax_2 = y_2.$$

Then for any $\alpha, \beta \in F$,

$$\alpha A x_1 = \alpha y_1,$$

$$\beta A x_2 = \beta y_2,$$

or

$$A(\alpha x_1) = \alpha y_1,$$

$$A(\beta x_2) = \beta y_1,$$

$$A(\alpha x_1 + \beta x_2) = \alpha y_1 + \beta y_2.$$

Hence,

$$x_1 = A^{-1} y_1,$$

$$x_2 = A^{-1} y_2,$$

$$\alpha x_1 = A^{-1}(\alpha y_1),$$

$$\beta x_2 = A^{-1}(\beta y_2),$$

$$\alpha x_1 + \beta x_2 = A^{-1}(\alpha y_1 + \beta y_2),$$

and

$$A^{-1}(\alpha y_1 + \beta y_2) = \alpha x_1 + \beta x_2$$

$$= \alpha A^{-1} y_1 + \beta A^{-1} y_2.$$

This completes the proof.

Theorem 1.4.3 *Let* $A : X \longmapsto Y$ *be a linear operator such that*

$$\|Ax\| \geq m\|x\| \tag{1.22}$$

for any $x \in X$ *and for some constant* $m > 0$. *Then* A *has inverse bounded operator* $A^{-1} : R(A) \longmapsto X$.

Proof 1.4.4 *Let* $x_1, x_2 \in X$ *be such that* $x_1 \neq x_2$ *and* $Ax_1 = Ax_2$. *Then, using* (1.22), *we get*

$$0 = \|Ax_1 - Ax_2\|$$

$$= \|A(x_1 - x_2)\|$$

$$\geq m\|x_1 - x_2\|,$$

which is a contradiction. Therefore, A *has inverse operator* $A^{-1} : R(A) \longmapsto X$. *Moreover, we have*

$$\|A^{-1}y\| \leq \frac{1}{m}\|AA^{-1}y\|$$

$$= \frac{1}{m}\|y\|$$

for any $y \in R(A)$. *This completes the proof.*

Theorem 1.4.5 *Let X be a Banach space and let a bounded linear operator A map X onto X with $\|A\| \leq q < 1$. Then, the operator $I + A$ has inverse which is a bounded linear operator.*

Proof 1.4.6 *Consider the series*

$$I - A + A^2 - \cdots + (-1)^n A^n + \cdots. \tag{1.23}$$

Let

$$S_n = I - A + A^2 - \cdots + (-1)^n A^n, \quad n \in \mathbb{N}.$$

Since $\|A^n\| \leq \|A\|^n$, it follows that

$$
\begin{aligned}
\left\| S_{n+p} - S_n \right\| &= \left\| (-1)^{n+1} A^{n+1} + \cdots + (-1)^{n+p} A^{n+p} \right\| \\[2mm]
&\leq \|A\|^{n+1} + \cdots + \|A\|^{n+p} \\[2mm]
&\leq q^{n+1} + \cdots + q^{n+p} \\[2mm]
&\rightarrow 0, \quad as \quad n \rightarrow \infty
\end{aligned}
$$

for any $p \in \mathbb{N}$. Therefore, the sequence $\{S_n\}_{n \in \mathbb{N}}$ is a Cauchy sequence in $\mathscr{L}(X)$. Since X is a Banach space, using Theorem 1.3.34, we have that $\mathscr{L}(X)$ is a Banach space. Consequently the sequence $\{S_n\}_{n \in \mathbb{N}}$ is convergent. Let

$$S = \lim_{n \to \infty} S_n.$$

Hence,

$$
\begin{aligned}
&S(I + A) \\
={}& \lim_{n \to \infty} S_n (I + A) \\[2mm]
={}& \lim_{n \to \infty} \left(I - A + A^2 - \cdots + (-1)^n A^n \right)(I + A) \\[2mm]
={}& \lim_{n \to \infty} \left(I - A + A^2 - \cdots + (-1)^n A^n + A - A^2 + \cdots + (-1)^{n+1} A^{n+1} \right)
\end{aligned}
$$

$$= \lim_{n\to\infty} \left(I - A^{n+1}\right)$$

$$= I.$$

As in above,

$$(I + A)S = I.$$

Therefore, $(I + A)^{-1} : X \longmapsto X$ *exists and*

$$(I + A)^{-1} = S.$$

Since $I + A$ *is a linear operator, by Theorem 1.4.1,* $(I + A)^{-1} : X \longmapsto X$ *is a linear operator. Besides,*

$$
\begin{aligned}
\|S\| &= \left\| \sum_{n=0}^{\infty} A^n \right\| \\
&\leq \sum_{n=0}^{\infty} \|A\|^n \\
&\leq \sum_{n=0}^{\infty} q^n \\
&= \frac{1}{1-q}.
\end{aligned}
$$

Therefore $(I + A)^{-1} : X \longmapsto X$ *is bounded. This completes the proof.*

Theorem 1.4.7 *Let* $A : X \longmapsto Y$ *have inverse* $A^{-1} : R(A) \longmapsto X$ *and there is an operator* $B : X \longmapsto Y$ *such that*

$$\|B\| < \left\| A^{-1} \right\|^{-1}.$$

Then, $C = A + B$ *has an inverse* $C^{-1} : R(A + B) \longmapsto X$ *and*

$$\left\| C^{-1} - A^{-1} \right\| \leq \frac{\|B\|}{1 - \|A^{-1}\| \|B\|} \left\| A^{-1} \right\|^2.$$

Proof 1.4.8 *We have*

$$A + B = A\left(I + A^{-1}B\right).$$

Since

$$\left\|A^{-1}B\right\| \leq \left\|A^{-1}\right\| \|B\|$$

$$< 1,$$

by Theorem 1.4.5, $I + A^{-1}B$ has an inverse and

$$\left(I + A^{-1}B\right)^{-1} = \sum_{n=0}^{\infty} \left(-A^{-1}B\right)^n.$$

Note that

$$I = A\left(I + A^{-1}B\right)\left(I + A^{-1}B\right)^{-1}A^{-1},$$

$$I = \left(I + A^{-1}B\right)^{-1}A^{-1}A\left(I + A^{-1}B\right).$$

Therefore,

$$\left(A\left(I + A^{-1}B\right)\right)^{-1} = \left(I + A^{-1}B\right)^{-1}A^{-1},$$

i.e.,

$$C^{-1} = \left(I + A^{-1}B\right)^{-1}A^{-1}.$$

Moreover,

$$\left\|(A+B)^{-1} - A^{-1}\right\| = \left\|\left(A\left(I + A^{-1}B\right)\right)^{-1} - A^{-1}\right\|$$

$$= \left\|\left(I + A^{-1}B\right)^{-1}A^{-1} - A^{-1}\right\|$$

$$= \left\|\left(\left(I + A^{-1}B\right)^{-1} - I\right)A^{-1}\right\|$$

$$\leq \left\|I - \left(I + A^{-1}B\right)^{-1}\right\| \left\|A^{-1}\right\|$$

$$= \left\| \sum_{n=1}^{\infty} \left(-A^{-1}B \right)^n \right\| \left\| A^{-1} \right\|$$

$$\leq \sum_{n=1}^{\infty} \left\| A^{-1}B \right\|^n \left\| A^{-1} \right\|$$

$$= \frac{\left\| A^{-1} \right\| \left\| A^{-1}B \right\|}{1 - \left\| A^{-1}B \right\|}$$

$$\leq \frac{\left\| A^{-1} \right\|^2 \left\| B \right\|}{1 - \left\| A^{-1}B \right\|}.$$

This completes the proof.

Theorem 1.4.9 *Let A be a bounded linear operator that maps the Banach space X onto the Banach space Y in a one-one fashion. Then there exists a bounded linear operator A^{-1}, inverse to the operator A, which maps Y onto X.*

Proof 1.4.10 *Since $A : X \longmapsto Y$ is onto and one-one, there exists A^{-1} that maps Y onto X. By Theorem 1.4.1, it follows that A^{-1} is a linear operator. By Theorem 1.3.14, the Banach space Y can be represented in the form $Y = \bigcup_{k=1}^{\infty} Y_k$, where $Y_k \subset Y$, $\left\| A^{-1}y \right\| \leq k \left\| y \right\|$ for any $y \in Y_k$ and at least one Y_l is everywhere dense in Y. We denote it by Y_n. Let $y \in Y$ be arbitrarily chosen and $\left\| y \right\| = a$. Because $S_a[0] \bigcap Y_n$ is everywhere dense in $S_1[0]$, there exists $y_1 \in Y_n$ such that*

$$\left\| y - y_1 \right\| \leq \frac{a}{2} \quad and \quad \left\| y_1 \right\| \leq a.$$

Since $S_{\frac{a}{2}}[0] \bigcap Y_n$ is everywhere dense in $S_{\frac{a}{2}}[0]$, there is $y_2 \in Y_n$ such that

$$\left\| (y - y_1) - y_2 \right\| \leq \frac{a}{2^2},$$

$$\left\| y_2 \right\| \leq \frac{a}{2},$$

and so on, there exists $y_m \in Y_n$ such that

$$\left\| (y - y_1 - \cdots - y_{m-1}) - y_m \right\| \leq \frac{a}{2^m},$$

$$\|y_m\| \leq \frac{a}{2^{m-1}}.$$

Thus,

$$y = \lim_{m\to\infty} \sum_{k=1}^{m} y_k.$$

We set

$$x_k = A^{-1} y_k.$$

Then

$$\|x_k\| \leq n\|y_k\|$$

$$\leq \frac{na}{2^{k-1}}.$$

Let

$$s_k = \sum_{m=1}^{k} x_m.$$

Then

$$\|s_{r+p} - s_r\| = \left\| \sum_{l=r+1}^{r+p} x_l \right\|$$

$$\leq \sum_{l=r+1}^{r+p} \|x_l\|$$

$$\leq na \sum_{l=r+1}^{r+p} \frac{1}{2^{l-1}}$$

$$< \frac{na}{2^{r-1}}.$$

Since X is a complete normed space, the sequence $\{s_k\}_{k\in\mathbb{N}}$ is convergent to some element $x \in X$. Consequently

$$x = \lim_{k\to\infty} \sum_{i=1}^{k} x_l$$

$$= \sum_{i=1}^{\infty} x_i.$$

We have

$$Ax = A \left(\lim_{k \to \infty} \sum_{i=1}^{k} x_i \right)$$

$$= \lim_{k \to \infty} A \left(\sum_{i=1}^{k} x_i \right)$$

$$= \lim_{k \to \infty} \sum_{i=1}^{k} A x_i$$

$$= \lim_{k \to \infty} \sum_{i=1}^{k} y_i$$

$$= y.$$

Hence,

$$\left\| A^{-1} y \right\| = \|x\|$$

$$= \lim_{k \to \infty} \left\| \sum_{i=1}^{k} x_i \right\|$$

$$\leq \lim_{k \to \infty} \sum_{i=1}^{k} \|x_i\|$$

$$\leq \lim_{k \to \infty} \sum_{i=1}^{k} \frac{na}{2^{k-1}}$$

$$= 2na$$

$$= 2n \|y\|.$$

Because $y \in Y$ was arbitrarily chosen, we conclude that A^{-1} is bounded. This completes the proof.

1.5 Measures of Noncompactness

The following abstract definition can be found in [1]:

Definition 1.5.1 *A function ψ defined on the set of all bounded subsets of a Banach space E with values in $\mathbb{R}^+$ is called a measure of noncompactness (MNC in short) on E if for any bounded subset M of E we have $\psi(\overline{co}M) = \psi(M)$, where $\overline{co}M$ stands for the closed convex hull of M. An MNC is said to be*
(a) Full: $\psi(M) = 0$ if M is a relatively compact set.
(b) Monotone: for all bounded subsets M_1 and M_2 of E, we have

$$M_1 \subset M_2 \implies \psi(M_1) \leq \psi(M_2).$$

(c) Nonsingular: $\psi(M \cup \{x\}) = \psi(M)$, for every bounded subset M of E and for all $x \in E$.

We are going to present two examples of the most important measures of noncompactness and investigate their properties, including properties (a)–(c). For more details, we refer the reader to [1, 5–7]

1.5.1 The general setting

Definition 1.5.2 *Let (E,d) be a metric space and $A \subset E$ a bounded subset. Define the sets:*
(a)

$$K(A) = \{D > 0 : \exists N \in \mathbb{N}, \ \exists (A_i)_{i=1}^N \subset E \text{ such that } A \subset \bigcup_{i=1}^N A_i \text{ with}$$

$$\operatorname{diam}(A_i) \leq D, \ \forall 1 \leq i \leq N\}.$$

(b)

$$H(A) = \{r > 0 : \exists N \in \mathbb{N}, \ \exists \{x_1, x_2, \cdots, x_N\} \subset E, \text{ such that}$$

$$A \subset \bigcup_{i=1}^N B(x_i, r)\}.$$

The following inclusions are immediate.

Lemma 1.5.3
$$2H(A) \subset K(A) \subset H(A).$$

Proof 1.5.4 *(a) Given $D \in K(A)$, we have*

$$\exists N \in \mathbb{N}, \ \exists (A_i)_{i=1}^{N} \subset E \text{ such that } A \subset \bigcup_{i=1}^{N} A_i \text{ with diam}(A_i) \leq D,$$

$$\forall 1 \leq i \leq N.$$

Since A_i is bounded for all $1 \leq i \leq N$, then

$$A_i \subset B(x_i, D), \text{ with } x_i \in A_i, \ \forall 1 \leq i \leq N.$$

Hence

$$\bigcup_{i=1}^{N} A_i \subset \bigcup_{i=1}^{N} B(x_i, D) \Rightarrow A \subset \bigcup_{i=1}^{N} B(x_i, D),$$

$$D \in H(A), \text{ and } K(A) \subset H(A).$$

(b) For $r \in H(A)$, there exists $N \in \mathbb{N}$ and $\{x_1, x_2, \cdots, x_N\} \subset E$ such that
$$A \subset \bigcup_{i=1}^{N} B(x_i, r) \text{ where diam}(B(x_i, r)) \leq 2r, \ \forall 1 \leq i \leq N. \text{ Therefore,}$$

$$2r \in K(A) \Rightarrow 2H(A) \subset K(A).$$

Definition 1.5.5 *The Kuratowski measure of noncompactness (KMNC) and the Hausdorff measure of noncompactness (HMNC) are respectively defined by:*

$$\alpha(A) := \inf(K(A)) \text{ and } \chi(A) := \inf(H(A)).$$

Immediately, we have

Lemma 1.5.6

(a) $\chi(A) \leq \alpha(A) \leq 2\chi(A)$.

(b) $0 \leq \chi(A) \leq \alpha(A) \leq \text{diam}(A)$.

Proof 1.5.7

(a) From Lemma 1.5.3, $2H(A) \subset K(A) \subset H(A)$. Hence

$$\inf(H(A)) \leq \inf(K(A)) \leq 2\inf(H(A)) \ \text{ and } \ \chi(A) \leq \alpha(A) \leq 2\chi(A).$$

(b) Taking $N = 1$ and $D = \text{diam}(A)$, we get $A \subset A$. Then

$$\text{diam}(A) \in K(A) \ \Rightarrow \ \inf(K(A)) \leq \text{diam}(A),$$

$$\Rightarrow \ \alpha(A) \leq \text{diam}(A),$$

$$\Rightarrow \ \chi(A) \leq \alpha(A) \leq \text{diam}(A).$$

Proposition 1.5.8

$$\alpha(A) = 0 \Leftrightarrow \chi(A) = 0 \Leftrightarrow A \text{ is totally bounded.}$$

Proof 1.5.9 *The first equivalence follows from Lemma 1.5.6. As for the second one, we have*

$$\chi(A) = 0 \ \Leftrightarrow \ \inf\{r > 0 : A \text{ has an } r - net\} = 0$$

$$\Leftrightarrow \ A \text{ has an } r - net, \ \forall r > 0$$

$$\Leftrightarrow \ A \text{ totally bounded.}$$

Recall that a subset

> *1. A is totally bounded if and only if A has an ε-net, for all $\varepsilon > 0$.*
>
> *2. A has an ε-net if there exists $N \in \mathbb{N}$ such that $A \subset \bigcup_{i=1}^{N} B(x_i, \varepsilon)$ and $\{x_1, x_2, \cdots, x_N\} \subset E$.*
>
> *3. $H(A) = \{r > 0 : A \text{ has an } r\text{-net}\}$.*

Lemma 1.5.10 *Let $A \subset B \subset E$ be bounded subsets and $\gamma = \alpha$ or $\gamma = \chi$. Then*

$$\gamma(A) \leq \gamma(B).$$

Proof 1.5.11 *We only check for $\gamma = \alpha$. Given $D \in K(B)$, we have*

$$\exists N \in \mathbb{N}, \; \exists (A_i)_{i=1}^{N} \subset E \text{ such that } B \subset \bigcup_{i=1}^{N} A_i$$

with $\operatorname{diam}(A_i) \leq D, \; \forall 1 \leq i \leq N.$

Since $A \subset B \subset \bigcup_{i=1}^{N} A_i$, we have

$$D \in K(A) \;\;\Rightarrow\;\; K(B) \subset K(A),$$

$$\Rightarrow\;\; \inf K(A) \leq \inf K(B),$$

$$\Rightarrow\;\; \alpha(A) \leq \alpha(B).$$

Lemma 1.5.12 *The sets A and its closure $\overline{A}$ have same measures.*

Proof 1.5.13

(a) Since $A \subset \overline{A}$, Lemma 1.5.10 implies $\alpha(A) \leq \alpha(\overline{A})$. Given $D \in K(A)$, there exist $N \in \mathbb{N}$ and $(A_i)_{i=1}^{N} \subset E$ such that

$$A \subset \bigcup_{i=1}^{N} A_i, \text{ where } \operatorname{diam}(A_i) \leq D, \; \forall 1 \leq i \leq N.$$

Hence

$$\overline{A} \subset \overline{\bigcup_{i=1}^{N} A_i} = \bigcup_{i=1}^{N} \overline{A_i},$$

where

$$\operatorname{diam}(\overline{A_i}) = \operatorname{diam}(A_i) \leq D, \; \forall 1 \leq i \leq N.$$

Finally

$$D \in K(\overline{A}), \; K(A) \subset K(\overline{A}), \; \alpha(\overline{A}) \leq \alpha(A), \text{ and } \alpha(A) = \alpha(\overline{A}).$$

(b) $\chi(A) = \chi(\overline{A})$. Indeed

$$A \subset \overline{A} \Rightarrow \chi(A) \leq \chi(\overline{A}). \tag{1.24}$$

For $r \in H(A)$, we have $A \subset \bigcup\limits_{i=1}^{N} B(x_i, r)$, where $\{x_1, x_2, \cdots, x_N\} \subset E$.

Hence,

$$\overline{A} \subset \overline{\bigcup\limits_{i=1}^{N} B(x_i, r)}$$

$$= \bigcup\limits_{i=1}^{N} \overline{B(x_i, r)}$$

$$= \bigcup\limits_{i=1}^{N} B(x_i, r)$$

$$\subset \bigcup\limits_{i=1}^{N} B(x_i, r+\varepsilon), \quad \forall \ \varepsilon > 0,$$

which yields

$$r + \varepsilon \in H(\overline{A}), \forall \varepsilon > 0 \Rightarrow \chi(\overline{A}) \leq r + \varepsilon, \ \forall \varepsilon > 0.$$

Therefore, $\chi(\overline{A}) \leq r$. As a consequence $\chi(\overline{A}) \leq \chi(A)$ and $\chi(A) = \chi(\overline{A})$, as claimed.

Corollary 1.5.14 *Let $A \subset E$ be a bounded subset. We have*

$$\alpha(A) = 0 \Leftrightarrow \chi(A) = 0 \Leftrightarrow A \text{ is relatively compact.}$$

Proof 1.5.15 *According to Proposition 1.5.8, if one MNC is zero, then $\overline{A}$ is totally bounded. Since $\overline{A}$ is a closed subset of the complete metric space E, then $\overline{A}$ is compact. The converse implication is clear.*

Remark 1.5.16 *Let $A \subset B \subset E$ be two bounded subsets of the metric space E. Then if B is relatively compact, then A is relatively compact. Moreover, if A is relatively compact subset of E, then $0 = \alpha(A) \leq \alpha(B)$, that is the farther B is from A, the larger is its measure. This justifies why α and χ are called measures of noncompactness (MNCs).*

Proposition 1.5.17 *Let A and B be bounded subsets of a metric space* (E,d) *and* $\gamma = \alpha$ *or* $\gamma = \chi$. *Then*
(1) $\gamma(A \cup B) = \max(\gamma(A), \gamma(B))$.
(2) $\gamma(A \cap B) \leq \min(\gamma(A), \gamma(B))$.

Proof 1.5.18

(1) $\gamma = \alpha$ *(a) We have* $A \subset A \cup B$ *and* $B \subset A \cup B$. *Then*

$$\alpha(A) \leq \alpha(A \cup B) \text{ and } \alpha(B) \leq \alpha(A \cup B).$$

Hence,

$$\max(\alpha(A), \alpha(B)) \leq \alpha(A \cup B). \tag{1.25}$$

By the infimum property, we have

$$\forall \varepsilon > 0, \ \exists D_\varepsilon, \ \exists N \in \mathbb{N}, \ \exists (A_i)_{i=1}^N \text{ such that } A \subset \bigcup_{i=1}^N A_i,$$

where

$$\text{diam}(A_i) \leq D_\varepsilon, \forall i \in [1, N] \text{ and } D_\varepsilon < \alpha(A) + \varepsilon \leq \max(\alpha(A), \alpha(B)) + \varepsilon.$$

Also there exist $D'_\varepsilon, M \in \mathbb{N}, (B_j)_{j=1}^M$ *such that* $B \subset \bigcup_{j=1}^M B_j$, $\text{diam}(B_j) \leq$
$D'_\varepsilon, \forall j \in [1, M]$, *and* $D'_\varepsilon < \alpha(B) + \varepsilon \leq \max(\alpha(A), \alpha(B)) + \varepsilon$. *Then*

$$A \cup B \subset \left(\bigcup_{i=1}^N A_i \right) \cup \left(\bigcup_{j=1}^M B_j \right)$$

$$= \bigcup_{K=1}^{N+M} C_k,$$

where

$$C_k = \begin{cases} A_i, & \forall k \in [1, N]; \\ B_j, & \forall k \in [N+1, M] \end{cases}$$

and

$$\text{diam}(C_k) < \max(\alpha(A), \alpha(B)) + \varepsilon, \forall \varepsilon > 0.$$

Then

$$\max\left(\alpha(A),\alpha(B)\right)+\varepsilon\in K(A\cup B),\ \forall\varepsilon>0$$

and

$$\alpha(A\cup B)\leq\max\left(\alpha(A),\alpha(B)\right)+\varepsilon,\ \forall\varepsilon>0.$$

Then

$$\alpha(A\cup B)\leq\max\left(\alpha(A),\alpha(B)\right).\tag{1.26}$$

From (1.25) and (1.26), we deduce that

$$\alpha(A\cup B)=\max\left(\alpha(A),\alpha(B)\right).$$

(b) Since $A\cap B\subset A$ and $A\cap B\subset B$, then

$$\chi(A\cap B)\leq\chi(A)\ \text{and}\ \chi(A\cap B)\leq\chi(B)\Rightarrow\chi(A\cap B)\leq\min\left(\chi(A),\chi(B)\right).$$

(2) $\gamma=\chi$ (a) We have

$$\max\left(\chi(A),\chi(B)\right)\leq\chi(A\cup B).\tag{1.27}$$

Let $d=\max(\chi(A),\chi(B))$. Again by the infimum property, we find

$$\forall\varepsilon>0,\ \exists r_\varepsilon>0,\ \exists\{x_1,x_2,\cdots,x_N\}\subset E\ \text{such that}\ A\subset\bigcup_{i=1}^{N}B_{r_\varepsilon}(x_i),$$

$$\exists r'_\varepsilon,\ \exists\{x_1,x_2,\cdots,x_M\}\subset E\ \text{such that}\ B\subset\bigcup_{j=1}^{M}B_{r'_\varepsilon}(x_j),$$

where

$$r_\varepsilon<\chi(A)+\varepsilon\leq d+\varepsilon\ \text{and}\ r'_\varepsilon<\chi(B)+\varepsilon\leq d+\varepsilon.$$

Hence $A\subset\bigcup_{i=1}^{N}B_{d+\varepsilon}(x_i)$ and $B\subset\bigcup_{j=1}^{M}B_{d+\varepsilon}(x_j)$. As a consequence

$$A\cup B\subset\bigcup_{k=1}^{N+M}B_{d+\varepsilon}(z_k),\ \text{where}\ z_k=\begin{cases}x_i,&\forall k\in[1,N];\\x_j,&\forall k\in[N+1,M].\end{cases}$$

Then

$$\chi(A\cup B)\leq d+\varepsilon,\ \forall\varepsilon>0$$

and

$$\chi(A\cup B)\leq d=\max\left(\chi(A),\chi(B)\right).\tag{1.28}$$

From 1.27 and 1.28, we infer $\chi(A\cup B)=\max\left(\chi(A),\chi(B)\right)$.

(b) Since $A \cap B \subset A$ and $A \cap B \subset B$, then

$$\alpha(A \cap B) \le \alpha(A) \text{ and } \alpha(A \cap B) \le \alpha(B).$$

$$\alpha(A \cap B) \le \min(\alpha(A), \alpha(B)).$$

Definition 1.5.19 *The r-neighborhood of a set A is defined as*

$$\mathcal{N}_r(A) = \{x \in E : d(x,A) < r\}.$$

Lemma 1.5.20 *We have*

$$\text{diam}(\mathcal{N}_r(A)) \le \text{diam}(A) + 2r.$$

Proof 1.5.21 *Let $x, y \in \mathcal{N}_r(A)$. Then $d(x,A) < r$ and $d(y,A) < r$. By the infimum property, we have*

$$\forall \varepsilon > 0, \exists x_0 \in A, \exists y_0 \in A : d(x,x_0) < r + \varepsilon, d(y,y_0) < r + \varepsilon.$$

So for all $x, y \in \mathcal{N}_r(A)$, we have

$$\begin{aligned} d(x,y) &\le d(x,x_0) + d(x_0,y_0) + d(y_0,y), \\ &< r + \varepsilon + \text{diam}(A) + r + \varepsilon \\ &= \text{diam}(A) + 2r + 2\varepsilon. \end{aligned}$$

Then

$$\begin{aligned} \text{diam}(\mathcal{N}_r(A)) &= \sup_{x,y \in \mathcal{N}_r(A)} d(x,y) \\ &\le \text{diam}(A) + 2r + 2\varepsilon, \forall \varepsilon > 0. \end{aligned}$$

Therefore,

$$\text{diam}(\mathcal{N}_r(A)) \le \text{diam}(A) + 2r.$$

Lemma 1.5.22 *Let $A = B(x_0, r_0)$. Then*

$$\mathcal{N}_r(A) \subset B(x_0, r_0 + r).$$

Proof 1.5.23 *Let $x \in \mathcal{N}_r(A)$. Then*

$$d(x,A) < r \Leftrightarrow \inf_{y \in A}(x,y) < r.$$

Since $y \in A$, then for all $y \in A$,

$$d(x,x_0) \leq d(x,y) + d(y,x_0) \leq d(x,y) + r_0.$$

Hence

$$d(x,x_0) \leq \inf_{y \in A}(x,y) < r_0,$$

$$\Rightarrow d(x,x_0) \leq d(x,A) + r_0,$$

$$\Rightarrow d(x,x_0) < r + r_0,$$

$$\Rightarrow x \in B(x_0, r_0 + r).$$

Lemma 1.5.24 *We have*

$$A \subset \bigcup_{i \in I} A_i \Rightarrow \mathcal{N}_r(A) \subset \bigcup_{i \in I} \mathcal{N}_r(A_i).$$

Proof 1.5.25 *Let $x \in \mathcal{N}_r(A)$. Then $d(x,A) < r$ and $A \subset B \Rightarrow d(x,B) \leq d(x,A)$. Hence*

$$d := d(x, \bigcup_{i \in I} A_i) < r,$$

$$\Rightarrow \quad \exists i_0 \in I : d(x, A_{i_0}) < r,$$

$$\Rightarrow \quad x \in \mathcal{N}_r(A_{i_0}) \Rightarrow x \in \bigcup_{i \in I} \mathcal{N}_r(A_i).$$

If by contradiction $\forall i \in I, d(x,A_i) \geq r$, then by the infimum property

$$\forall \varepsilon > 0, \ \exists y_\varepsilon \in \bigcup_{i \in I} A_i : d \leq d(x,y_\varepsilon) < d + \varepsilon.$$

Hence

$$\forall \varepsilon > 0, \ \exists i_0 \in I : y_\varepsilon \in A_{i_0},$$

where $r \le d(x, A_{i_0}) \le d(x, y_\varepsilon) < d + \varepsilon$. Finally

$$r < d + \varepsilon, \forall \varepsilon > 0 \Rightarrow r < d,$$

a contradiction.

Proposition 1.5.26

$$\chi(\mathcal{N}_r(A)) \le \chi(A) + r \text{ and } \alpha(\mathcal{N}_r(A)) \le \alpha(A) + 2r.$$

Proof 1.5.27

(1) Given $\varepsilon > 0$, we have

$$\exists r_0 > 0 : A \subset \bigcup_{i=1}^{N} B(x_i, r_0) \text{ with } \{x_1, x_2, \cdots, x_N\} \subset E.$$

By the infimum property,

$$\chi(A) \le r_0 < \chi(A) + \varepsilon.$$

Lemma 1.5.24 yields

$$\mathcal{N}_r(A) \subset \bigcup_{i=1}^{N} \mathcal{N}_r(B(x_i, r_0))$$

while Lemma 1.5.22 entails

$$\mathcal{N}_r(B(x_0, r_0)) \subset B(x_0, r_0 + r), \forall i \in [1, N].$$

Therefore, $\mathcal{N}_r(A) \subset \bigcup_{i=1}^{N} B(x_0, r_0 + r)$. Hence

$$r_0 + r \in H(\mathcal{N}_r(A)) \Rightarrow \chi(\mathcal{N}_r(A)) \le r_0 + r < \chi(A) + \varepsilon + r, \forall \varepsilon > 0.$$

We conclude that

$$\chi(\mathcal{N}_r(A)) \le \chi(A) + r.$$

(2) Given $\varepsilon > 0$, there exist $D_\varepsilon > 0$, $N \in \mathbb{N}$, $(A_i)_{i=1}^N \subset E$ such that

$$A \subset \bigcup_{i=1}^N A_i,$$

where

$$\mathrm{diam}\,(A_i) \leq D_\varepsilon, \; \forall i \in [1,N].$$

Then

$$\alpha(A) \leq D_\varepsilon < \alpha(A) + \varepsilon.$$

By Lemma 1.5.24,

$$\mathcal{N}_r(A) \subset \bigcup_{i=1}^N \mathcal{N}_r(A_i)$$

and by Lemma 1.5.20,

$$\mathrm{diam}\,(\mathcal{N}_r(A_i)) \leq \mathrm{diam}\,(A_i) + 2r \leq D_\varepsilon + 2r < \alpha(A) + \varepsilon + 2r, \; \forall \varepsilon > 0.$$

It follows that

$$\forall \varepsilon > 0, \; \alpha(A) + 2r + \varepsilon \in K(\mathcal{N}_r(A)).$$

Then

$$\alpha(\mathcal{N}_r(A)) \leq \alpha(A) + 2r + \varepsilon, \; \forall \varepsilon > 0.$$

As a consequence

$$\alpha(\mathcal{N}_r(A)) \leq \alpha(A) + 2r.$$

Proposition 1.5.28 *[Cantor's Intersection Theorem] Let (E,d) be a complete space and $(F_n)_n$ a decreasing sequence of closed nonempty subsets of E such that $\lim_{n \to \infty} \alpha(F_n) = 0$. Then $F_\infty := \bigcap_n F_n$ is nonempty and compact.*

Proof 1.5.29

(a) We show that $F_\infty \neq \phi$. Let $(x_n)_n \subset E$ be a sequence such that $x_n \in F_n$, $\forall n$ and $\mathcal{A}_n = \{x_n, x_{n+1}, \cdots\}$. Since $\mathcal{A}_n \subset F_n$, $\forall n$, then $\mathcal{A}_1 \subset \{x_1, \ldots, x_n\} \bigcup F_n$. Hence

$$\alpha(\mathcal{A}_1) \leq \alpha(F_n), \; \forall n.$$

As $n \to \infty$, we get $\alpha(\mathscr{A}_1) = 0$. Hence $\mathscr{A}_1$ is relatively compact. This implies that the sequence $(x_n)_n$ has a limit point $x \in \bigcap_n \overline{A}_n$. Since

$\bigcap_n \overline{A}_n \subset \bigcap_n \overline{F}_n$, then $x \in \bigcap_n F_n = F_\infty$.

(b) We show that F_∞ is compact. Since $F_\infty \subset F_n \, \forall n$, then

$$0 \leq \alpha(F_\infty) \leq \alpha(F_n), \, \forall n.$$

This proves $\alpha(F_\infty) = 0$, as $n \to \infty$, that is F_∞ is compact.

1.5.2 Case of normed spaces

Lemma 1.5.30 *Let $(E, \|\cdot\|)$ be a normed space, A, B two bounded subsets of E, and $\lambda \in \mathbb{R}$. Then,*

(a) $\operatorname{diam}(A + B) \leq \operatorname{diam}(A) + \operatorname{diam}(B)$.

(b) $\operatorname{diam}(\lambda A) = |\lambda| \operatorname{diam}(A)$.

Proof 1.5.31

(a) *Let $x, y \in A + B$. Then there exist $a_1, a_2 \in A$, $b_1, b_2 \in B$ such that $x = a_1 + b_1$ and $y = a_2 + b_2$. Hence*

$$
\begin{aligned}
\|x - y\| &= \|a_1 + b_1 - a_2 - b_2\| \\[2mm]
&\leq \|a_1 - a_2\| + \|b_1 - b_2\| \\[2mm]
&\leq \operatorname{diam}(A) + \operatorname{diam}(B),
\end{aligned}
$$

which entails

$$\operatorname{diam}(A + B) \leq \operatorname{diam}(A) + \operatorname{diam}(B).$$

(b) *Let $x, y \in \lambda A$ with $\lambda \neq 0$. Then*

$$\exists a_1, a_2 \in A : x = \lambda a_1 \text{ and } y = \lambda a_2.$$

Hence

$$\|x - y\| = \|\lambda a_1 - \lambda a_2\| = |\lambda| \|a_1 - a_2\| \leq |\lambda| \operatorname{diam}(A),$$

which yields

$$\text{diam}(\lambda A) \leq |\lambda| \text{diam}(A).$$

Conversely, let $a_1, a_2 \in A$. Then $\lambda a_1, \lambda a_2 \in \lambda A$.

$$\|a_1 - a_2\| = \|\frac{\lambda}{\lambda} a_1 - \frac{\lambda}{\lambda} a_2\| = \frac{1}{|\lambda|} \|\lambda a_1 - \lambda a_2\| \leq \frac{1}{|\lambda|} \text{diam}(\lambda A).$$

Hence

$$\text{diam}(A) \leq \frac{1}{|\lambda|} \text{diam}(\lambda A).$$

Proposition 1.5.32 *Let γ be α or χ an MNC in a normed space $(E, \|\cdot\|)$. Then for all bounded subsets $A, C \subset E$ and for all $\lambda \in \mathbb{R}^*$,*

(a) $\gamma(A+C) \leq \gamma(A) + \gamma(C)$ (sub-additivity),

(b) $\gamma(A + \{x\}) = \gamma(A)$ (invariance under shifting),

(c) $\gamma(\lambda A) = |\lambda| \gamma(A)$ (homogeneity).

Proof 1.5.33

(a) $\alpha(A+C) \leq \alpha(A) + \alpha(C)$. Let $(A_i)_{i=1}^N$ and $(C_j)_{j=1}^M$ cover A and C respectively. Then $(A_i)_{i=1}^N + (C_j)_{j=1}^M$ cover $A+C$. For $\varepsilon > 0$, there exist

$$D_1, D_2 > 0, (A_i)_{i=1}^N, (C_j)_{j=1}^M : A \subset \bigcup_{i=1}^N A_i, C \subset \bigcup_{j=1}^M C_j,$$

with $\text{diam}(A_i) \leq D_1$, $\forall i \in [1,N]$, $\text{diam}(C_i) \leq D_2$, $\forall j \in [1,N]$, and

$$\alpha(A) \leq D_1 < \alpha(A) + \frac{\varepsilon}{2} \text{ and } \alpha(C) \leq D_2 < \alpha(C) + \frac{\varepsilon}{2}.$$

By Lemma 1.5.30, we have

$$\begin{aligned}
\text{diam}(A_i + C_j) &\leq \text{diam}(A_i) + \text{diam}(C_j), \forall i \in [1,N], j \in [1,j] \\
&\leq D_1 + D_2, \\
&< \alpha(A) + \frac{\varepsilon}{2} + \alpha(C) + \frac{\varepsilon}{2}, \forall \varepsilon > 0 \\
&= \alpha(A) + \alpha(C) + \varepsilon, \forall \varepsilon > 0.
\end{aligned}$$

Hence

$$\alpha(A+C) < \alpha(A) + \alpha(C) + \varepsilon, \ \forall \varepsilon > 0,$$

that is $\alpha(A+C) \le \alpha(A) + \alpha(C)$. *We claim that* $\chi(A+C) \le \chi(A) + \chi(C)$. *For some* $\varepsilon > 0$, *there exist* $r_1, r_2 > 0$, $\{a_1, a_2 \ldots, a_N\} \subset E \{c_1, c_2 \ldots, c_j\} \subset E$ *such that*

$$A \subset \bigcup_{i=1}^{N} B_{r_1}(a_i) \ \text{and} \ C \subset \bigcup_{j=1}^{M} B_{r_2}(c_j),$$

such that

$$\chi(A) \le r_1 < \chi(A) + \frac{\varepsilon}{2} \ \text{and} \ \chi(C) \le r_1 < \chi(C) + \frac{\varepsilon}{2}.$$

Since $(B_{r_1}(a_i))_{i=1}^{N}$ *cover A and* $(B_{r_2}(c_j))_{j=1}^{M}$ *cover C, then*

$$A + C \subset \bigcup_{i=1}^{N} B_{r_1}(a_i) + \bigcup_{j=1}^{M} B_{r_2}(c_j).$$

Let us show that

$$\bigcup_{i=1}^{N} B_{r_1}(a_i) + \bigcup_{j=1}^{M} B_{r_2}(c_j) \subset \bigcup_{k=1}^{N+M} B_{r_1+r_2}(a_k + c_k),$$

where $a_k = 0$, $\forall k > N$ *and* $c_j = 0$, $\forall k \le N$. *Let* $x_0 \in \bigcup_{i=1}^{N} B_{r_1}(a_i) + \bigcup_{j=1}^{M} B_{r_2}(c_j)$. *Then there exist* $i_0 \in [1, N]$, $j_0 \in [1, M]$ *such that* $x_0 \in B_{r_1}(a_{i_0}) + B_{r_2}(c_{j_0})$. *Hence*

$$\exists t_1 \in B_{r_1}(a_{i_0}) \ \text{and} \ \exists t_2 \in B_{r_2}(c_{j_0}) : x_0 = t_1 + t_2,$$

which implies

$$\|t_1 - a_{i_0}\| < r_1 \ \text{and} \ \|t_2 - c_{j_0}\| < r_2.$$

Then

$$\|x_0 - (a_{i_0} + c_{j_0})\| = \|t_1 + t_2 - a_{i_0} - c_{j_0}\|,$$

$$\leq \ \|t_1 - a_{i_0}\| + \|t_2 - c_{j_0}\|,$$

$$< \ r_1 + r_2.$$

Thus,

$$x_0 \in B_{r_1+r_2}(a_{i_0} + c_{j_0}) \Rightarrow x_0 \in \bigcup_{k=1}^{N+M} B_{r_1+r_2}(a_k + c_k).$$

We obtain

$$A + C \subset \bigcup_{k=1}^{N+M} B_{r_1+r_2}(a_k + c_k)$$

and then

$$\chi(A+C) \ \leq \ r_1 + r_2$$

$$< \ \chi(A) + \frac{\varepsilon}{2} + \chi(C) + \frac{\varepsilon}{2}, \ \forall \varepsilon > 0,$$

which entails

$$\chi(A+C) \leq \chi(A) + \chi(C).$$

(b) Part (a) yields $\gamma(A + \{x\}) \leq \gamma(A) + \gamma(\{x\}) = \gamma(A)$. Note that

$$\gamma(\{x\}) \leq \mathrm{diam}\,(\{x\}) = 0 \Rightarrow \gamma(\{x\}) = 0.$$

Hence

$$A = A + \{x\} + \{-x\} \Rightarrow \gamma(A) \ = \ \gamma(A + \{x\} + \{-x\})$$

$$\leq \ \gamma(A + \{x\}) + \gamma(\{-x\})$$

$$\Rightarrow \ \gamma(A) \leq \gamma(A + \{x\}).$$

Hence $\gamma(A + \{x\}) = \gamma(A)$.

(c) Let $(A_i)_{i=1}^N$ cover A. Then $(\lambda A_i)_{i=1}^N$ cover λA. Indeed

$$\lambda A \subset \lambda \bigcup_{i=1}^N A_i = \bigcup_{i=1}^N \lambda A_i.$$

Let $\varepsilon > 0$, then $\exists D > 0$, $\exists (A_i)_{i=1}^N$, such that $A \subset \bigcup_{i=1}^N A_i$ with $\mathrm{diam}\,(A_i) \leq D$, $\forall i \in [1,N]$, such that $\alpha(A) \leq D < \alpha(A) + \varepsilon$. Then, for all $\varepsilon > 0$

$$\mathrm{diam}\,(\lambda A_i) = |\lambda|\,\mathrm{diam}\,(A_i) \leq |\lambda|\,D < |\lambda|\,(\alpha(A) + \varepsilon) = |\lambda|\,\alpha(A) + |\lambda|\,\varepsilon.$$

Then, for all $\varepsilon > 0$

$$\alpha(\lambda A) \leq |\lambda|\,\alpha(A) + |\lambda|\,\varepsilon,$$

which implies

$$\alpha(\lambda A) \leq |\lambda|\,\alpha(A). \tag{1.29}$$

For any $\varepsilon > 0$, there exists $D' > 0$, and $(\lambda A_i)_{i=1}^M$ such that $\lambda A \subset \bigcup_{i=1}^M \lambda A_i$ with $\mathrm{diam}\,(\lambda A_i) \leq D'$, $\forall i \in [1,M]$, and

$$\alpha(\lambda A) \ \leq \ D'$$

$$< \ \alpha(\lambda A) + \varepsilon.$$

Then

$$\mathrm{diam}\,(A_i) \ = \ \frac{\mathrm{diam}\,(\lambda A_i)}{|\lambda|}$$

$$\leq \ \frac{D'}{|\lambda|}$$

$$< \ \frac{\alpha(\lambda A)}{|\lambda|} + \frac{\varepsilon}{|\lambda|}, \ \forall \varepsilon > 0.$$

Thus,

$$\alpha(A) \ \leq \ \frac{\alpha(\lambda A)}{|\lambda|} + \frac{\varepsilon}{|\lambda|} \Rightarrow$$

$$\alpha(A) \;\leq\; \frac{\alpha(\lambda A)}{|\lambda|}, \; \forall \varepsilon > 0.$$

Hence,

$$|\lambda|\,\alpha(A) \leq \alpha(\lambda A). \tag{1.30}$$

From (1.29) and (2.28), we deduce that

$$\alpha(\lambda A) = |\lambda|\,\alpha(A), \; \forall \lambda \in \mathbb{R}.$$

We claim that $\chi(\lambda A) = |\lambda|\,\chi(A)$. Firstly, we will show that

(a) $\displaystyle\bigcup_{i=1}^{N} \lambda\,B(a_i,r) = \bigcup_{i=1}^{N} B_{|\lambda|r}(\lambda\,a_i),$ *where $(a_i)_{i=1}^{N} \subset E$.*

*Let $x \in \displaystyle\bigcup_{i=1}^{N} \lambda\,B(a_i,r)$. Then there exist $i_0 \in [1,N] : x \in \lambda\,B(a_i,r)$. Hence
there exists $y \in B(a_i,r) : x = \lambda\,y$ and $\|y - a_{i_0}\| < r$. As a consequence*

$$
\begin{aligned}
\|x - \lambda\,a_{i_0}\| &= \|\lambda\,y - \lambda\,a_{i_0}\| \\[2mm]
&= |\lambda|\,\|y - a_{i_0}\| \\[2mm]
&< |\lambda|\,r.
\end{aligned}
$$

Hence,

$$
\begin{aligned}
x &\in B(\lambda\,a_{i_0}, |\lambda|\,r) \Rightarrow \\[2mm]
x &\in \bigcup_{i=1}^{N} B(\lambda\,a_i, |\lambda|\,r).
\end{aligned}
$$

*Conversely, let $x \in \displaystyle\bigcup_{i=1}^{N} B(\lambda\,a_i, |\lambda|\,r)$. Then there exists $i_0 \in [1,N]$ such
that $x \in B_{|\lambda|r}(\lambda\,a_{i_0})$. Hence,*

$$\|x - \lambda\,a_{i_0}\| < r|\lambda| \Rightarrow |\lambda|\,\left\|\frac{x}{\lambda} - a_{i_0}\right\| < r|\lambda|$$

$$\Rightarrow \quad \left\| \frac{x}{\lambda} - a_{i_0} \right\| < r$$

$$\Rightarrow \quad \frac{x}{\lambda} \in B(a_{i_0}, r) \Rightarrow x \in \lambda B(a_{i_0}, r).$$

(b) $r \in H(A) \Rightarrow |\lambda| r \in H(\lambda A).$

$$r \in H(A) \quad \Rightarrow \quad \exists N \in \mathbb{N}, \exists \{a_1, a_2, \cdots, a_N\} \subset E : A \subset \bigcup_{i=1}^{N} B(a_i, r)$$

$$\Rightarrow \quad \lambda A \subset \bigcup_{i=1}^{N} \lambda B(a_i, r) \subset \bigcup_{i=1}^{N} B_{|\lambda| r}(\lambda a_i), \; (by \; (a))$$

$$\Rightarrow \quad r|\lambda| \in H(\lambda A).$$

(c) If $r \in H(\lambda A)$, then

$$\exists N \in \mathbb{N}, \exists \{\lambda a_1, \lambda a_2, \cdots, \lambda a_N\} \subset E : \lambda A \subset \bigcup_{i=1}^{N} B(\lambda a_i, r)$$

$$\Rightarrow \quad A \subset \bigcup_{i=1}^{N} \frac{1}{\lambda} B(\lambda a_i, r) \subset \bigcup_{i=1}^{N} B_{\frac{r}{|\lambda|}}(a_i), \; (by \; (a))$$

$$\Rightarrow \quad \frac{r}{|\lambda|} \in H(A).$$

Secondly, let $\varepsilon > 0$. Then there exist $r_1, > 0$ and $\{a_1, a_2, \cdots, a_N\} \subset$
E such that $A \subset \bigcup_{i=1}^{N} B_{r_1}(a_i)$ and

$$\chi(A) \leq r_1 < \chi(A) + \varepsilon.$$

Using (b), we get

$$r_1 \in H(A) \quad \Rightarrow \quad |\lambda| r_1 \in H(\lambda A)$$

$$\Rightarrow \quad \chi(\lambda A)$$

$$\leq \quad |\lambda| r_1$$

$$< \quad |\lambda| \chi(A) + |\lambda| \varepsilon, \ \forall \varepsilon > 0.$$

$$\Rightarrow \quad \chi(\lambda A) \leq |\lambda| \chi(A).$$

Conversely, let $\varepsilon > 0$. Then there exist $r_2 > 0$, $\{\lambda a_1, \lambda a_2, \cdots, \lambda a_N\} \subset$
E such that $\lambda A \subset \bigcup\limits_{i=1}^{N} B_{r_1}(\lambda a_i)$ and

$$\chi(\lambda A) \quad \leq \quad r_2$$

$$< \quad \chi(\lambda A) + \varepsilon.$$

By using (c), we get

$$r_2 \in H(\lambda A) \quad \Rightarrow \quad \frac{r_2}{|\lambda|} \in H(A)$$

$$\Rightarrow \quad \chi(A)$$

$$\leq \quad \frac{r_2}{|\lambda|}$$

$$< \quad \frac{\chi(\lambda A) + \varepsilon}{|\lambda|}$$

$$< \quad \frac{\chi(\lambda A)}{|\lambda|} + \frac{\varepsilon}{|\lambda|}, \ \forall \varepsilon > 0.$$

$$\Rightarrow \quad |\lambda| \chi(A) < \chi(\lambda A).$$

Finally, $\chi(\lambda A) = |\lambda| \chi(A)$.

Definition 1.5.34 *The convex hull of A, denoted $\mathrm{conv}(A)$, is the small-est convex set that contains A (i.e. $\mathrm{conv}(A)$ is the intersection of all convex sets containing A).*

Proposition 1.5.35 *[1,5,7] (Invariance under the convex hull). Let A be a subset of a normed space and $\gamma = \alpha$ or χ. Then*

$$\gamma(A) = \gamma(\text{conv}(A)).$$

The proof follows from the following facts:
(a) $\text{diam}(A) = \text{diam}(\text{conv}(A))$,
(b) $A \subset \text{conv } A \Rightarrow \gamma(A) \le \gamma(\text{conv}(A))$,
and uses the following Carathéodory's characterization of the convex hull (see [8]):

$$\text{conv}(A) = \left\{ \sum_{i=1}^{n} \lambda_i a_i,\ a_i \in A,\ n \in \mathbb{N}^*,\ \lambda_i \ge 0,\ \sum_{i=1}^{n} \lambda_i = 1 \right\}.$$

Recall the classical result from functional analysis.

Lemma 1.5.36 *(Riesz Theorem) [40] A normed linear space is finite-dimensional if and only if the closed unit ball is compact.*

Proposition 1.5.37 *Let $B = B(0,1)$ be the unit ball in a normed space $(E, \|\cdot\|)$. Then*

$$\chi(B) = \begin{cases} 0, & if \quad dim(E) < \infty, \\ 1, & if \quad dim(E) = \infty. \end{cases}$$

Proof 1.5.38 *By Riesz Lemma, we have, since E is complete*

$$dim(E) < \infty \quad \Leftrightarrow \quad B(0,1) \text{ relatively compact},$$

$$\Leftrightarrow \quad \chi(B) = 0.$$

Assume now that $\text{diam}(E) = \infty$. *Then*

$$B(0,1) \subset B(0,1) \Rightarrow 1 \in H(B) \Rightarrow \chi(B) \le 1.$$

To prove that $\chi(B) = 1$, we proceed by contradiction and assume that $\chi(B) < 1$ and let $0 < \varepsilon < 1 - \chi(B)$. Then there exit $r > 0, N \in \mathbb{N}, \{x_1, x_2, \cdots, x_N\} \subset E$ such that $B \subset \bigcup_{i=1}^{N} B(x_i, r)$ and

$$\chi(B) \le r < \chi(B) + \varepsilon < 1.$$

Since $B \subset \bigcup_{i=1}^{N} B(x_i, r)$*, then*

$$\chi(B) \leq \max_{1 \leq i \leq N} \chi(B(x_i, r))$$

$$= \max_{1 \leq i \leq N} \chi(\{x_i\} + rB(0, 1))$$

$$\leq \chi(rB(0, 1))$$

$$= r\chi(B).$$

By Riesz Theorem, $\chi(B) \neq 0$*, which is a contradiction with* $1 > r$*, and so* $\chi(B) = 1$*.*

Corollary 1.5.39 *Let* $(E, \|\cdot\|)$ *be a normed space and* $B = B(x_0, r) \subset E$*. Then*

$$\chi(B) = \begin{cases} 0, & if \quad dim(E) < \infty, \\ r, & if \quad dim(E) = \infty. \end{cases}$$

Proof 1.5.40 *Since* $B(x_0, r) = \{x_0\} + rB(0, 1)$*, then*

$$\chi(B(x_0, r)) = \chi(\{x_0\} + rB(0, 1))$$

$$= \chi(rB(0, 1))$$

$$= r\chi(B(0, 1))$$

$$= \begin{cases} 0, & if \quad dim(E) < \infty, \\ r, & if \quad dim(E) = \infty. \end{cases}$$

Lemma 1.5.41 *Let* $(E, \|\cdot\|)$ *be a normed space and* S *the unit sphere. Then*

$$\mathrm{conv}\,(S) = \overline{B}(0, 1).$$

Proof 1.5.42 *Clearly* $S \subset B(0,1)$. *So* $B(0,1)$ *convex implies that, by definition,* $\mathrm{conv}\,(S) \subset \bar{B}(0,1)$. *Indeed, let* $x,y \in \bar{B}(0,1)$ *and* $\lambda \in [0,1]$. *Then*

$$\|\lambda y + (1-\lambda)x\| \leq \|\lambda y\| + \|(1-\lambda)x\|$$

$$= |\lambda|\,\|y\| + |1-\lambda|\,\|x\|$$

$$\leq \lambda + 1 - \lambda$$

$$= 1.$$

So $\lambda y + (1-\lambda)x \in B(0,1) \Rightarrow B(0,1)$ *is convex. Let us show that* $\bar{B}(0,1) \subset \mathrm{conv}\,(S)$. *Let* $x \in B(0,1)$ *and* $\lambda = \dfrac{1+\|x\|}{2}$. *Then* $\lambda \in (0,1]$ *and* $x = \lambda\dfrac{-x}{\|x\|} + (1-\lambda)\dfrac{x}{\|x\|}$, *where* $\dfrac{\pm x}{\|x\|} \in S$. *Hence,* $x \in \mathrm{conv}\,(S) \Rightarrow$ $\bar{B}(0,1) \subset \mathrm{conv}\,(S)$. *We conclude that*

$$\mathrm{conv}\,(S) = \bar{B}(0,1).$$

Remark 1.5.43 *By Lemma 1.5.41 and Proposition 1.5.35, we deduce that*

$$\gamma(S) = \gamma(\mathrm{conv}\,(S)) = \gamma(\bar{B}(0,1)) = \gamma(B(0,1)).$$

However, in order to compute $\alpha(B)$, we need

Lemma 1.5.44 *(Ljusternik-Schrinelman-Borsuk Theorem) [48] Let S be the sphere in a normed space E with* $dim(E) = n$. *Then, for every covering* $(A_i)_{i=1}^n$ *by closed sets, there exists at least one set* A_{i_0} *that contains two antipodal points of the sphere S.*

Proposition 1.5.45 *Let* $(E, \|\cdot\|)$ *be a normed space and* $B = B(0,1)$ *be the unit ball in E. Then*

$$\alpha(B) = \begin{cases} 0, & \text{if } dim(E) < \infty, \\ 2, & \text{if } dim(E) = \infty. \end{cases}$$

Proof 1.5.46 *By Riesz Lemma, we have*

$$dim(E) \leq \infty \quad \Rightarrow \quad B(0,1) \text{ is relatively compact,}$$
$$\Rightarrow \quad \alpha(B) = 0.$$

Assume that $dim(E) = \infty$. *Then by Proposition 1.5.37*

$$\chi(B) \leq \alpha(B) \leq 2\chi(B) \Rightarrow \alpha(B) \leq 2.$$

By contradiction, assume that $\alpha(S) = \alpha(B) < 2$ *(by Remark 1.5.43).*
Then $\forall \varepsilon \in (0, 2 - \alpha(S))$, $\exists D > 0$, $\exists (A_i)_{i=1}^{N}$ *(chosen closed):* $S \subset \bigcup_{i=1}^{N} A_i$
with $\mathrm{diam}(A_i) < \alpha(S) + \varepsilon < 2$, $\forall i \in [1, N]$. *Let* $L = \{x_1, x_2, \cdots, x_N\}$ *be*
a linearly independent subset of E *and* $E = [L]$. *Then* $\mathrm{diam}(E) = N$.
Let $S_N = \{x \in E : \|x\| = 1\}$. *Then* $S \cap S_N = S_N \subset \bigcup_{i=1}^{N} (S_N \cap A_i)$ *with*
$\mathrm{diam}(S_N \cap A_i) \leq \mathrm{diam}(A_i) < 2$, $\forall i \in [1, N]$. *This is a contradiction*
with Lemma 1.5.44. So $\alpha(B) = 2$.

1.6   Related Maps

1.6.1   *k*-set contractions

Let (E, d), (E', d') be two metric spaces, $f : E \to E'$ a continuous and
bounded map. $\mathscr{P}_B(E)$ will denote the family of all bounded subsets of
E.

Remark 1.6.1 *Lipschitz maps can be characterized by:*

$$f \text{ is } k - Lipschitz \iff \forall A \in \mathscr{P}_B(E), \ \mathrm{diam}(f(A)) \leq k\,\mathrm{diam}(A).$$

Indeed

$$f \text{ is } k - Lipschitz \Rightarrow \exists k \geq 0 : d'(f(x), f(y)) \leq k\,d(x, y), \ \forall x, y \in A.$$
$$\Rightarrow d'(f(x), f(y)) \leq k \sup_{x, y \in A} d(x, y) = k\,\mathrm{diam}(A).$$
$$\Rightarrow \mathrm{diam}(f(A)) \leq k\,\mathrm{diam}(A).$$

Conversely, let $A = \{x, y\} \in \mathscr{P}_B(E)$. Then

$$\operatorname{diam}(f(A)) \leq k\operatorname{diam}(A) \quad \Rightarrow \quad d'(f(x), f(y)) \leq kd(x, y)$$
$$\Rightarrow \quad f \text{ is } k - Lipschitz.$$

The observation in Remark 1.6.1 suggests to introduce k-set Lipschitz maps for some MNC γ:

Definition 1.6.2

(a) f is called a k-set contraction, if there exists $k \geq 0$, such that

$$\gamma(f(A)) \leq k\gamma(A), \; \forall A \in \mathscr{P}_B(E).$$

(b) f is called a 1-set contraction, if $k = 1$.

(c) f is called a strict k-set contraction if $0 \leq k < 1$.

(d) f is called a condensing, if $\forall A \in \mathscr{P}_B(E)$ with $\gamma(A) > 0$, we have $\gamma(f(A)) < \gamma(A)$.

Example 1.6.3 *Let E be a normed space and $\Omega \subset E$ a nonempty sub-set. The Minkowski functional $g_\Omega = g : X \longrightarrow [0, +\infty)$ is defined by*

$$g(x) \quad = \quad \inf\{\lambda > 0 : \lambda^{-1} x \in \overline{\Omega}\}$$

$$= \quad \inf\{\lambda > 0 : x \in \lambda\overline{\Omega}\}.$$

It satisfies (see [42, Lemma 4.2.5]) $0 \leq g(x) \leq 1$, if $x \in \Omega$ and $g(\lambda x) = |\lambda| g(x)$, for $\lambda \in \mathbb{R}$ and $x \in E$.

Let $C \subset E$ be a bounded closed convex subset containing the origin. Then the map r defined by $r(x) = \dfrac{x}{\max\{1, g(x)\}}$ is a retraction. Then $r(x) = x$ for every $x \in C$. In addition, since C is convex, $g(x) \geq 1$ for every $x \notin C$ and then $g(r(x)) = g\left(\dfrac{x}{g(x)}\right) = 1$, which yields $r(x) \in \partial C$, that is, $r(X \setminus C) \subset \partial C$. Finally for every bounded subset $A \subset X$, we have

$$r(A) \subset \overline{conv}(A \cup \{0\}) \implies \alpha(r(A)) \leq \alpha(A),$$

proving our claim.

Definition 1.6.4 *Let $f\colon D \subset E \to E'$ be a continuous map. f is said to be*

>   *1. compact if the set $f(D)$ is relatively compact.*
>
>   *2. completely continuous if it maps bounded sets into relatively compact sets.*

Remark 1.6.5

(a) If f is a strict k-set contraction, then f is condensing, hence a 1-set contraction, where f is continuous and E is complete. Indeed, let $A \in \mathscr{P}_B(E)$ with $\gamma(A) > 0$. Then, since f is a strict k-set contraction, there exists $0 \le k < 1$ such that $\gamma(f(A)) \le k\gamma(A) < \gamma(A)$, that is f is condensing.

(b) Suppose that f is condensing, and E is complete. Then
(i) if $\gamma(A) > 0$, then $\gamma(f(A)) \le \gamma(A) \Rightarrow f$ 1-set contraction,
(ii) if $\gamma(A) = 0$, then $\overline{A}$ is compact for E is complete. Hence $f(\overline{A})$ is compact for f is continuous. As a consequence $\gamma(f(A)) = 0 \le \gamma(A)$ for $\overline{f(A)} \subset f(\overline{A})$ and $\gamma(f(\overline{A})) = 0$).

(c) Compact maps are 0-set contractions, whenever (E', d') is complete. Indeed,

$$f \text{ is compact} \quad \Rightarrow \quad \overline{f(A)} \text{ compact}, \forall A \in \mathscr{P}_B(E)$$

$$\Rightarrow \quad \gamma(f(A))$$

$$= \quad \gamma(\overline{f(A)})$$

$$= \quad 0$$

$$\Rightarrow \quad f \text{ is } 0-\text{set contraction}.$$

Conversely,

$$f \text{ is } 0-\text{set contraction} \quad \Rightarrow \quad \gamma(\overline{f(A)})$$

$$= \gamma(f(A))$$

$$= 0, \forall A \in \mathscr{P}_B(E)$$

$$\Rightarrow \overline{f(A))} \text{ compact, (since } E' \text{ is complete)}$$

$$\Rightarrow f \text{ is compact.}$$

(d) Let $f : (E, \|\cdot\|_E) \to (F, \|\cdot\|_F)$ *be a k-set contraction, and* $g : (E, \|\cdot\|_E) \to (F, \|\cdot\|_F)$ *be a compact function. Then* $f + g$ *is a k-set contraction. Indeed let* $A \in \mathscr{P}_B(E)$. *We have*

$$\gamma((f+g)(A)) = \gamma(f(A) + g(A))$$

$$\leq \gamma(f(A)) + \gamma(g(A))$$

$$= \gamma(f(A)) + 0,$$

$$\leq k\gamma(A).$$

Hence $f + g$ *is a k-set contraction.*

Proposition 1.6.6 *Every k-Lipschitz map is a k-set contraction (with respect to the Kuratowski MNC).*

Proof 1.6.7 *Let* $A \in \mathscr{P}_B(E)$. *Then*

$$\forall \varepsilon > 0, \exists D_\varepsilon > 0, \exists N \in \mathbb{N}, \exists \{A_1, A_2, \cdots, A_N\} \subset E : A \subset \bigcup_{i=1}^{N} A_i,$$

with $\operatorname{diam}(A_i) \leq D_\varepsilon$, $\forall i \in [1, N]$ *such that* $\alpha(A) \leq D_\varepsilon < \alpha(A) + \varepsilon$. *We have*

$$f(A) \subset f\left(\bigcup_{i=1}^{N} A_i\right) \subset \bigcup_{i=1}^{N} f(A_i).$$

Then

$$\alpha(f(A)) \ \leq \ \alpha\left(\bigcup_{i=1}^{N} f(A_i)\right)$$

$$\leq \ \max_{1\leq i\leq N} \alpha(f(A_i))$$

$$\leq \ \max_{1\leq i\leq N} \text{diam}\,(f(A_i)).$$

By Remark 1.6.1, we have

$$\alpha(f(A)) \ \leq \ \max_{1\leq i\leq N} \text{diam}\,(f(A_i))$$

$$\leq \ \max_{1\leq i\leq N} k\,\text{diam}\,(A_i), \ (\textit{for } f \textit{ is Lipschitz}).$$

$$\leq \ kD_\varepsilon$$

$$< \ k(\alpha(A)+\varepsilon), \ \forall \varepsilon > 0$$

Hence $\alpha(f(A)) \leq k\,\alpha(A)$.

Remark 1.6.8 *In case of the Hausdorff MNC, we can show in a similar manner that every k-Lipschitz map is 2k-set contraction. Thus, according to Propositions 1.5.37 and 1.5.45, we can say that every k-Lipschitz map is $\gamma(B)k$-set contraction, where γ is either α or χ and B the unit ball.*

Proposition 1.6.9 *Let* $f : (E_1,d_1) \rightarrow (E_2,d_2)$ *and* $g : (E_2,d_2) \rightarrow (E_3,d_3)$ *be* k_1-set *and* k_2-set *contraction, respectively. Then* $g \circ f : (E_1,d_1) \rightarrow (E_3,d_3)$ *is a* $k_1.k_2$-set *contraction.*

Proof 1.6.10 *Let* $A \in \mathscr{P}_B(E_1)$. *Then*

$$\gamma(g(f(A))) \ \leq \ k_2\,\gamma(f(A)) \ (\textit{for } g \textit{ is } k_2\textit{-set contraction}).$$

$$\leq \ k_2.k_1\,\gamma(A) \ (\textit{for } f \textit{ is } k_1\textit{-set contraction}).$$

Proposition 1.6.11 *Let* $f : (E_1, \| \cdot \|_{E_1}) \to (E_2, \| \cdot \|_{E_2})$ *and* $g : (E_1, \| \cdot \|_{E_1}) \to (E_2, \| \cdot \|_{E_2})$ *be* k_1-*set and* k_2-*set contraction, respectively. Then* $f + g : (E_1, \| \cdot \|_{E_1}) \to (E_2, \| \cdot \|_{E_2})$ *is a* $(k_1 + k_2)$-*set contraction.*

Proof 1.6.12 *Given* $A \in \mathscr{P}_B(E_1)$, *we have*

$$\gamma(f(A) + g(A)) \leq \gamma(f(A)) + \gamma(g(A))$$

$$\leq k_1 \gamma(A) + k_2 \gamma(A)$$

$$= (k_1 + k_2) \gamma(A).$$

Next we prove that MNCs are Lipschitz maps with respect to a more general metric. First recall.

Definition 1.6.13 *Let* $A, B \subset E$ *be bounded subsets. The Hausdorff distance between* A *and* B *is defined by*

$$H_d(A, B) = \max\{\sup_{a \in A} d(a, B), \sup_{b \in B} d(b, A)\},$$

where $d(a, B) = \inf_{b \in B} d(a, b)$ *and* $d(b, A) = \inf_{a \in A} d(b, a)$.

Proposition 1.6.14

$$H_d(A, B) = \inf \{r > 0 : A \subset \mathscr{N}_r(B) \text{ and } B \subset \mathscr{N}_r(A)\}.$$

Proof 1.6.15 *Let* $D = H_d(A, B)$ *and* $F = \{r > 0 : A \subset \mathscr{N}_r(B) \text{ and } A \subset \mathscr{N}_r(A)\}$. *By definition of the Hausdorff distance, for all* $a \in A$, *we have*

$$d(a, B) \leq \sup_{a \in A} d(a, B) \leq D < D + \varepsilon, \ \forall \varepsilon > 0.$$

Then

$$a \in \mathscr{N}_{D+\varepsilon}(B), \ \forall a \in A \Rightarrow A \subset \mathscr{N}_{D+\varepsilon}(B).$$

Also $B \subset \mathscr{N}_{D+\varepsilon}(A)$. *Hence* $D + \varepsilon \in F$, $\forall \varepsilon > 0$ *and* $\inf(F) \leq D + \varepsilon$, $\forall \varepsilon > 0$. *Consequently*

$$\inf(F) \leq D. \tag{1.31}$$

Let $r \in F$. Then

$$\forall a \in A \quad \Rightarrow \quad a \in \mathcal{N}_r(B) \tag{1.32}$$

$$\Rightarrow \quad d(a,B) < r, \ \forall a \in A \tag{1.33}$$

$$\Rightarrow \quad \sup_{a \in A} d(a,B) < r, \ \forall r \in F \tag{1.34}$$

$$\Rightarrow \quad \sup_{a \in A} d(a,B) \le \inf(F). \tag{1.35}$$

In the same manner,

$$\sup_{b \in B} d(b,A) \le \inf(F). \tag{1.36}$$

Combining (1.32) and (1.36), we find

$$D = \max \left\{ \sup_{a \in A} d(a,B), \sup_{b \in B} d(b,A) \right\} \le \inf(F). \tag{1.37}$$

By (1.31) and (1.37), we conclude that $D = \inf(F)$.

It is easy to check that H_d is a distance over the family of all closed subsets of E, denoted $\mathscr{P}_{cl}(E)$. Our result on the Lipschitz character of the MNCs is

Proposition 1.6.16 *Let $A, B \subset E$ be such that A and B are bounded and $\gamma = \alpha$ or $\gamma = \chi$. Then*

$$|\gamma(A) - \gamma(B)| \le \gamma(B(0,1)).H_d(A,B).$$

Proof 1.6.17

(a) $\gamma = \alpha$. By Proposition 1.6.14, we have

$$H_d(A,B) = \inf \{ r > 0 : A \subset \mathcal{N}_r(B) \text{ and } B \subset \mathcal{N}_r(A) \}.$$

Let $r \in F := \{ r > 0 : A \subset \mathcal{N}_r(B) \text{ and } B \subset \mathcal{N}_r(A) \}$. Then $A \subset \mathcal{N}_r(B)$ and $B \subset \mathcal{N}_r(A)$, which implies that

$$\alpha(A) \le \alpha(\mathcal{N}_r(B)) \text{ and } \alpha(B) \le \alpha(\mathcal{N}_r(A)).$$

Using Proposition 1.5.26, we get

$$\alpha(A) \le \alpha(B) + 2r \ \text{and} \ \alpha(B) \le \alpha(\dot{A}) + 2r.$$

Then

$$|\alpha(A) - \alpha(B)| \le 2r, \forall \varepsilon \in F.$$

Hence,

$$|\alpha(A) - \alpha(B)| \ \le \ 2\inf(F)$$

$$= \ 2H_d(A, B).$$

(b) $\gamma = \chi$, From the proof of part (a), we get $\chi(A) \le \chi(\mathcal{N}_r(B))$ and $\chi(B) \le \chi(\mathcal{N}_r(A))$. Hence

$$\chi(A) \le \chi(B) + r \ \text{and} \ \chi(B) \le \chi(A) + r,$$

that is

$$|\chi(A) - \chi(B)| \ \le \ r, \forall r \in F$$

$$|\chi(A) - \chi(B)| \ \le \ \inf(F)$$

$$= \ H_d(A, B).$$

1.6.2 Nonexpansive and expansive maps

Definition 1.6.18 *Let (E, d) be a metric space and D be a subset of E. A map $T : D \to E$ is said to be expansive if there exists a constant $h > 1$ such that*

$$d(Tx, Ty) \ge hd(x, y), \quad \text{for all} \ x, y \in D.$$

Suppose E is a Banach space and $T : \mathscr{D}(T) \to E$, where $\mathscr{D}(T)$ refers to the domain of T and $\mathscr{R}(T)$ is its range.

Definition 1.6.19

(a) T is called nonexpansive if

$$\|Tx - Ty\| \leq \|x - y\|, \ \forall x, y \in \mathscr{D}(T).$$

(b) T is nonlinear expansive (or ψ-expansive) if

$$\|Tx - Ty\| \geq \psi(\|x - y\|), \ \forall x, y \in \mathscr{D}(T),$$

where $\psi : [0, \infty) \rightarrow [0, \infty)$ satisfies $\psi(0) = 0$ and $\psi(t) > t$, for all $t \geq 0$.

(c) T is called a nonlinear contraction (or a ϕ-contraction) if

$$\|Tx - Ty\| \leq \phi(\|x - y\|), \ \forall x, y \in \mathscr{D}(T),$$

where $\phi : [0, \infty) \rightarrow [0, \infty)$ satisfies $\phi(0) = 0$ and $\phi(t) < t$, for all $t \geq 0$.

Proposition 1.6.20

(a) If T is ϕ-contractive, then $(I - T)$ is ψ-expansive, invertible, and $(I - T)^{-1}$ is a continuous mapping.

(b) If T is ψ-expansive, then $(I - T)$ is $(\psi - 1)$-expansive, invertible, and $(I - T)^{-1}$ is a continuous mapping.

Proof 1.6.21

(a)

$$
\begin{aligned}
\|(I - T)x - (I - T)y\| \ &\geq \ \|x - y\| - \|Tx - Ty\| \\[2mm]
&\geq \ \|x - y\| - \phi(\|x - y\|) \\[2mm]
&= \ \psi(\|x - y\|),
\end{aligned}
$$

where $\psi(s) = s - \phi(s)$, for $s > 0$.

(b)

$$\|(I-T)x-(I-T)y\| \geq \|Tx-Ty\| - \|x-y\|$$

$$\geq \psi(\|x-y\|) - \|x-y\|$$

$$= \widetilde{\psi}(\|x-y\|),$$

where $\widetilde{\psi}(s) = \psi(s) - s$, for $s > 0$. In particular, if $\psi(s) = hs$ with $h > 1$, then $(I-T)^{-1}$ is $(h-1)^{-1}$-Lipschitz.

Remark 1.6.22 *Let $T : \mathbb{R}^2 \to \mathbb{R}^2$ be defined by $T(x,y) = (y,-x)$. Then $(I-T)$ is ψ-expansive with $\psi(t) = \sqrt{2}t$ but T is not a nonlinear contraction, showing that the converse in part (a) is not true.*

Clearly if $T : \Omega \to E$ is a ψ-expansive mapping, then T is injective and T^{-1} is uniformly continuous on the image set.

Remark 1.6.23 *The class of mappings T such that T is nonexpansive and $(I-T)$ is ψ-expansive encompasses the so-called separate contractions:*
(a) $\|Tx-Ty\| \leq \phi(\|x-y\|), \forall x,y \in \mathscr{D}(T)$,
(b) $\phi(r) \leq r - \psi(r), \forall r > 0$,
(c) $\psi(0) = 0$ and ψ is non-decreasing.
As a consequence, if T is a separate contraction, then $I-T$ is a homeomorphism on $\mathscr{D}(T)$.

1.7   Ascoli-Arzelà Theorem

1.7.1   Ascoli-Arzelà theorem: a first version

Let (X,τ) be a topological space, (Y,d) a metric space, and $\mathscr{C}(X,Y)$ denotes the space of continuous functions from X to Y. Let $\mathscr{H} \subset \mathscr{C}(X,Y)$.

Definition 1.7.1 *$\mathscr{H}$ is said to be equi-continuous at a point $x_0 \in X$ if*

$$\forall \varepsilon > 0, \exists U_\varepsilon \in \mathscr{V}(x_0), \forall x \in X,$$
$$(x \in U_\varepsilon \Longrightarrow f(x) \in B(f(x_0),\varepsilon)), \forall f \in \mathscr{H}.$$

$\mathscr{H}$ is equi-continuous if it is equi-continuous at every point $x_0 \in X$.

Remark 1.7.2 *When (X,d) is a compact metric space, then $\mathscr{H}$ is equi-continuous if and only if (it is uniformly equi-continuous):*

$$\forall \varepsilon > 0, \exists \delta, \forall x, y \in X,$$

$$(d(x,y) < \delta \Longrightarrow d(f(x), f(y)) < \varepsilon), \forall f \in \mathscr{H}.$$

Proof 1.7.3 *Since uniformly equi-continuity is a stronger condition, we only prove necessity. So, let $\mathscr{H}$ be an equi-continuous family of functions and let $\varepsilon > 0$. By assumption, for every $x \in X$, there exists $\delta = \delta(\varepsilon) > 0$ such that $d(f(x), f(y)) < \varepsilon$ for all $f \in \mathscr{H}$ and $d(x,y) < \delta$. Since X is compact, it can be covered by a finite number of balls $B(x_i, \delta_{x_i})$ $(1 \leq i \leq m)$. Let $\delta = \min_{1 \leq i \leq m} \{\delta_{x_i}\}$ and let $x, y \in X$ be such that $d(x,y) < \delta$. Then there exists $i_0 \in \{1, \ldots, m\}$ such that $x \in B(x_{i_0}, \delta_{x_{i_0}}/2)$. Hence, $y \in B(x_{i_0}, \delta_{x_{i_0}}/2)$ and for all $f \in \mathscr{H}$,*

$$d(f(x), f(y)) \quad \leq \quad d(f(x), f(x_{i_0})) + d(f(y), f(x_{i_0}))$$

$$< \quad \varepsilon.$$

Remark 1.7.4 *Let $D \subset X$ be a dense subset of a space X. It can be easily proved that if $\mathscr{H}$ is equi-continuous in $\mathscr{C}(D,Y)$, then it is equi-continuous in $\mathscr{C}(X,Y)$. In addition, the closure of an equi-continuous family $\mathscr{H} \subset \mathscr{C}(X,Y)$ is also equi-continuous.*

Remark 1.7.5 *A subset $\mathscr{H} \subset \mathscr{C}(X,Y)$ is said to be evenly continuous if it satisfies:*

$$\forall x \in X, \forall y \in Y, \forall V \in \mathscr{V}(y),$$

$$\exists U \in \mathscr{V}(x), \exists W \in \mathscr{V}(y),$$

$$(f(x) \in W \Longrightarrow f(U) \subset V), \forall f \in \mathscr{H}.$$

To prove Ascoli-Arzelà Theorem, we consider, for the sake of simplicity, a special situation in which (X,d) is a compact metric space and

$(Y, \|\cdot\|)$ a Banach space. The space $E = \mathscr{C}(X,Y)$ is endowed with the norm:

$$\|f\| = \sup_{x \in X} \|f(x)\|_Y.$$

Theorem 1.7.6 (Ascoli-Arzelà Theorem) *A subset $\mathscr{H} \subset E$ is relatively compact if and only if*

(a) $\mathscr{H}$ is equi-continuous.

(b) $\forall x \in X$, the set $\mathscr{H}(x) = \{f(x), f \in \mathscr{H}\}$ is relatively compact in Y.

Proof 1.7.7

(a) The condition is necessary. If $\mathscr{H}$ is relatively compact, then for all $\varepsilon > 0$, there exist a finite number of elements $\{f_i\}_{1 \le i \le n}$ in E such that $\mathscr{H} \subset \bigcup\limits_{i=1}^{n} B(f_i, \varepsilon/3)$, i.e.,

$$\forall f \in \mathscr{H}, \exists i \in \{1,n\}, \|f - f_i\|_E \le \varepsilon/3.$$

Hence,

$$\forall f \in \mathscr{H}, \forall x \in X, \exists i \in \{1,n\}, \|f(x) - f_i(x)\| \le \varepsilon/3.$$

Therefore,

$$\mathscr{H}(x) \subset \bigcup\limits_{i=1}^{n} B\left(f_i(x), \varepsilon/3\right),$$

which implies that $\mathscr{H}(x)$ is relatively compact in Y.

Now we prove that $\mathscr{H}$ is equi-continuous. For all $i = 1, 2, \ldots, n$, the function f_i is continuous. Then for all positive ε, there exists $\delta_i > 0$, such that $\forall x, y \in X$, we have

$$d(x,y) \le \delta_i \implies \|f_i(x) - f_i(y)\| \le \varepsilon/3.$$

Let $\delta = \min_{1 \le i \le n} \delta_i$ *and* $f \in \mathcal{H}$. *There exists* $i \in \{1,n\}$ *such that* $f \in B(f_i, \varepsilon/3)$ *and for all* $x, y \in X$, *we have*

$$d(x,y) \le \delta \Longrightarrow \|f(x) - f(y)\| \le \|f(x) - f_i(x)\| + \|f(y) - f_i(y)\|$$

$$+ \|f_i(x) - f_i(y)\|$$

$$\le \frac{\varepsilon}{3} + \frac{\varepsilon}{3} + \frac{\varepsilon}{3}$$

$$= \varepsilon,$$

we get the equi-continuity of $\mathcal{H}$.

(b) The condition is sufficient. Since $E = \mathscr{C}(X,Y)$ *is complete, it is sufficient to prove that* $\mathcal{H}$ *is totally bounded. Let* $\varepsilon > 0$. *Since* $\mathcal{H}$ *is equi-continuous, for every* $x \in X$, *there exists some* $\delta > 0$ *such that for all* $y \in X$ *and* $f \in \mathcal{H}$ *we have*

$$d(x,y) \le \delta \Longrightarrow \|f(x) - f(y)\| \le \varepsilon/4.$$

The space X being compact, can be covered by a finite number of balls $B_{x_i} = B(x_i, r)$, $1 \le i \le m$. *By assumption, each subset* $\mathcal{H}(x)$ *is relatively compact in Y, then the same holds for their finite union* $\mathcal{H} = \bigcup_{i=1}^{m} \mathcal{H}(x_i)$. *Therefore, we can cover* $\mathcal{H}$ *by a finite number of balls centered at* c_j ($1 \le j \le p$) *and with radius* $\varepsilon/4$. *Let* $I = \{1,2,\ldots,m\}$, $J = \{1,2,\ldots,p\}$, *and let* Φ *be the set of all mappings* $\varphi : I \longrightarrow J$. *For all* $\varphi \in \Phi$, *denote by* L_φ *the set of all mappings* $f \in \mathcal{H}$ *such that*

$$\forall i \in I, \ \|f(x_i) - c_{\varphi(i)}\| \le \varepsilon/4.$$

Some of the sets L_φ *may be empty, but* $\mathcal{H}$ *is covered by the union of* L_φ. *It remains to prove that the diameter of each* L_φ *is less than or equal to* ε. *Let* $f, g \in L_\varphi$. *For every* $y \in X$, *there exists* $i \in I$ *such that* $y \in B_{x_i}$. *Hence,*

$$\|f(y) - f(x_i)\| \le \varepsilon/4 \ \text{and} \ \|g(y) - g(x_i)\| \le \varepsilon/4.$$

For all $y \in X$, we have

$$\|f(y) - g(y)\| \leq \|f(y) - f(x_i)\| + \|g(y) - g(x_i)\|$$

$$+ \|f(x_i) - c_{\varphi(i)}\| + \|g(x_i) - c_{\varphi(i)}\|$$

$$\leq \frac{\varepsilon}{4} + \frac{\varepsilon}{4} + \frac{\varepsilon}{4} + \frac{\varepsilon}{4} = \varepsilon.$$

Hence, $\|f - g\|_{\mathscr{C}(X,Y)} \leq \varepsilon$, and our claim follows.

Next, we give a simple proof of Ascoli-Arzelà Theorem using the notion of a measure of noncompactness, (MNC). Let $J = [a,b]$, Y be a Banach space, and α some measure of noncompactness in Y. The following technical lemma is needed, where α_E denotes the MNC in the space E.

Lemma 1.7.8 *If $\mathscr{H} \subset E = \mathscr{C}(J,Y)$ is bounded and equicontinuous, then*

$$\alpha_E(\mathscr{H}) = \alpha(\mathscr{H}(J)) = \max_{t \in J} \alpha(\mathscr{H}(t)),$$

where

$$\mathscr{H}(J) = \{x(t) : x \in \mathscr{H}, t \in J\}$$

$$= \bigcup_{t \in J} \mathscr{H}(t).$$

Then, when Y is finite dimensional, we have

Corollary 1.7.9 (Ascoli-Arzelà Theorem) *If $\mathscr{H} \subset \mathscr{C}(J,Y)$ is bounded and equicontinuous, then $\mathscr{H}$ is relatively compact.*

Proof 1.7.10

(a) First proof (direct). The result follows from Lemma 1.7.8.

(b) Second proof. Let ω be the modulus of continuity:

$$\omega(x, \delta) = \sup\{|x(t) - x(s)| : |t - s| \leq \delta\}$$

and

$$\omega_0(\mathcal{H}, \delta) = \sup_{x \in \mathcal{H}} \omega(x, \delta).$$

Then it can be proved that

$$\omega_0(\mathcal{H}) = \lim_{\delta \to 0} \omega(\mathcal{H}, \delta).$$

is a regular measure of noncompactness. Using Remark 1.7.2, we then deduce that

$$\mathcal{H} \text{ equicontinuous } \Leftrightarrow \omega_0(\mathcal{H}) = 0 \Leftrightarrow \overline{\mathcal{H}} \text{ compact.}$$

This completes the proof.

1.7.2 Applications

Corollary 1.7.11 *For all $k \in \mathbb{N}$, the space $\mathscr{C}^{k+1}([a,b], \mathbb{R}^n)$ is embedded compactly in $\mathscr{C}^k([a,b], \mathbb{R}^n)$.*

Corollary 1.7.12 *Let $\mathcal{M} \subset \mathscr{C}^1([a,b], \mathbb{R})$ satisfy the following conditions:*

(a) The exists $L > 0$ such that for all $t \in [a,b]$ and $u \in \mathcal{M}$,

$$|u(t)| \leq L \text{ and } |u'(t)| \leq L$$

(b) For every positive real number $\varepsilon > 0$, there exists $\delta(\varepsilon) > 0$ such that for all $t_1, t_2 \in [a,b]$ with $|t_1 - t_2| < \delta(\varepsilon)$ and for all $u \in \mathcal{M}$,

$$|u(t_1) - u(t_2)| \leq \varepsilon \text{ and } |u'(t_1) - u'(t_2)| \leq \varepsilon.$$

Then, the set $\mathcal{M}$ is relatively compact in $\mathscr{C}^1([a,b], \mathbb{R})$.

Proof 1.7.13 *Let $\{u_n\}_{n \in \mathbb{N}}$ be a sequence of $\mathcal{M} \subset \mathscr{C}^1([a,b], \mathbb{R})$. To prove that $\mathcal{M}$ is relatively compact in $\mathscr{C}^1([a,b], \mathbb{R})$, it is equivalent to show that $\{u_n\}_{n \in \mathbb{N}}$ has a subsequent converging in $\mathscr{C}^1([a,b], \mathbb{R})$. Since $\{u_n\}_{n \in \mathbb{N}}$ is a sequence of $\mathcal{M} \subset \mathscr{C}^1([a,b], \mathbb{R})$, $\{u'_n\}_{n \in \mathbb{N}}$ (resp. $\{u_n\}_{n \in \mathbb{N}}$) is a sequence of $\mathscr{C}([a,b], \mathbb{R})$. Ascoli-Arzelà and Assumptions (a)–(b) guarantee that the sequence of derivatives $\{u'_n\}_{n \in \mathbb{N}}$ (resp. $\{u_n\}_{n \in \mathbb{N}}$) is relatively compact in $\mathscr{C}([a,b], \mathbb{R})$. As a consequence, there exists a* *subsequence, also denoted $\{u_n\}_{n \in \mathbb{N}}$ which converges in $\mathscr{C}([a,b],\mathbb{R})$ to a limit $u \in \mathscr{C}([a,b],\mathbb{R})$, and a subsequence of $\{u'_n\}_{n \in \mathbb{N}}$, also denoted $\{u'_n\}_{n \in \mathbb{N}}$, converging in $\mathscr{C}([a,b],\mathbb{R})$ to a limit $v \in \mathscr{C}([a,b],\mathbb{R})$. Using the integral representation of u_n, we find that for all $t, t_0 \in [a,b]$,*

$$u_n(t) = u(t_0) + \int_{t_0}^{t} u'_n(s)ds$$

$$\rightarrow u(t_0) + \int_{t_0}^{t} v(s)ds,$$

as $n \rightarrow \infty$. Then for all $t \in [a,b]$, $\lim_{n \to \infty} u_n(t) = u(t)$ and the uniqueness of the limit yields that $u(t) = u(t_0) + \int_{t_0}^{t} v(s)ds$. Hence, $u \in \mathscr{C}^1([a,b],\mathbb{R})$ and $u' = v$.

Example 1. Set $f_n(x) = \sin nx$, $x \in [0, 2\pi]$ and $\mathscr{H} = \{f_n(.) : n \in \mathbb{N}\}$. Then $\mathscr{H}$ is bounded. However, it is not equi-continuous in $\mathscr{C}([0, 2\pi], \mathbb{R})$ (for this, consider the sequence $x_n = \dfrac{\pi}{2n}$). Hence, $\mathscr{H}$ is not relatively compact, i.e., we cannot extract a convergent subsequence.

Example 2. Let $G : [a,b] \times [a,b] \longrightarrow \mathbb{R}$ be a continuous function and $T : \mathscr{C}([a,b], \mathbb{R}) \longrightarrow \mathscr{C}([a,b], \mathbb{R})$ be the linear operator defined by

$$Tx(t) = \int_{a}^{b} G(t,s)x(s)ds.$$

Then T is compact.

Example 3. The set F of functions f on $[a,b]$ that are uniformly bounded and satisfy the Hölder condition of order $0 < a \le 1$ with a fixed constant M, i.e.,

$$|f(x) - f(y)| \le M|x - y|^{\alpha}, \ x, y \in [a,b]$$

is relatively compact in $\mathscr{C}([a,b], \mathbb{R})$. In particular, the unit ball of the Hölder space $\mathscr{C}^{0,\alpha}([a,b], \mathbb{R})$ is compact in $\mathscr{C}([a,b], \mathbb{R})$.

Example 4. Let T be a compact linear operator from a Banach space X to a Banach space Y. Then its transpose T^* is compact from the continuous dual Y^* to X^*. Indeed, the image $T(B)$ of the closed unit ball of X is contained in a compact subset $K \in Y$. The unit ball B^* of Y^* defines, by restricting Y to K, a set F of linear continuous functions on K that is bounded and equi-continuous. Ascoli-Arzelà Theorem guarantees that, for every sequence $\{y_n^*\}_{n \in \mathbb{N}}$ in B^*, there is a subsequence that converges uniformly on K, and this implies that the image $T^*(\{y_{n_k}^*\}_{n \in \mathbb{N}})$ of that subsequence is a Cauchy sequence in X^*.

Example 5. Let f be a p-integrable function on $[0,1]$ ($1 < p \le \infty$) and define F by

$$F(x) = \int_0^x f(t)dt.$$

Let $\mathscr{H}$ be the set of functions F corresponding to functions f in the unit ball of the space $L^p([0,1])$. If q is the Hölder conjugate of p, then Hölder's inequality implies that all functions in $\mathscr{H}$ satisfy the Hölder condition with $\alpha = 1/q$ and constant $M = 1$. Hence, $\mathscr{H}$ is compact in $\mathscr{C}([0,1])$, that is the correspondence $f \longmapsto F$ is a linear compact operator from $L^p([0,1])$ to $\mathscr{C}([0,1])$. Composing with the injection of $\mathscr{C}([0,1])$ into $L^p([0,1])$, we find that T acts compactly from $L^p([0,1])$ to itself.

1.7.3   General forms of Ascoli-Arzelà theorem

Let X, Y be two topological spaces. Notice that in the space $E = \mathscr{C}(X,Y)$, we can as well as consider the topology of pointwise convergence (simple convergence), which is not metrizable, and we have the following result.

Theorem 1.7.14 *A subset $\mathscr{H} \subset E$ is relatively compact if and only if*

(a) $\mathscr{H}$ is pointwise closed.

(b) $\forall x \in X$, the set $\mathscr{H}(x) = \{f(x) : f \in H\}$ is relatively compact in Y.

The proof follows from Tychonov's Theorem because the topology of pointwise convergence is inherited from the product topology on $Y^X = \mathscr{F}(X,Y)$. This topology is given by the family of all sets of the form $T(\{x\},U) = \{f : f(x) \subset U\}$, where $x \in X$ and $U \subset Y$ is an open set. This topology is finer (larger) than the compact-open topology which is introduced in next theorems.

Now, we give two general forms of Ascoli-Arzelà Theorem and then one remark.

Theorem 1.7.15 *[41, T. 2, XX, 4; 1, page 312] Let X be a topological space, Y a metric space and $\mathscr{H} \subset \mathscr{C}(X,Y)$. Suppose that:*

(a) $\mathscr{H}$ is equi-continuous on every compact subsets of X,

(b) for every $x \in X$, $\mathscr{H}(x)$ is compact.

Then $\mathscr{H}$ is relatively compact for the topology of the compact convergence. If, further X is locally compact topological space, the conditions (a)-(b) are also necessary.

Remark 1.7.16 *The compact-open topology is the topology in which a subbase is given by the family of all sets of the form $F(K,U) = \{f : f(K) \subset U\}$, where $K \subset X$ is compact and $U \subset Y$ is an open set, while a base consists of finite intersection of such sets. When X is compact and Y a metric space, the compact-open topology coincides with the D-topology induced by the sup-metric on $\mathscr{C}(X,Y)$:*

$$D(f,g) = \sup\{d(f(x),g(x)) : x \in X\}.$$

Recall that a locally compact topological space is regular. For the sake of completeness, we recall the following relation connecting special structures of topological spaces.

Theorem 1.7.17 *[33, Theorem 21, page 236] Let X be a locally compact (hence regular) topological space, Y a Hausdorff regular topological space and $\mathscr{H} \subset \mathscr{C}(X,Y)$ be endowed with the compact-open topology. Then $\mathscr{H}$ is compact if and only if*

(a) $\mathscr{H}$ is closed in $\mathscr{C}(X,Y)$.

(b) $\mathscr{H}$ is evenly continuous on every compact subsets of X.

(c) For every $x \in X$, $\mathscr{H}(x)$ is compact.

1.8   Corduneanu-Avramescu Compactness Criterion in $\mathscr{C}_\ell$

1.8.1   Main results

Let $I = \mathbb{R}$, $\mathscr{C}_b = \mathscr{C}_b(I, \mathbb{R}^n)$ denote the vector topological space of all bounded and continuous functions defined on I and having values in $\mathbb{R}^n$, and

$$
\begin{aligned}
\mathscr{C}_\ell &= \{x \in \mathscr{C}(I, \mathbb{R}^n), \quad \lim_{t \to \pm\infty} x(t) \text{ exist}\}, \\[6pt]
\mathscr{C}_{\ell\ell} &= \{x \in \mathscr{C}(I, \mathbb{R}^n), \quad \lim_{t \to +\infty} x(t) = \lim_{t \to -\infty} x(t)\}, \\[6pt]
\mathscr{C}_{(a,b)} &= \{x\colon [a,b] \to \mathbb{R}^n,\ x \text{ continuous}\}, \\[6pt]
\mathscr{C}_{[a,b]} &= \{x \in \mathscr{C}([a,b], \mathbb{R}^n),\ x(a) = x(b)\}.
\end{aligned}
\tag{1.38}
$$

These are Banach spaces with respect to the norm

$$
\|x\|_\infty = \sup_{t \in I} \|x(t)\|.
$$

The proof that $\mathscr{C}_\ell$ is complete is easy.

Lemma 1.8.1 *The spaces $\mathscr{C}_\ell$ and $\mathscr{C}_{(a,b)}$ (respectively $\mathscr{C}_{\ell\ell}$ and $\mathscr{C}_{[a,b]}$) are isomorphic.*

Proof 1.8.2 *Let $\varphi\colon (a,b) \to \mathbb{R}$ be a continuous, strictly nondecreasing function[1] with $\lim_{t \to a^+} \varphi(t) = -\infty$ and $\lim_{t \to b^-} \varphi(t) = +\infty$. Define the mapping*

$$
(\Phi x)(t) = \begin{cases} x(\varphi(t)), & \text{if } t \in (a,b) \\ x(-\infty), & \text{if } t = a \\ x(+\infty), & \text{if } t = b. \end{cases}
\tag{1.39}
$$

Then, it is clear that Φ is an isometric isomorphism between $\mathscr{C}_\ell$ and $\mathscr{C}_{(a,b)}$ in one hand and between $\mathscr{C}_{\ell\ell}$ and $\mathscr{C}_{[a,b]}$ in the other one. This completes the proof.

---

[1] For instance $\varphi(t) = \tan\left(\dfrac{\pi(t-a)}{b-a} - \dfrac{\pi}{2}\right)$.

Before stating a compactness criterion in $\mathscr{C}_\ell$, we give

Definition 1.8.3 *A family* $\mathscr{A} \subset \mathscr{C}_\ell$ *is called equi-convergent if*

$$\forall \varepsilon > 0, \ \exists T = T(\varepsilon) > 0, \ \forall t_1, t_2 \in \mathbb{R},$$
$$|t_1| > T, |t_2| > T \ \Rightarrow \ \|x(t_1) - x(t_2)\| < \varepsilon, \ \forall x \in \mathscr{A}.$$

Remark 1.8.4 *Equivalently,* $\mathscr{A}$ *is equi-convergent if for any* $\varepsilon > 0$, *there exists some* $T = T(\varepsilon) > 0$ *such that* $\|x(t) - l_x^+\| \leq \varepsilon$ *and* $\|x(t) - l_x^-\| \leq \varepsilon$ *for all* $|t| \geq T$ *and for all* $x \in \mathscr{A}$. *Here,* $l_x^+ : = \lim\limits_{t \to +\infty} x(t)$ *and* $l_x^- : = \lim\limits_{t \to -\infty} x(t)$ *(Cauchy criterion for limits).*

Remark 1.8.5 *For a sufficient condition for* $\mathscr{A}$ *be equi-convergent is the existence of some function* $\gamma \in \mathscr{C}_\ell$ *such that*

$$\|x(t) - x(\infty)\| \leq \|\gamma(t) - \gamma(\infty)\|, \ \forall x \in \mathscr{A}.$$

Now, we state and prove Avramescu-Corduneanu compactness criterion in $\mathscr{C}_\ell$ [4].

Theorem 1.8.6 *A family* $\mathscr{A} \subset \mathscr{C}_\ell$ *is relatively compact if and only if the following conditions are satisfied:*

(a) $\mathscr{A}$ *is uniformly bounded in* $\mathscr{C}_\ell$.

(b) $\mathscr{A}$ *is equi-continuous on every compact interval of* $\mathbb{R}$ *(we say that* $\mathscr{A}$ *is almost equi-continuous).*

(c) $\mathscr{A}$ *is equi-convergent.*

Proof 1.8.7 *We only prove sufficiency.*

(a) First proof. From the conditions (b) and (c), it follows $\mathscr{A}$ *is equi-continuous in* $\mathscr{C}_\ell$. *By Lemma 1.8.1,* $\Phi(\mathscr{A})$ *is equi-continuous and uniformly bounded in* $\mathscr{C}_{(a,b)}$. *Ascoli-Arzelà Theorem guarantees that* $\Phi(\mathscr{A})$ *is relatively compact in* $\mathscr{C}_{(a,b)}$. *Therefore,* $\mathscr{A}$ *is compact as the compact image of a compact set.*

(b) Direct proof, detailed. For the sake of simplicity of presentation, we take $I = [0, +\infty)$. *Let* $\{x_n\}_{n \in \mathbb{N}} \subset \mathscr{A}$ *be a bounded sequence. The proof is split into three steps:*

Step 1. For every $t \in I$, the real sequence $\{x_n(t)\}_{n\in\mathbb{N}}$, including $\{x_n(+\infty)\}_{n\in\mathbb{N}}$, is bounded. By the Bolzano-Weirstrass theorem, there exists a subsequence $\{x_{n_k}(t)\}_{k\in\mathbb{N}}$ converging for every $t \in I$; also $\{x_{n_k}(+\infty)\}_{k\in\mathbb{N}}$ satisfies the Cauchy criterion, i.e., for all positive $\varepsilon > 0$, there is some $N_1(\varepsilon) \in \mathbb{N}$ such that

$$p,q \geq N_1 \implies \|x_{n_p}(+\infty) - x_{n_q}(+\infty)\| \leq \varepsilon/3. \tag{1.40}$$

Step 2. The sequence $\{x_{n_k}\}_{k\in\mathbb{N}}$ being equi-convergent, for all positive $\varepsilon > 0$, there exists $T = T(\varepsilon) > 0$ such that

$$\|x_{n_k}(t) - x_{n_k}(+\infty)\| \leq \varepsilon/3, \ \forall t \geq T, \ \forall k \in \mathbb{N}. \tag{1.41}$$

As a consequence of (1.40)–(1.41), we obtain that for all positive $\varepsilon > 0$, there exist $N_1(\varepsilon) \in \mathbb{N}$ and a real number $T = T(\varepsilon)$ such that for all $p,q \geq N_1$ and $t \geq T$, we have

$$\begin{aligned}
\|x_{n_p}(t) - x_{n_q}(t)\| &\leq \|x_{n_p}(t) - x_{n_p}(+\infty)\| \\
&\quad + \|x_{n_p}(+\infty) - x_{n_q}(+\infty)\| + \|x_{n_q}(t) - x_{n_q}(+\infty)\| \\
&\leq \frac{\varepsilon}{3} + \frac{\varepsilon}{3} + \frac{\varepsilon}{3} = \varepsilon.
\end{aligned}$$
$$\tag{1.42}$$

Step 3. The sequence $\{x_{n_k}\}_{k\in\mathbb{N}}$ being bounded and equi-continuous on $[0,T]$, admits by the Ascoli-Arzelà Theorem, a subsequence $\{x_{n_l}\}_{l\in\mathbb{N}}$ converging uniformly to some limit x, as $l \to +\infty$, i.e.,

$$\sup_{t\in[0,T]} \|x_{n_l}(t) - x(t)\| \to 0, \ as \ l \to +\infty.$$

$\{x_{n_l}\}_{l\in\mathbb{N}}$ is then a Cauchy sequence in $\mathscr{C}([0,T],\mathbb{R})$. Therefore, there exists an integer $N_2 = N_2(\varepsilon)$ such that

$$p,q \geq N_2 \implies \|x_{n_p}(t) - x_{n_q}(t)\| \leq \varepsilon, \ \forall t \in [0,T]. \tag{1.43}$$

Finally (1.42) and (1.43) lead to that for all positive $\varepsilon > 0$, there exists $N_0(\varepsilon) = \max(N_1, N_2)$ such that for all $t \in \mathbb{R}$ and all $p,q \geq N_0$,

$$\|x_{n_p}(t) - x_{n_q}(t)\| \leq \varepsilon,$$

i.e., $\|x_{n_p} - x_{n_q}\| \leq \varepsilon$, proving that $\{x_{n_k}\}_{k\in\mathbb{N}}$ is a Cauchy sequence in the Banach space $\mathscr{C}_\ell(I,\mathbb{R})$, hence converges in this space.

1.8.2　Application 1

Let $\theta > 0$ be some real parameter and consider the space

$$E \;=\; \mathscr{C}^1_\infty([0,\infty),\mathbb{R})$$

$$=\; \left\{ x \in \mathscr{C}^1([0,\infty),\mathbb{R}): \; \lim_{t\to+\infty}\frac{x(t)}{e^{\theta t}} \text{ and } \lim_{t\to+\infty}\frac{x'(t)}{e^{\theta t}} \text{ exist} \right\}.$$

normed by

$$\|x\|_\theta = \max\{\|x\|_1, \|x\|_2\},$$

where

$$\|x\|_1 \;=\; \sup_{t\in[0,\infty)}\frac{|x(t)|}{e^{\theta t}} \text{ and}$$

$$\|x\|_2 \;=\; \sup_{t\in[0,\infty)}\frac{|x'(t)|}{e^{\theta t}}.$$

Lemma 1.8.8　*E is a Banach space.*

Proof 1.8.9　*We prove that E is complete. Let $\{x_n\}_{n\in\mathbb{N}} \subset X$ be a Cauchy sequence; then*

$$\left\{ y_n : y_n(t) = \frac{x_n}{e^{\theta t}} \right\} \text{ and } \left\{ z_n : z_n(t) = \frac{x'_n(t)}{e^{\theta t}} \right\}$$

are Cauchy sequences in the Banach space $\mathscr{C}_\ell$. Thus there exist $y_0, z_0 \in \mathscr{C}_\ell$ such that $\lim_{n\to+\infty}\|y_n - y_0\|_\infty = 0$ and $\lim_{n\to+\infty}\|z_n - z_0\|_\infty = 0$. Let $x_0(t) = y_0(t)e^{\theta t}$, $t \in \mathbb{R}^+$. On every compact interval $[0,T]$, $\{x_n\}$ converges uniformly to x_0 and $\{x'_n\}_{n\in\mathbb{N}}$ converges to the function $t \mapsto z_0(t)e^{\theta t}$. Then x_0 is differentiable on $[0,T]$ and $x'_0(t) = z_0(t)e^{\theta t}$ for $t \in [0,T]$. It follows that x_0 is differentiable on $\mathbb{R}^+$ and $x'_0(t) = z_0(t)e^{\theta t}$, $\forall t \geq 0$. Finally, $\lim_{n\to+\infty}\|x_n - x_0\|_\theta = \lim_{n\to+\infty}\max\{\|y_n - y_0\|_\infty, \|z_n - z_0\|_\infty\} = 0$, ending the proof of our claim.

Lemma 1.8.10　*Let $M \subset \mathscr{C}^1_\infty(\mathbb{R}^+,\mathbb{R})$. Then M is relatively compact in $\mathscr{C}^1_\infty(\mathbb{R}^+,\mathbb{R})$ if the following conditions hold:*

(a) M *is uniformly bounded in* $\mathscr{C}^1_\infty(\mathbb{R}^+,\mathbb{R})$.

(b) *The functions belonging to the sets*

$$\mathscr{A} = \{y : y(t) = \frac{x(t)}{e^{\theta t}}, x \in M\} \text{ and } \mathscr{B} = \{z : z(t) = \frac{x'(t)}{e^{\theta t}}, x \in M\}$$

are almost equi-continuous on $\mathbb{R}^+$.

(c) *The functions from* $\mathscr{A}$ *and* $\mathscr{B}$ *are equi-convergent at* $+\infty$.

Proof 1.8.11 *Let* $\{x_n\}_{n \in \mathbb{N}} \subset M$ *and* $y_n(t) = e^{-\theta t} x_n(t)$, $n \geq 1$. *Since the set* $\mathscr{A}$ *satisfies the conditions of Theorem 1.8.6, there exists* $y_0 \in \mathscr{C}_\ell$ *such that* $\lim_{n \to +\infty} \|y_n - y_0\|_l = 0$. *If we set* $\mathscr{B}_n = \{z_n : z_n(t) = \frac{x'_n(t)}{e^{\theta t}}, t \in \mathbb{R}^+\}$, *then again from Theorem 1.8.6, there exists a subsequence* $\{z_{n_j}\}_{j \in \mathbb{N}} \subset \{z_n\}_{n \in \mathbb{N}}$ *and* $z_0 \in \mathscr{C}_\ell$ *such that* $\lim_{j \to +\infty} \|z_{n_j} - z_0\|_l = 0$. *Moreover, for any* $T > 0$, $\{x_{n_j}\}$ *converges to* $x_0 = y_0 e^{\theta t}$ *uniformly on* $[0, T]$ *and* $\{x'_{n_j}\}_{j \in \mathbb{N}}$ *converges to the function* $t \mapsto z_0(t) e^{\theta t}$ *uniformly on* $[0, T]$. *Writing the integral representation of* x_n, *we find that* x_0 *is differentiable on* $[0, T]$ *and* $x'_0(t) = z_0(t) e^{\theta t}$. *Since T is arbitrary, it follows that* x_0 *is differentiable on* $\mathbb{R}^+$ *and that* $x'_0(t) = z_0(t) e^{\theta t}$, $\forall t \geq 0$. *Finally,* $\lim_{j \to +\infty} \|x_{n_j} - x_0\|_\theta = \lim_{j \to +\infty} \max\{\|y_{n_j} - y_0\|_l, \|z_{n_j} - z_0\|_l\} = 0$. *Therefore, the sequence* $\{x_{n_j}\}_{j \in \mathbb{N}} \subset M$ *is convergent, ending the proof of the lemma.*

1.8.3   Application 2

Let $t_0 > 0$ and consider the functional spaces

$$V(t_0) \quad = \quad \{u \in \mathscr{C}^1([t_0, +\infty), \mathbb{R}) :$$

$$\lim_{t \to +\infty} u'(t) = a_u < +\infty\}$$

and

$$W(t_0) \quad = \quad \{u \in \mathscr{C}^1([t_0, +\infty), \mathbb{R}) :$$

$$\lim_{t \to +\infty} u'(t) = a_u, \lim_{t \to +\infty} (u(t) - a_u t) = b_u\},$$

where a_u, b_u are positive constants depending on u.

Proposition 1.8.12 $V(t_0)$ *and* $W(t_0)$ *are complete when equipped with the respective norms*

$$\|u\|_V = \sup_{t \geq t_0} \frac{|u(t)|}{t} + \sup_{t \geq t_0} |u'(t)|$$

and

$$\|u\|_W = \sup_{t \geq t_0} \frac{|u(t)|}{t} + \sup_{t \geq t_0} |u'(t)| + \sup_{t \geq t_0} |u(t) - a_u t|.$$

Proof 1.8.13

(a) *Let* $\{u_n\}_{n \in \mathbb{N}}$ *be a Cauchy sequence in* $V(t_0)$. *Then* $\{u_n'\}_{n \in \mathbb{N}}$ *is a Cauchy sequence in* $\mathscr{C}_\ell$ *which is complete. Let* v *be the uniform limit of* $\{u_n'\}_{n \in \mathbb{N}}$ *in* $\mathscr{C}_\ell$. *Since, for every* $T > t_0$ *and* $t \in [t_0, T]$,

$$|u_p(t) - u_q(t)| \leq T \left| \frac{u_p(t)}{t} - \frac{u_q(t)}{t} \right|,$$

then $(u_n)_{n \in \mathbb{N}}$ *is also a Cauchy sequence in* $\mathscr{C}([t_0, T])$. *Hence, it converges to some limit* u *on every compact subset of* $[t_0, \infty)$. *Passing to the limit in the integral representation*

$$u_n(t) = u_n(t_0) + \int_{t_0}^{t} u_n'(s)\,ds, \ t \geq t_0,$$

we deduce that $u \in \mathscr{C}^1([t_0, +\infty))$ *with* $u' = v \in \mathscr{C}_\ell$. *Finally*

$$\left| \frac{u_n(t)}{t} - \frac{u(t)}{t} \right| \leq \frac{|u_n(t_0) - u(t_0)|}{t} + \|u_n' - u'\|_0 \frac{t - t_0}{t}, \ \forall t \geq t_0,$$

where $\|u\|_0 = \sup_{t \geq t_0} |u(t)|$. *Hence,* $\|u_n - u\|_{V(t_0)} \to 0$, *as* $n \to \infty$, *as claimed.*

(b) *First, note that* $\|u\|_{W(t_0)} = \|u\|_{V(t_0)} + \sup_{t \geq t_0} |u(t) - t a_u|$, *where* $a_u = \lim_{t \to +\infty} u'(t)$. *Now, let* $(u_n)_{n \geq 1}$ *be a Cauchy sequence in* $W(t_0)$. *Then* $\{u_n\}_{n \in \mathbb{N}}$ *is a Cauchy sequence in* $V(t_0)$, *which is complete by part (a). Hence,* u_n *converges uniformly in* $V(t_0)$ *to some limit* $u \in \mathscr{C}_\ell$. *In particular,* u_n' *converges uniformly to* u' *and so, we can invert the limits* *as $t \to \infty$ and $n \to \infty$ to get $a_u = \lim\limits_{n \to \infty} a_{u_n}$. Also $v_n(t) = u_n(t) - t a_n$ is a Cauchy sequence in $\mathscr{C}_\ell$ which is complete. Let v be the uniform limit of v_n in $\mathscr{C}_\ell$. In addition, u_n converges to u on every compact subset of $[t_0, +\infty)$. We deduce that for all $t \geq t_0$, $v_n(t)$ converges to $u(t) - t a_u$ and even $\|v_n - v\|_0 \to$, as $n \to \infty$. As a consequence,*

$$\|u_n - u\|_{W(t_0)} = \|u_n - u\|_{V(t_0)} + \|v_n - v\|_{\mathscr{C}_\ell}$$

$$\to 0, \text{ as } n \to \infty,$$

proving the second part of the proposition.

Now, we can enunciate two compactness criteria derived from Avramescu's compactness criterion.

Proposition 1.8.14 *Let $\mathscr{M} \subset V(t_0)$ satisfy:*

(a) there exists $L > 0$ such that for all $t \geq t_0$ and $u \in \mathscr{M}$,

$$|u'(t)| \leq L \text{ and } \frac{|u(t)|}{t} \leq L.$$

(b) For each $\varepsilon > 0$, there exists $\delta(\varepsilon) > 0$ such that for all $t_1, t_2 \geq t_0$ with $|t_1 - t_2| \leq \delta(\varepsilon)$ and all $u \in \mathscr{M}$,

$$|u'(t_1) - u'(t_2)| < \varepsilon \text{ and } \left| \frac{u(t_1)}{t_1} - \frac{u(t_2)}{t_2} \right| < \varepsilon.$$

Then $\mathscr{M}$ is relatively compact in $V(t_0)$. Conversely, if the set $\mathscr{M}$ is relatively compact, then all conditions (a), (b), (c) are satisfied.

Proof 1.8.15 *We restrict our attention to the proof of sufficiency. Let $\{u_n\}_{n \in \mathbb{N}}$ be a sequence in $\mathscr{M} \subset V(t_0)$. To prove that $\mathscr{M}$ is relatively compact in the Banach space $V(t_0)$, it is equivalent to show that $\{u_n\}_{n \in \mathbb{N}}$ admits a subsequence converging in $V(t_0)$. From the conditions (a)–(b) and Avramescu's compactness criterion, we find that $u'_n(t)$ and $\dfrac{u_n(t)}{t}$ have subsequences, denoted the same, which converge to $v(t)$ and $\dfrac{u(t)}{t}$, respectively. By the integral representation*

$$\frac{u_n(t)}{t} = \frac{u_n(t_0)}{t} + \frac{1}{t} \int_{t_0}^{t} u'_n(s) ds,$$

we deduce that

$$\frac{u(t)}{t} = \frac{u(t_0)}{t} + \frac{1}{t}\int_{t_0}^{t} v(s)ds.$$

Hence $v = u'$ *and* $u \in V(t_0)$. *We have used the fact that under the boundedness of* u' *and* $\dfrac{u(t)}{t}$, *we have that* $\lim\limits_{t\to\infty} u'(t) = l$ *implies*

$$\lim_{t\to\infty} \frac{u(t)}{t} = l.$$

Proposition 1.8.16 *Assume that a subset* $\mathscr{M} \subset W(t_0)$ *satisfy*

(a) there exists a constant $L > 0$ *such that for all* $t \geq t_0$ *and all* $u \in \mathscr{M}$,

$$|u'(t)| \leq L \text{ and } |u(t) - a_u t| \leq L.$$

(b) For each $\varepsilon > 0$, *there exists* $\delta(\varepsilon) > 0$ *such that for all* $t_1, t_2 \geq t_0$ *with* $|t_1 - t_2| \leq \delta(\varepsilon)$ *and all* $u \in \mathscr{M}$,

$$|u'(t_1) - u'(t_2)| < \varepsilon \text{ and } |u(t_1) - u(t_2) - a_u(t_1 - t_2)| < \varepsilon.$$

(c) For each $\varepsilon > 0$, *there exists* $T\varepsilon) > t_0$ *such that for all* $t \geq T(\varepsilon)$ *and all* $u \in \mathscr{M}$,

$$|u'(t) - a_u| < \varepsilon \text{ and } |u(t) - a_u t - b_u| < \varepsilon.$$

Then $\mathscr{M}$ *is relatively compact in* $W(t_0)$. *Conversely, if* $\mathscr{M}$ *is relatively compact, then all conditions* (a), (b), *and* (c) *are satisfied.*

Proof 1.8.17 *Again we only check sufficiency. Let* $\{u_n\}_{n\in\mathbb{N}}$ *be a sequence in* $\mathscr{M} \subset W(t_0)$. *We prove that it has a subsequence converging in* $W(t_0)$. *Notice first that,* $\{u'_n\}_{n\in\mathbb{N}}$ *has a subsequence which converges to some limit* $v \in \mathscr{C}_\ell[t_0, +\infty)$ *(uniformly on* $[t_0, +\infty)$*) with* $l_v = \lim\limits_{n\to\infty} a_{u_n}$. *In addition,* $v_n(t) = u_n(t) - ta_{u_n}$ *is relatively compact in* $\mathscr{C}_\ell[t_0, +\infty)$. *Hence, converges in this space to some limit* $v(t)$ *(uniformly on* $[t_0, +\infty)$*). Then, on every compact subset of* $[t_0, +\infty)$, $u_n(t)$ *converges to* $b + tl_v : = u$. *However, the integral representation of* u_n *guarantees that*

$$u(t) = u(t_0) + \int_{t_0}^{t} v(s)ds, \ t \geq t_0.$$

Therefore, $v = b' = u'$ *and thus,* $l_v = a_b = a_u$. *Hence,* $\{u_n\}_{n\in\mathbb{N}}$ *converges uniformly to* u *in* $W(t_0)$. *This completes the proof.*

Before ending this section, notice that Avramescu-Corduneanu compactness criterion can be generalized to the sets of functions taking values in a general Banach space Y. For this we appeal to a measure of noncompactness to get the following result.

Theorem 1.8.18 *A family $\mathscr{A} \subset \mathscr{C}_\ell([0,+\infty),Y)$ is relatively compact if and only if the following conditions are satisfied:*

(a) $\mathscr{A}$ is uniformly bounded.

(b) $\mathscr{A}$ is almost equi-continuous.

(c) $\mathscr{A}$ is equi-convergent.

(d) For any $t \in [0,+\infty)$, the set $\mathscr{A}(t)$ is relatively compact in Y.

The proof can be modeled on that of Theorem 1.12.1 and uses the following lemma which is a generalization of Lemma 1.12.2 to unbounded intervals.

Lemma 1.8.19 *Let $\mathscr{H} \subset \mathscr{C}(J,Y)$ be countable and there exists $\rho \in L^1(J,\mathbb{R})$ such that $\|x(t)\| \le \rho(t)$ for $t \in J$ and $x \in \mathscr{H}$. Then $\alpha(\{u(t) : u \in \mathscr{H}\}$ is integrable on J and*

$$\alpha\left(\left\{\int_0^\infty x(t)dt : x \in \mathscr{H}\right\}\right) \le 2\int_0^\infty \alpha(\{u(t) : u \in \mathscr{H}\})dt.$$

1.9   Przeradzki's Compactness Criterion in $\mathscr{BC}(\mathbb{R},Y)$

1.9.1   Main result

Let Y be a Banach space and $\mathscr{BC}(\mathbb{R},Y)$ denotes the space of all bounded continuous functions $x : \mathbb{R} \longrightarrow Y$ with the norm:

$$\|x\|_\infty = \sup_{t \in I} \|x(t)\|.$$

$(\mathscr{CB}(\mathbb{R},Y), \|.\|)$ is a Banach space. We have now the following compactness criterion [39] where equi-convergent is replaced by the stability condition near infinity.

Theorem 1.9.1 *For a subset $H \subset \mathscr{BC}(\mathbb{R},Y)$ to be relatively compact, it is necessary and sufficient that*

(a) the set $H(t)$ is relatively compact in Y, for every $t \in \mathbb{R}$,

(b) for each $a > 0$, the family $H_a = \{x_{|[-a,a]} : x \in H\}$ is equi-continuous (almost-equi-continuous),

(c) H is stable at $\pm\infty$, i.e., for any $\varepsilon > 0$, there exist $T > 0$ and $\delta > 0$ such that for all $x, y \in H$, we have

$$\|x(T) - y(T)\| \le \delta \Longrightarrow \|x(t) - y(t)\| \le \varepsilon, \ \text{for } t \ge T,$$

and

$$\|x(-T) - y(-T)\| \le \delta \Longrightarrow \|x(t) - y(t)\| \le \varepsilon, \ \text{for } t \le -T.$$

Proof 1.9.2

(a) Necessity. Suppose that conditions (a)–(c) hold and let $\varepsilon > 0$. We shall construct a finite ε-net for H. Let T, δ be positive constants chosen for ε given by (c) and $\delta \le \varepsilon$. By (a), we can find finite δ-nets: $x_1(T), \ldots x_p(T)$ in $\{x(T) : x \in D\}$ and $y_1(-T), \ldots y_r(-T)$ in $\{y(-T) : y \in D\}$. Repeating the proof of the general Ascoli-Arzelà, we get a finite δ-net of the relatively compact set $H_T \subset \mathscr{C}([-T,T],Y)$ consisting of functions $z_1, \ldots, z_s$ which take one of the values $y_j(-T)$, $j = 1, \ldots, r$, at $-T$, and one of the values $x_i(T)$, $i = 1, \ldots, p$, at T. It is then clear that the set of all continuous functions $\mathbb{R} \longrightarrow Y$, which are obtained by "gluing" y_j, z_k, and x_i together, is a finite ε-net for H.

(b) Sufficiency. If H is relatively compact, then (a) and (b) hold. Indeed, since H_a, $a > 0$ are relatively compact in $\mathscr{C}([-a,a],Y)$, this follows from the general Ascoli-Arzelà Theorem. Now, suppose that H is not stable at $+\infty$ (the proof is the same at $-\infty$). Then there exists $\varepsilon_0 > 0$ and sequences $\{x_n\}_{n \in \mathbb{N}}, \{y_n\}_{n \in \mathbb{N}} \subset H$, and $\{t_n\}_{n \in \mathbb{N}} \subset \mathbb{R}$ such that for any $n \in \mathbb{N}$,

$$\|x_n(n) - y_n(n)\| \le 1/n,$$

$$\|x_n(t_n) - y_n(t_n)\| \le 1/n, \ t_n \ge n.$$

Since D is relatively compact, it has a finite $\varepsilon/4$-net $z_1, \ldots, z_p$. Define $w_1, \ldots, w_p \subset \mathscr{BC}(\mathbb{R}, Y)$ such that

$$\|w_i(t) - z_i(t)\| \leq \varepsilon_0/4, \ t \in \mathbb{R}, \tag{1.44}$$

for $i \leq p$ and

$$\|w_i(n) - w_j(n)\| \geq \varepsilon_0/8p, \tag{1.45}$$

for $n \in \mathbb{N}$ and $i \neq j$. Let $w_1 = z_1$ and assume that $w_1, \ldots, w_k$ have already been defined. Fix $n \in \mathbb{N}$. If (1.45) is satisfied for $i \leq k$ and w_j is replaced by z_{k+1}, we put $h_{k+1,n} = z_{k+1}(n)$. If not, we choose $h_{k+1,n}$ arbitrarily from the set

$$\overline{B}(z_{k+1}(n), \varepsilon_0/4) \setminus \bigcup_{i=1}^{k} B(w_i(n), \varepsilon_0/8p).$$

Then, we define a multi-valued mapping $\Phi_{k+1} : \mathbb{R} \longrightarrow 2^Y$ by

$$\Phi_{k+1}(t) = \begin{cases} \overline{B}(z_{k+1}(t), \varepsilon_0/4), & t \notin \mathbb{N} \\ h_{k+1,n}, & t \in \mathbb{N}. \end{cases}$$

Simple calculations show that Φ_{k+1} is lower semicontinuous. By the Michael selection theorem, it has a continuous selection $w_{k+1} : \mathbb{R} \longrightarrow Y$. By induction, we obtain some $w_i, \ldots, w_p$ having properties (2.13) and (1.45). Clearly, $\{w_1, \ldots, w_p\}$ is an $\varepsilon_0/2$-net for H. Let us return to the sequences $\{x_n\}_{n \in \mathbb{N}}$ and $\{y_n\}_{n \in \mathbb{N}}$. We can choose a subsequence $\{x_{n_m}\}_{m \in \mathbb{N}}$ contained in an $\varepsilon_0/2$-neighborhood of one element w_j. By (1.45) and $x_n(n) - y_n(n) \to 0$, as $n \to +\infty$ y_{n_m} is in the same ball for sufficiently large m. Hence,

$$\|x_{n_m}(t) - y_{n_m}(t)\| \ \leq \ 2\varepsilon_0/2$$

$$= \ \varepsilon_0,$$

for each $t \in \mathbb{R}$, which is impossible for $t = t_{n_m}$. This completes the proof.

1.9.2 Application

Consider the nonlinear integral operator

$$S : \mathscr{BC}(\mathbb{R}, Y) \longrightarrow \mathscr{BC}(\mathbb{R}, Y)$$

defined by

$$Sx(t) = \int_{-\infty}^{+\infty} G(t,s)f(x(s),s)ds,$$

where the Green function $G : \mathbb{R}^2 \longrightarrow L(Y)$ satisfies

$$\exists \alpha, N > 0, \; \|G(t,s)\| \le N e^{-\alpha|t-s|}$$

and $f : Y \times \mathbb{R} \longrightarrow Y$ is continuous. We have

Theorem 1.9.3 *S is completely continuous if, in addition, f satisfies the properties*

(a) $f(.,t) : Y \longrightarrow Y$ is completely continuous for any $t \in \mathbb{R}$,

(b) there exists a bounded continuous function $b : \mathbb{R} \longrightarrow Y$ such that for any M and $\varepsilon > 0$, there is $T > 0$ such that $\|f(x,t) - b(t)\| \le \varepsilon$ where $\|x\| \le M$ and $|t| \ge T$.

It is noteworthy to observe how condition (b) is essential to guarantee the stability condition in Theorem 1.9.1.

1.10   Compactness Criterion in $\mathscr{C}([0,+\infty),\mathbb{R}^n)$

The space $\mathscr{C}([0,+\infty),\mathbb{R}^n)$ is not a normed space but it can be organized as a Fréchet space, i.e., a topological linear space which is metrizable and complete. For instance, if (X,τ) is a topological linear space with a countable family of semi-norms $\{|x_k| : k > 1\}$, then (X,d) is a metric space with

$$d(x,y) = \sum_{k=1}^{\infty} \frac{1}{2^k} \frac{|x-y|_k}{1+|x-y|_k}.$$

If, in addition (X,τ) is complete, then it is a Fréchet space. This is exactly the situation with $\mathscr{C}([0,+\infty),\mathbb{R}^n)$, where

$$|x|_n = \sup\{\|x(t)\| : t \in [0,t_n]\}, \; n \ge 1,$$

and $\{t_n\}_{n\in\mathbb{N}}$ is some sequence increasing to $+\infty$. Then it is easy to check the following statement.

Proposition 1.10.1 *The distance d defines uniform convergence on any compact interval of* $[0, +\infty)$.

Yet, of particular interest for our purpose is the following proposition.

Proposition 1.10.2 *A subset* $\mathscr{H} \subset \mathscr{C}([0,+\infty), \mathbb{R}^n)$ *is relatively compact (for the topology of uniform convergence) if and only if it is bounded and equi-continuous on each compact interval of* $[0, +\infty)$.

Proof 1.10.3 *Indeed, from Proposition 1.10.1, we can define the relative compactness in* $\mathscr{C}([0,+\infty), \mathbb{R}^n)$ *as follows: the set* $\mathscr{H} \subset \mathscr{C}([0,+\infty), \mathbb{R}^n)$ *is relatively compact if from any sequence* $\{x^n, n \geq 1\} \subset \mathscr{H}$, *one can extract a subsequence, also denoted* $\{x^n, n \geq 1\}$, *which is uniformly convergent on any compact interval of* $[0,+\infty)$. *If we consider the restriction of the terms of the sequence* $\{x^n, n \geq 1\}$ *to some* $[0, t_m]$, *then these restrictions satisfy the classical Ascoli-Arzelà Theorem. This leads to a compactness criterion in* $\mathscr{C}([0,+\infty), \mathbb{R}^n)$ *for the compact-open topology. This completes the proof.*

1.11   Compactness Criteria in $\mathscr{BC}(X, \mathbb{R})$

1.11.1   The Stone-Čech compactification

Let X be a completely regular topological space (not necessarily compact). We know that there exist a compact space $\check{X}$ and an embedding $\rho : X \longrightarrow \check{X}$ such that $\rho : X \longrightarrow \rho(X)$ is a homeomorphism and $\rho(X)$ is dense in $\check{X}$. Moreover, $\mathscr{BC}(X, \mathbb{R})$ is isomorphic (even isometric) to $\mathscr{C}(\check{X}, \mathbb{R})$. The pair $(\check{X}, \rho)$ is called the Stone-Čech compactification of the space X. For instance, the Stone-Čech compactification of the positive real line $[0, +\infty)$ is the unit compact interval $[0, 1]$ with homeomorphism $\rho(t) = \dfrac{t}{1+t}$, generating an isometry between $\mathscr{BC}([0,+\infty), \mathbb{R})$ and $\mathscr{C}([0, 1], \mathbb{R})$. This is the main idea used in the proof of Theorem 1.8.6 (first proof) where such an isometry is explicitly defined.

As a consequence of this observation, the Ascoli-Arzelà theorem is valid in $\mathscr{BC}([0,+\infty), \mathbb{R})$ and the compactness is understood as the topology of uniform convergence on compact subsets. However, this does not mean in any way that the compactness criteria in the two spaces $\mathscr{BC}([0,+\infty),\mathbb{R})$ and $\mathscr{C}([0,1],\mathbb{R})$ are the same. The convergence on compact subsets of $[0,+\infty)$ does not imply the strong convergence on $[0,+\infty)$ as shown by the following two counter-examples.

(a) The sequence of functions defined by $f_n(t) = 1 - e^{-t/n}$, $n \geq 1$ lies in $\mathscr{BC}([0,+\infty),\mathbb{R})$ for $\|f_n\|_{\mathscr{BC}([0,+\infty)} = 1$. Hence, $\mathscr{H} = \{f_n\}_{n\in\mathbb{N}}$ is bounded and even equi-continuous for f_n is nonexpansive. Notice further that for every compact interval $[a,b]$ and all $t \in [a,b]$, $\lim_{n\to+\infty} f_n(t) = 0$. However, $\{f_n\}_{n\in\mathbb{N}}$ does not converge to $f = 0$ for the sup-norm of $\mathscr{BC}([0,+\infty),\mathbb{R})$. Otherwise, $1 = \sup_{t\geq 0}|f_n(t) - f(t)| \to$ 0, as $n \to +\infty$, which is a contradiction.

(b) The sequence of functions defined by

$$f_n(t) = \begin{cases} 1, & t \leq n \\ n/t, & t \geq n \end{cases}$$

lies in $\mathscr{BC}([0,+\infty),\mathbb{R})$ for $\|f_n\|_{\mathscr{BC}([0,+\infty)} = 1$. It converges to 1 on compact subsets but $1 = \sup_{t\geq 0}|f_n(t) - 1| \not\to 0$, as $n \to +\infty$.

1.11.2 Bartles's compactness criterion and consequences

The following result was proved in 1955 [9].

Theorem 1.11.1 *Let X be a topological space and $\mathscr{H} \subset \mathscr{BC}(X,\mathbb{R})$ a bounded subset. Then the following statements are equivalent:*

(1) $\mathscr{H}$ is compact (understood in the compact-open topology).

(2) $\mathscr{H}$ is equi-continuous.

(3) For any positive ε, there is a finite partition $X = \bigcup_{i=1}^{n} A_i$ such that if x,y belong to the same A_i, then

$$|f(x) - f(y)| < \varepsilon, \ \forall f \in \mathscr{H}.$$

From Bartle's Theorem, some consequences have been recently obtained. However, since the compactness is not given in the sup-norm in $\mathscr{BC}(X,\mathbb{R})$, then the scope of these results is somewhat limited.

Proposition 1.11.2 *Let $\mathscr{H} \subset \mathscr{BC}(X, \mathbb{R})$ be a bounded subset and ϕ : $X \longrightarrow \mathbb{R}$ be a bounded function for which*

$$|f(t) - f(s)| \leq |\phi(t) - \phi(s)|, \text{ for all } f \in \mathscr{H} \text{ and } t, s \in X.$$

Then $\mathscr{H}$ is relatively compact in $\mathscr{BC}(X, \mathbb{R})$.

Proof 1.11.3 *Let $\varepsilon > 0$ be fixed and $Im(\phi) \subset (-M, M)$. Put $A_i = \phi^{-1}(((i-1)\varepsilon, i\varepsilon))$, $i = -n, -(n-1), \dots, 0, 1, \dots, n$, where $n \geq M/\varepsilon$. Then $\{A_i, i = -n, \dots, n\}$ is a finite partition of the space X which satisfies condition (3) of Bartles's Theorem.*

Corollary 1.11.4 *Let $h \in \mathscr{BC}([0, +\infty), \mathbb{R})$ be a differentiable function with $h'(t) \neq 0$ for any $t \geq 0$. Let $A = \{f \in \mathscr{BC}([0, +\infty), \mathbb{R}) : 0 \leq f(t) \leq M$, there exist f'_- and f'_+ for $t \geq 0\}$. If F is a subset of A such that*

$$\sup \left\{ \frac{\max\{|f'_-|, |f'_+|\}}{|h'(t)|} : t \geq 0, f \in F \right\} = K < \infty,$$

then F is relatively compact.

Corollary 1.11.5 *Let $\mathscr{H} \subset \mathscr{BC}(X, \mathbb{R})$ be a bounded subset and ϕ_i : $X \longrightarrow \mathbb{R}$ $(1 \leq i \leq k)$ be k bounded functions for which*

$$|f(t) - f(s)| \leq \Sigma_{i=1}^{k} |\phi_i(t) - \phi_i(s)|, \text{ for all } f \in \mathscr{H} \text{ and } t, s \in X.$$

Then $\mathscr{H}$ is relatively compact in $\mathscr{BC}(X, \mathbb{R})$.

Corollary 1.11.6 *Let $K : \mathscr{BC}(X, \mathbb{R}^n) \longrightarrow \mathscr{BC}(X, \mathbb{R}^n)$ be a continuous operator. Suppose that for any bounded set $\mathscr{H} \subset \mathscr{BC}(X, \mathbb{R}^n)$, $K(\mathscr{H})$ is a bounded set and there exist k bounded functions $\phi_i : X \longrightarrow \mathbb{R}^n$ $(1 \leq i \leq k)$ such that*

$$|(Kf)(t) - (Kf)(s)| \leq \Sigma_{i=1}^{k} |\phi_i(t) - \phi_i(s)|, \text{ for all } f \in \mathscr{H} \text{ and } t, s \in X.$$

Then K is a compact operator.

1.12 Higher-order Derivative Spaces

Let $\mathscr{C}^m(J,Y)$ denote the space of continuously differentiable functions defined on some interval $J \subset \mathbb{R}$, which take values in a Banach space Y. For $\mathscr{H} \subset \mathscr{C}^m(J,Y)$ and $k = 1,2,\ldots,m$, denote by $\mathscr{H}^{(k)}$, the space of functions $\mathscr{H}^{(k)} = \{x^{(k)} : x \in \mathscr{H}\}$ and $\mathscr{H}^{(k)}(t) = \{x^{(k)}(t) : x \in \mathscr{H}\}$.

1.12.1 The compact case

First, we present a generalization of Ascoli-Arzelà Theorem to the space $\mathscr{C}^m(J,Y)$ when $J = [a,b]$ is compact. We have

Theorem 1.12.1 $\mathscr{H} \subset \mathscr{C}^m(J,Y)$ *is relatively compact if and only if*

(a) $\mathscr{H}^{(m)}$ *is equi-continuous and, for any $t \in J$, $\mathscr{H}^{(m)}(t)$ is relatively compact in Y,*

(b) for each $k \in \{0,1,\ldots,m\}$, there exists $t_k \in J$ such that $\mathscr{H}^{(k)}(t_k)$ is relatively compact in Y.

Note that the relative compactness of $\mathscr{H}$ is equivalent to that of the sets $\mathscr{H}, \mathscr{H}', \ldots, \mathscr{H}^{(m)}$. To prove Theorem 1.12.1, we need two auxiliary results regarding the computation of a measure of noncompactness α.

Lemma 1.12.2 *Let $\mathscr{H} \subset \mathscr{C}^m(J,Y)$ be bounded and equi-continuous. Then $\alpha(\mathscr{H}(t))$ is continuous on J and*

$$\alpha\left(\left\{\int_J x(t)dt : x \in \mathscr{H}\right\}\right) \le \int_J \alpha(\mathscr{H}(t))dt.$$

Now $\alpha_{(m)}$ will refer to the Kuratowski measure of noncompactness in the space $\mathscr{C}^m(J,Y)$. The following lemma is a generalization of Lemma 1.7.8 to higher-order derivatives.

Lemma 1.12.3 *Let $\mathscr{H} \subset \mathscr{C}^m(J,Y)$ be such that $\mathscr{H}^{(m)}$ is equi-continuous. Then*

$$\alpha_{(m)}(\mathscr{H}) = \max\{\max_{t\in J} \alpha(\mathscr{H}(t)), \max_{t\in J} \alpha(\mathscr{H}'(t)), \ldots, \max_{t\in J} \alpha(\mathscr{H}^{(m)}(t))\}.$$

Proof 1.12.4 (Proof of Theorem 1.12.1) *Since necessity is obvious, we only prove sufficiency. Because $\mathscr{H}^{(m)}$ is equi-continuous and $\mathscr{H}^{(m)}(a)$ is bounded in Y, then $\mathscr{H}^{(m)}$ is bounded in $\mathscr{C}^m(J,Y)$. By the mean value theorem, $\mathscr{H}^{(m-1)}$ is bounded in $\mathscr{C}^m(J,Y)$. Lemma 1.12.2 implies that for all $t,t' \in J$:*

$$\alpha(\{x^{(m-1)}(t) - x^{(m-1)}(t') : x \in \mathscr{H}\}) = \alpha(\{\int_{t'}^{t} x^{(m)}(s)ds : x \in \mathscr{H}\})$$

$$\leq \int_{t'}^{t} \alpha(\mathscr{H}^{(m)}(s))ds = 0.$$

Using the additivity property of the MNC, we deduce that

$$\left|\alpha(\mathscr{H}^{(m-1)})(t)) - \alpha(\mathscr{H}^{(m-1)})(t'))\right|$$

$$\leq \alpha(\{x^{(m-1)}(t) - x^{(m-1)}(t') :$$

$$x \in \mathscr{H}\})$$

$$= 0.$$

Hence $\alpha(\mathscr{H}^{(m-1)}(t)) = \alpha(\mathscr{H}^{(m-1)}(t'))$ for any $t,t' \in J$. Now, condition (b) guarantees that $\alpha(\mathscr{H}^{(m-1)}(t_{m-1})) = 0$ so that $\alpha(\mathscr{H}^{(m-1)}(t)) = 0$, for all $t \in J$. Repeating this procedure for all the derivatives $\mathscr{H}^k$ $(0 \leq k \leq m-2)$ and using Lemma 2.2.13, we deduce that $\alpha_{(m)}(\mathscr{H}) = 0$, ending the proof of the theorem.

1.12.2 The noncompact case

Now, we consider the case of unbounded intervals and we first present some compactness results for continuously differentiable functions with derivatives having some limits at infinity. For some positive integer m, let $p_i : [0,\infty) \longrightarrow \mathbb{R}$ $(1 \leq i \leq m)$ be some continuous functions such that $p_i(t) > 0$, $\forall t \geq 0$ for all $i \in \{1,\ldots,m\}$. Set

$$E = \mathscr{C}_\infty^m([0,\infty),\mathbb{R})$$

$$= \left\{ x \in \mathscr{C}^m([0,\infty),\mathbb{R}) : \lim_{t \to +\infty} p_i(t)x^{(i)}(t), \text{ exist} \right.$$

$$\left. \text{for all } i = 1,\dots,m \right\}$$

which becomes a Banach space when endowed with the norm

$$\|x\|_E = \max_{1 \le i \le m} \sup_{t \ge 0} |p_i(t)x^{(i)}(t)|.$$

Arguing exactly as in Lemma 1.8.10, we can prove the following compact result.

Theorem 1.12.5 *Let $M \subset E$. Then M is relatively compact if the following conditions hold:*

(a) M is uniformly bounded in E.

(b) the functions belonging to the sets

$$\mathscr{A}_i = \{ y_i : y_i(t) = p_i(t)x^{(i)}(t), \ x \in M \}$$

are almost equi-continuous on $\mathbb{R}^+$, for every $i = 1,\dots,m$.

(c) The functions from $\mathscr{A}_i$ are equi-convergent at $+\infty$, for every $i = 1,\dots,m$.

If all derivatives have the same limit at positive infinity but functions take values in any Banach space, a compactness criterion is presented. Let

$$\mathscr{C}_0^m([0,+\infty),E) = \{x \in \mathscr{BC}([0,+\infty),E) : \lim_{t \to +\infty} x^{(k)}(t) = \theta, k = 0,1,\dots,m\},$$

where θ is the zero element. The following result has been proved in 2006. The first three conditions are the same as in Theorem 1.12.1.

Theorem 1.12.6 *A subset $\mathscr{H} \subset \mathscr{C}_0^m([0,+\infty),E)$ is relatively compact if and only if the following conditions are satisfied:*

(1) $\mathscr{H}^{(m)}$ is almost equi-continuous.

(2) For any $t \in [0,+\infty)$, $\mathscr{H}^{(m)}(t)$ is relatively compact in E.

(3) For any $k = 0,\dots,m-1$, there exists $t_k \in [0,+\infty)$ such that $\mathscr{H}^{(k)}(t_k)$ is relatively compact in E.

(4) If t → ∞, then $x^{(k)}(t)$ converges to θ uniformly for all $x \in \mathcal{H}$ and any $k = 0, 1, \ldots, m$.

We end this section with a compactness criterion in $\mathscr{BC}^m([0,+\infty), Y)$, the space of all functions that are continuous and bounded together with their derivatives up to the m-th ($m \geq 0$) order from $[0,+\infty)$ to a Banach space Y:

$$\mathscr{BC}^m([0,+\infty), Y) = \{f \in \mathscr{C}([0,+\infty), Y) : \forall k = 0, \ldots m,$$
$$f^{(k)} \text{ is bounded and continuous}\}.$$

Equipped with the norm

$$\|y\|_{\mathscr{BC}^m} = \begin{cases} \|y\|_\infty = \sup_{t \geq 0} \|y(t)\|, & \text{if } m = 0 \\ \sum_{k=0}^m \|y^{(k)}\|_0 & \text{if } m > 0, \end{cases}$$

it becomes a Banach space. A recent compactness criterion has been proved in 2013.

Theorem 1.12.7 *A subset $\mathcal{H} \subset \mathscr{BC}^m([0,+\infty), Y)$ is relatively compact if and only if the following conditions are satisfied:*

(1) $\mathcal{H}^{(m)}$ is relatively compact in $\mathscr{BC}([0,+\infty), Y)$.

(2) For all $k = 0, \ldots, m-1$ and for all $\varepsilon > 0$, there exists $M = M(\varepsilon) > 0$ such that

$$\alpha_{|\mathscr{BC}([0,+\infty),E)} \left(\mathcal{H}^{(k)}_{|[M,+\infty)} \right) < \varepsilon,$$

where α is some measure of noncompactness.

An application to the study of the following second-order boundary value problem can be found in [47]

$$\begin{cases} x'' + m^2 x' = f(t,x), \ t \geq 0 \\ x(0) = x_0, \ x(\infty) = \theta, \end{cases}$$

where $f : [0,+\infty) \times Y \longrightarrow Y$, Y is some Banach space, $\theta \in Y$ the zero element, $x_0 \in Y$, and $m \neq 1$. The function f is continuous and satisfies some sublinear growth conditions.

1.13 Zima's Compactness Criterion

1.13.1 Main result

Let $I: = [0,+\infty[$ and $p: I \longrightarrow (0,+\infty)$ be a continuous function. Denote by $E = \mathscr{C}_p([0,+\infty),\mathbb{R})$ the Banach space consisting of all weighted functions u continuous on I such that

$$\sup_{x\in I}\{|u(x)|p(x)\} < \infty.$$

Equipped with the norm $\|u\|_p = \sup_{x\in I}\{|u(x)|p(x)\}$, this is a Banach space. We have the following result.

Theorem 1.13.1 *If the functions $u \in \Omega$ are almost equi-continuous on I and uniformly bounded in the sense of the norm*

$$\|u\|_q = \sup_{x\in I}\{|u(x)|q(x)\},$$

where the function q is positive, continuous on I and satisfies

$$\lim_{x\to+\infty} \frac{p(x)}{q(x)} = 0 \ (q \text{ dominates } p \text{ at } +\infty),$$

then Ω is relatively compact in E.

Proof 1.13.2 *If $\{u_n\}_{n\in\mathbb{N}}$ is a sequence in Ω, uniformly bounded for the norm $\|.\|_q$, then there exists some $k > 0$ such that*

$$\forall n \in \mathbb{N}, \ \forall x \in I, \ |u_n(x)| \leq \frac{k}{q(x)}.$$

The functions $\{u_n\}_{n\in\mathbb{N}}$ are then uniformly bounded on any subinterval of I. In addition, these functions are, by assumption, equi-continuous on subintervals of I. By the Ascoli-Arzelà Theorem and a diagonal procedure, we get some subsequence $\{u_{n_k}\}_{k\in\mathbb{N}}$ converging almost uniformly to a limit function u which satisfies $|u(x)|.q(x) \leq k$. Let us prove that the sequence $\{u_{n_k}\}_{k\in\mathbb{N}}$, which we shorten to u_n, converges in E for the p-weighted norm. Indeed, for any $a > 0$, we have

$$\|u_n - u\|_p = \sup_{x\in I}\{|u_n(x) - u(x)|p(x)\}$$

$$\leq \sup_{x\in(0,a)} \{|u_n(x)-u(x)|p(x)\}$$

$$+\sup_{x>a}\{|u_n(x)-u(x)|q(x)\frac{p(x)}{q(x)}\},$$

and thus,

$$\|u_n-u\|_p \leq \sup_{x\in(0,a)} \{|u_n(x)-u(x)|\} \sup_{x\in(0,a)} p(x)+2k\sup_{x>a}\{p(x)/q(x)\}.$$

Since, $\{u_n\}_{n\in\mathbb{N}}$ converges almost uniformly to u (on compact subsets) and $\lim\limits_{x\to+\infty}\dfrac{p(x)}{q(x)}=0$, we infer that $\lim\limits_{n\to\infty}\|u_n-u\|_p=0$, proving our claim.

Remark 1.13.3 *In fact, Zima's compactness criterion (Theorem 1.13.1) is a direct consequence of Avramescu-Corduneanu compact-nees criterion with weight p (Theorem 1.12.5). Indeed, under the assumptions of Theorem 1.13.1, we have $p(t)u(t)=q(t)u(t)r(t)$, where $r(t)=p(t)/q(t)$ and thus,*

$$\lim_{t\to+\infty} p(t)u(t)=0,$$

that is $u\in\mathscr{C}_p^0$. As for equi-convergence condition, it is also satisfied. Since $\lim\limits_{t\to+\infty} r(t)=0$, for every $\varepsilon>0$, there is some $T(\varepsilon)>0$ such that for all $t,s\geq T(\varepsilon)$, we have

$$|p(t)u(t)-p(s)u(s)| \leq |q(t)u(t)|r(t)-r(s)|$$

$$+|r(s)||q(t)u(t)-q(s)u(s)|$$

$$\leq \|qu\|_0(|r(t)-r(s)|+2|r(s)|)$$

$$\leq k\left(\frac{\frac{\varepsilon}{k}}{2}+2\frac{\frac{\varepsilon}{k}}{4}\right)$$

$$= \varepsilon,$$

where $\|qu\|_0\leq k$, for all $u\in\mathscr{C}_p$.

1.13.2 Application 1

Let $E = \mathscr{C}_p^1$ be the set of the functions $x \in \mathscr{C}^1([0,\infty)$ such that

$$\sup_{x \in I}\{[|u(x)| + |u'(x)|]p(x)\} < \infty.$$

Equipped with the norm $\|u\|_{p,1} = \sup_{x \in I}\{[|u(x)| + |u'(x)|]p(x\}$, it is also a Banach space. Analogously to Theorem 1.13.1, we can prove the following result.

Corollary 1.13.4 *If the functions $u \in \Omega$ and their derivatives are almost equi-continuous on I and uniformly bounded in the sense of the norm*

$$\|u\|_{p,1} = \sup_{x \in I}\{[|u(x)| + |u'(x)|]p(x\},$$

where the function q is positive, continuous on I and satisfies

$$\lim_{x \to +\infty} \frac{p(x)}{q(x)} = 0,$$

then Ω is relatively compact in E.

1.13.3 Application 2

Let $p : I \longrightarrow I$ be a continuous function and let $E = \mathscr{C}_p^1$ be the set of the functions $x \in \mathscr{C}^3([0,\infty), \mathbb{R})$ such that

$$\sup_{t \in I}\{[|x(t)| + |x'(t)| + |x''(t)| + |x'''(t)|]p(t)\} < \infty.$$

Equipped with the norm $\|x\|_p = \sup_{t \in I}\{p(t)\sum_{k=0}^{k=3}|x^{(k)}(t)|\}$, E is a Banach space.

Corollary 1.13.5 *Let the functions $\{x \in \Omega \subset E\}$ and their derivatives x', x'', and x''' be almost equi-continuous on I and uniformly bounded in the sense of the norm $\|x\|_q = \sup_{t \in I}\{[\sum_{k=0}^{k=3}|x^{(k)}(t)|]q(t)\}$, where q is a positive continuous dominant function on I, that is $\lim_{t \to +\infty} \frac{p(t)}{q(t)} = 0$. Then Ω is relatively compact in E.*

Proof 1.13.6 *If $\{x_n\}_{n\in\mathbb{N}}$ is a sequence in Ω, uniformly bounded for the norm $\|.\|_q$, then there exists some $M > 0$ such that*

$$\forall n \in \mathbb{N}, \forall t \in I, \sum_{k=0}^{k=3} |x_n^{(k)}(t)| \leq \frac{M}{q(t)}.$$

The functions $\{x_n^{(k)}\}_{n\in\mathbb{N}}$, $k = 0,1,2,3$ are then uniformly bounded on any subinterval of I. In addition, these functions are, by assumption, equicontinuous on subintervals of I. By the Ascoli-Arzelà Theorem and a diagonal procedure, there exists some subsequence $\{x_{n_j}^{(k)}\}_{j\in\mathbb{N}}$ converging uniformly to some limit function $x^{(k)}$ on every compact subset of I. Moreover, $\sum_{k=0}^{k=3} |x^{(k)}(t)| \leq \frac{M}{q(t)}$. Let us prove that the sequence $\{x_{n_j}\}_{j\in\mathbb{N}}$, which we shorten to (x_n), converges in E for the p-weighted norm. Indeed, for any $T > 0$, we have

$$\|x_n - x\|_p = \sup_{t\in I} \sum_{k=0}^{k=3} |x_n^{(k)}(t) - x^{(k)}(t)|p(t)$$

$$\leq \sup_{t\in[0,T]} \sum_{k=0}^{k=3} |x_n^{(k)}(t) - x^{(k)}(t)|p(t)$$

$$+ \sup_{t>T} \sum_{k=0}^{k=3} |x_n^{(k)}(t) - x^{(k)}(t)|q(t)\frac{p(t)}{q(t)}$$

$$\leq \sum_{k=0}^{k=3} \sup_{t\in[0,T]} |x_n^{(k)}(t) - x^{(k)}(t)| \sup_{t\in[0,T]} p(t)$$

$$+ 2M \sup_{t>T} \frac{p(t)}{q(t)}.$$

Since $\{x_n\}_{n\in\mathbb{N}}$ converges uniformly to x on every compact subset of I and $\lim_{x\to+\infty} \frac{p(t)}{q(t)} = 0$, we infer that $\lim_{n\to\infty} \|x_n - x\|_p = 0$, proving our claim.

1.14 Cones and Partial Order Relations

Definition 1.14.1 *A nonempty subset $\mathscr{P}$ of a Banach space X is called a cone if $\mathscr{P}$ is convex, closed, and satisfies the conditions:*

(i) $\alpha x \in \mathscr{P}$ for all $x \in \mathscr{P}$ and $\alpha \geq 0$,

(ii) $x, -x \in \mathscr{P}$ implies $x = 0$.

Example 1.14.2 *Let $X = \mathbb{R}^n$. We set*

$$\mathscr{P} = \{(x_1,\ldots,x_n) : x_j \geq 0, \quad j \in \{1,\ldots,n\}\}.$$

Then $\mathscr{P}$ is a cone and

$$(x_1,\ldots,x_n) \leq (y_1,\ldots,y_n)$$

if and only if $x_j \leq y_j$, $j \in \{1,\ldots,n\}$.

Example 1.14.3 *Let $X = \mathscr{C}(\overline{\Omega})$ for a set $\Omega \subset \mathbb{R}^n$. We set*

$$\mathscr{P} = \{f \in \mathscr{C}(\overline{\Omega}) : f(x) \geq 0, \quad x \in \overline{\Omega}\}.$$

We have that $f \leq g$ if an only if $f(x) \leq g(x)$ for all $x \in \overline{\Omega}$.

Remark 1.14.4 *Condition (i) both with $\mathscr{P}$ convex is equivalent to*

$$\alpha, \beta \geq 0 \text{ and } x,y \in \mathscr{P} \text{ imply } \alpha x + \beta y \in \mathscr{P}.$$

Definition 1.14.5 *A cone $\mathscr{P}$ is called solid if it contains interior points, i.e., $\mathscr{P}^o \neq \emptyset$.*

Example 1.14.6 *Let $X = \mathbb{R}^n$. The cone $\mathscr{P}$, defined in Example 1.14.2, is a solid cone.*

Every cone $\mathscr{P}$ in a Banach space X defines a partial order relation $\leq$ in X as follows

$$x \leq y \iff y - x \in \mathscr{P}.$$

If $x \leq y$ and $x \neq y$, we write $x < y$; if $\mathscr{P}$ is a solid cone and $y - x \in \mathscr{P}^o$, we write $x \ll y$.

Definition 1.14.7 *By an ordered Banach space we mean a real Banach space with a cone.*

Definition 1.14.8 *An order interval in a Banach space with a cone $\mathscr{P}$ is an interval of the form*

$$[x,y] = \{x \in \mathscr{P} : x \leq z \leq y\}.$$

Definition 1.14.9 *A cone $\mathscr{P} \subset X$ is said to be normal if there exists a constant $N > 0$ such that*

$$0 \leq x \leq y \Rightarrow \|x\| \leq N\|y\|, \quad \forall x,y \in \mathscr{P}. \tag{1.46}$$

i.e., the norm $\|.\|$ is semi-monotone. The least positive number N, if exists, is called a normal constant. Clearly, $N \geq 1$. In fact, taking $y = x \neq 0$ in (3.23), we have $N \geq 1$.

Example 1.14.10 *Let $X = \mathbb{R}^n$. Then the cone $\mathscr{P}$, defined in Example 1.14.2, is a normal cone.*

Below with θ we will denote the zero element in a Banach space X.

Theorem 1.14.11 *Let $\mathscr{P}$ be a cone in a Banach space X. Then the following assertions are equivalent.*

(i) $\mathscr{P}$ *is normal.*

(ii) *There exists a equivalent norm $\|.\|_1$ on X such that*

$$\theta \leq x \leq y \Rightarrow \|x\|_1 \leq \|y\|_1, \quad \forall x,y \in \mathscr{P},$$

that is, the norm $\|.\|_1$ is monotone.

(iii) $x_n \leq z_n \leq y_n \ (n \in \mathbb{N})$ *and* $\|x_n - x\| \to 0, \|y_n - x\| \to 0$, *as* $n \to \infty$, *imply* $\|z_n - x\| \to 0$, *as* $n \to \infty$.

(iv) *The set $(B + \mathscr{P}) \cap (B - \mathscr{P})$ is bounded, where $B = \{x \in X : \|x\| \leq 1\}$.*

(v) *Any order interval $[x,y] = \{z \in X : x \leq z \leq y\}$ is bounded.*

(vi) *There exists* $\delta > 0$ *such that* $\|x+y\| \geq \delta$ *for all* $x,y \in \mathscr{P}$ *with* $\|x\| = \|y\| = 1$.

(vii) *There exists* $\gamma > 0$ *such that*

$$\|x+y\| \geq \gamma \max\{\|x\|, \|y\|\} \quad \text{for all } x,y \in \mathscr{P}.$$

Proof 1.14.12

1. $(i) \Longrightarrow (ii)$ *Let* $N \geq 1$ *be the normal constant for the cone* $\mathscr{P}$. *For an element* $x \in X$, *define*

$$\|x\|_1 = \inf_{u \leq x} \|u\| + \inf_{v \geq x} \|v\|. \tag{1.47}$$

We will prove that $\| \cdot \|_1$ *satisfies all axioms for a norm. Firstly, note that*

$$\|\theta\|_1 = \inf_{u \leq \theta} \|u\| + \inf_{v \geq \theta} \|v\|$$

$$= 0.$$

Assume that $\|x\|_1 = 0$ *for some* $x \in X$. *By* (1.47) *of* $\| \cdot \|_1$, *it follows that for any* $\varepsilon > 0$ *there exist* $u_1, v_1 \in X$ *such that*

$$u_1 \leq x \leq v_1, \quad \|u_1\| < \varepsilon \quad \text{and} \quad \|v_1\| < \varepsilon.$$

Because $\mathscr{P}$ *is a normal cone with normal constant* N *and* $x - u_1 \leq v_1 - u_1$, *we find*

$$\|x\| \leq \|x - u_1\| + \|u_1\|$$

$$\leq N\|v_1 - u_1\| + \|u_1\|$$

$$\leq N(\|v_1\| + \|u_1\|) + \varepsilon$$

$$< (2N+1)\varepsilon.$$

Since the last inequality is true for any $\varepsilon > 0$, *we conclude that* $x = \theta$. *Next, for* $\alpha \in R$ *and* $x \in X$, *we have*

$$\|\alpha x\|_1 = |\alpha|\|x\|_1.$$

Now, we take $x, y \in X$ and $\varepsilon > 0$ arbitrarily. Then there exist $u_1, u_2, v_1, v_2 \in X$ so that

$$u_1 \leq x \leq v_1 \quad and \quad u_2 \leq y \leq v_2$$

and

$$\|u_1\| + \|v_1\| \leq \|x\|_1 + \frac{\varepsilon}{2}, \quad \|u_2\| + \|v_2\| \leq \|y\|_1 + \frac{\varepsilon}{2}.$$

Note that

$$u_1 + u_2 \leq x + y$$
$$\leq v_1 + v_2$$

and from here,

$$\|u_1 + u_2\| + \|v_1 + v_2\| \geq \|x + y\|_1$$

and then

$$\|x + y\|_1 \leq \|u_1\| + \|v_1\| + \|u_2\| + \|v_2\|$$
$$\leq \|x\|_1 + \|y\|_1 + \varepsilon.$$

Now, from the arbitrariness of ε, we get

$$\|x + y\|_1 \leq \|x\|_1 + \|y\|_1.$$

Therefore $\|\cdot\|_1$, defined by (1.47), satisfies all axioms for a norm. Take $\theta \leq x \leq y$, $x, y \in X$. Then, we have

$$\inf_{u \leq x} \|u\| + \inf_{u \leq y} \|u\| = 0$$

and

$$\|x\|_1 = \inf_{v \geq x} \|v\|$$
$$\leq \inf_{v \geq y} \|v\|$$

$$\leq \quad \|y\|_1.$$

Next, we will show that the norms $\|\cdot\|$ and $\|\cdot\|_1$ are equivalent. Clearly, for $x \in X$, we have

$$\|x\|_1 \quad = \quad \inf_{u \leq x} \|u\| + \inf_{u \geq x} \|u\|$$

$$\leq \quad \|x\| + \|x\|$$

$$= \quad 2\|x\|.$$

Let now, $u, v, x \in X$ be such that $u \leq x \leq v$. Ten, using that $\mathscr{P}$ is a normal cone with normal constant N, we get

$$\|x\| \quad \leq \quad \|x - u\| + \|u\|$$

$$\leq \quad N\|v - u\| + \|u\|$$

$$\leq \quad N\|v\| + N\|u\| + \|u\|$$

$$\leq \quad (N+1)(\|u\| + \|v\|)$$

and from here,

$$\|x\| \quad \leq \quad (N+1)\left(\inf_{u \leq x} \|u\| + \inf_{v \geq x} \|v\|\right)$$

$$= \quad (N+1)\|x\|_1.$$

Consequently

$$\frac{1}{N+1}\|x\| \leq \|x\|_1 \leq 2\|x\|.$$

2. $(ii) \Longrightarrow (iii)$ *Suppose that $x, x_n, y_n, z_n \in X$, $n \in \mathbb{N}$, be such that*

$$\|x_n - x\| \to 0, \quad \|y_n - x\| \to 0, \quad as \quad n \to \infty.$$

Then $z_n - x_n \leq y_n - x_n$, $n \in \mathbb{N}$, and since $\|\cdot\|_1$ is monotone, we get

$$\|z_n - x_n\|_1 \leq \|y_n - x_n\|_1, \quad n \in \mathbb{N}.$$

Because $\|\cdot\|$ and $\|\cdot\|_1$ are equivalent, there exist positive constants $M > m$ and

$$m\|\cdot\| \leq \|\cdot\|_1 \leq M\|\cdot\|.$$

Hence,

$$
\begin{aligned}
\|z_n - x_n\| &\leq \frac{1}{m}\|z_n - x_n\|_1 \\[2mm]
&\leq \frac{1}{m}\|y_n - x_n\|_1 \\[2mm]
&\leq \frac{M}{m}\|y_n - x_n\| \\[2mm]
&\to 0, \quad as \quad n \to \infty.
\end{aligned}
$$

3. $(iii) \implies (iv)$ Assume that $(B + \mathscr{P}) \cap (B - \mathscr{P})$ is unbounded. Then there exists a sequence $\{z_n\}_{n \in \mathbb{N}} \subset (B + \mathscr{P}) \cap (B - \mathscr{P})$ such that $\|z_n\| \to \infty$, as $n \to \infty$, and there exist sequences $\{x_n\}_{n \in \mathbb{N}}$, $\{y_n\}_{n \in \mathbb{N}} \subset B$ such that

$$x_n \leq z_n \leq y_n, \quad n \in \mathbb{N}.$$

Set

$$
\begin{aligned}
u_n &= \frac{x_n}{\|z_n\|}, \\[2mm]
v_n &= \frac{y_n}{\|z_n\|}, \\[2mm]
w_n &= \frac{z_n}{\|z_n\|}, \quad n \in \mathbb{N}.
\end{aligned}
$$

Then

$$u_n \leq w_n \leq v_n, \quad n \in \mathbb{N},$$

and

$$\|u_n\|, \quad \|v_n\| \to 0, \quad as \quad n \to \infty.$$

Hence and 2), we get $\|w_n\| \to 0$, as $n \to \infty$. This is a contradiction because $\|w_n\| = 1$, $n \in \mathbb{N}$. Consequently $(B + \mathscr{P}) \cap (B - \mathscr{P})$ is bounded.

4. $(iv) \Longrightarrow (v)$ *Since* $(B + \mathscr{P}) \cap (B - \mathscr{P})$ *is bounded, there is a* $\beta > 0$ *so that* $(B + \mathscr{P}) \cap (B - \mathscr{P}) \subset \beta B$. *Set* $\gamma = \max\{\|x\|, \|y\|\}$. *Take* $z \in [x, y]$ *arbitrarily. Then* $x \leq z \leq y$ *and*

$$\frac{z}{\gamma} \leq \frac{y}{\|x\|} \quad and \quad \frac{z}{\gamma} \leq \frac{y}{\|y\|}$$

and then $\frac{z}{\gamma} \in (B + \mathscr{P}) \cap (B - \mathscr{P})$. *Therefore,* $z \in \beta \gamma B$ *and* $[x, y] \subset \beta \gamma B$.

5. $(v) \Longrightarrow (vi)$ *Suppose the contrary. Then there exist sequences* $\{x_n\}_{n \in \mathbb{N}}, \{y_n\}_{n \in \mathbb{N}} \subset \mathscr{P}$ *such that*

$$\|x_n\| = \|y_n\| = 1 \quad and \quad \|x_n + y_n\| < \frac{1}{4^n}, \quad n \in \mathbb{N}.$$

Set

$$u_n = \frac{x_n}{\sqrt{\|x_n + y_n\|}},$$

$$v_n = \frac{x_n + y_n}{\sqrt{\|x_n + y_n\|}}, \quad n \in \mathbb{N}.$$

Then $\theta \leq u_n \leq v_n$, $n \in \mathbb{N}$, *and*

$$\sum_{n=1}^{\infty} \|v_n\| = \sum_{n=1}^{\infty} \sqrt{\|x_n + y_n\|}$$

$$\leq \sum_{n=1}^{\infty} \frac{1}{2^n}$$

$$< \infty.$$

Thus, $\sum_{n=1}^{\infty} v_n$ *converges to a point* $v \in X$. *We have*

$$0 \leq u_n \leq v_n \leq v, \quad n \in \mathbb{N}.$$

Since

$$\|u_n\| = \frac{\|x_n\|}{\sqrt{\|x_n + y_n\|}}$$

$$= \frac{1}{\sqrt{\|x_n + y_n\|}}$$

$$> 2^n, \quad n \in \mathbb{N},$$

we conclude that the interval $[\theta, v]$ *is unbounded, which is a contradiction.*

6. $(vi) \Longrightarrow (vii)$ *We have that there exists a* $\delta > 0$ *so that*

$$\|x + y\| > \delta, \quad x, y \in \mathscr{P}, \quad \|x\| = \|y\| = 1.$$

Let $\gamma = \dfrac{\delta}{2}$. *Take* $x, y \in \mathscr{P}$. *If* $x = \theta$ *or* $y = \theta$, *the assertion is evident. Let* $x \neq \theta$ *and* $y \neq \theta$. *Without loss of generality, suppose that* $\|x\| = 1$ *and* $0 < \|y\| \leq 1$. *Otherwise, take* $x_1 = \dfrac{x}{\|x\|}$ *and* $y_1 = \dfrac{y}{\beta}$, *where* $\beta \geq \|y\|$. *Then*

$$1 = \|x\|$$

$$\leq \|x + y\| + \|y\|.$$

Hence,

$$1 - \|y\| \leq \|x + y\|.$$

On the other hand,

$$\left\| x + \frac{y}{\|y\|} \right\| > \delta$$

and

$$\|x + y\| = \left\| x + \frac{y}{\|y\|} - \frac{1 - \|y\|}{\|y\|} y \right\|$$

$$\geq \left\| x + \frac{y}{\|y\|} \right\| - (1 - \|y\|)$$

$$> \delta - \|x + y\|,$$

whereupon

$$\|x + y\| \geq \gamma \{\|x\|, \|y\|\}.$$

7. $(vii) \Longrightarrow (i)$ *Assume that the cone $\mathscr{P}$ is not normal. Then there exist sequences $\{x_n\}_{n\in\mathbb{N}}, \{y_n\}_{n\in\mathbb{N}} \subset \mathscr{P}$ such that*

$$\theta \le x_n \le y_n, \quad \|x_n\| > n\|y_n\|, \quad \|y_n\| > 0, \quad n \in \mathbb{N}.$$

Let

$$u_n = \frac{x_n}{\|x_n\|} + \frac{y_n}{n\|y_n\|},$$

$$v_n = -\frac{x_n}{\|x_n\|} + \frac{y_n}{n\|y_n\|}, \quad n \in \mathbb{N}.$$

We have that

$$u_n \ge \theta,$$

$$v_n \ge -\frac{x_n}{n\|y_n\|} + \frac{y_n}{n\|y_n\|}$$

$$\ge \theta, \quad n \in \mathbb{N}.$$

Next,

$$\|u_n\| = \left\| \frac{x_n}{\|x_n\|} + \frac{y_n}{n\|y_n\|} \right\|$$

$$\ge 1 - \frac{1}{n},$$

$$\|v_n\| = \left\| -\frac{x_n}{\|x_n\|} + \frac{y_n}{n\|y_n\|} \right\|$$

$$\ge 1 - \frac{1}{n}, \quad n \in \mathbb{N}.$$

Thus,

$$\|u_n + v_n\| = \frac{2}{n}, \quad n \in \mathbb{N},$$

and

$$\|u_n + v_n\| \ge \gamma \max\{\|u_n\|, \|v_n\|\}$$

$$\geq \gamma\left(1 - \frac{1}{n}\right), \quad n \in \mathbb{N}.$$

Hence,

$$\frac{2}{n} \geq \gamma\left(1 - \frac{1}{n}\right), \quad n \in \mathbb{N}.$$

Now, letting $n \to \infty$, *we obtain* $\gamma = 0$. *This is contradiction. Consequently* $\mathscr{P}$ *is a normal cone. This completes the proof.*

Example 1.14.13 *Let G be a closed and bounded subset of $\mathbb{R}^n$ and $X = \mathscr{C}(G)$, the space of the continuous functions on G equipped with the sup norm. Let*

$$\mathscr{P}_1 = \{x \in \mathscr{C}(G) : x(t) \geq 0, t \in G\}.$$

It is easy to show that $\mathscr{P}_1$ is a solid cone in $\mathscr{C}(G)$ and $\mathscr{P}_1^o = \{x \in \mathscr{C}(G) : x(t) > 0, t \in G\}$. From $x \leq y \Leftrightarrow x(t) \leq y(t)$ for all $t \in G$, its norm is monotone. Then $\mathscr{P}_1$ is normal and the normal constant $N = 1$. Let

$$\mathscr{P}_2 = \{x \in \mathscr{C}(G) : x(t) \geq 0, t \in G \text{ and } \int_{G_0} x(t)\,dt \geq \varepsilon_0 \|x\|\},$$

$$\mathscr{P}_3 = \{x \in \mathscr{C}(G) : x(t) \geq 0, t \in G \text{ and } \min_{t \in G_1} x(t)\,dt \geq \varepsilon_1 \|x\|\},$$

where G_0, G_1 are two closed subset of G and $\varepsilon_0, \varepsilon_1$ are constants with $0 < \varepsilon_0 < mesG_0$ and $0 < \varepsilon_1 < 1$. It is easy to see that $\mathscr{P}_2 \subset \mathscr{P}_1$ and $\mathscr{P}_3 \subset \mathscr{P}_1$, both $\mathscr{P}_2$ and $\mathscr{P}_3$ are solid cone in $\mathscr{C}(G)$.

Example 1.14.14 *Let $X = L^p(\Omega)$, where $\Omega \in R^n$, $0 < \text{meas}\,\Omega < +\infty$ and $1 \leq p < +\infty$, and $\mathscr{P}_4 = \{x \in L^p(\Omega) : x(t) \geq 0, a.e.t \in \Omega\}$. It is easy to know that $\mathscr{P}_4$ is a normal cone and its normal constant $N = 1$. Clearly, $\mathscr{P}_4^o = \emptyset$. Thus $\mathscr{P}_4$ is not a solid cone.*

Example 1.14.15 *Let $X = \mathscr{C}^1[a,b]$, formed by*

$$\|x\| = \max_{a \leq t \leq b} |x(t)| + \max_{a \leq t \leq b} |x'(t)|.$$

Let $\mathscr{P}_5 = \{x \in \mathscr{C}^1[a,b] : x(t) \geq 0, a \leq t \leq b\}$. Clearly, $\mathscr{P}_5$ is a solid cone in $\mathscr{C}^1[a,b]$ but is not normal. In fact, if $\mathscr{P}_5$ is normal, then there would exists a constant $N > 0$ such that, if $0 \le x \le y$, then $\|x\| \le N\|y\|$. Let $x_n(t) = 1 - \cos(nt)$ and $y_n(t) = 2$ for $n = 1, 2, \ldots$. So we have

$$0 \le x_n \le y_n, \quad \|x_n\| = 2 + n, \quad \|y_n\| = 2.$$

Then $2 + n \le 2N$ for $n = 1, 2, \ldots$, which is impossible.

Definition 1.14.16 *Let $\mathscr{P}$ be a cone in a Banach space X. If for any sequence $\{x_n\}_{n \in \mathbb{N}} \subset X$ that satisfies the condition*

$$x_1 \le x_2 \le \ldots \le x_n \le \ldots \le y$$

for some $y \in X$, there exists an element $x \in X$ so that $\|x_n - x\| \to 0$, as $n \to \infty$, then the cone $\mathscr{P}$ is said to be regular.

Example 1.14.17 *The cone $\mathscr{P}$ in Example 1.14.2 is regular.*

Example 1.14.18 *The cone $\mathscr{P}$ in Example 1.14.3 is not regular.*

Theorem 1.14.19 *Let $\mathscr{P}$ be a regular cone in a Banach space X. Then $\mathscr{P}$ is a normal cone.*

Proof 1.14.20 *Suppose that $\mathscr{P}$ is not a normal cone. Then there exist sequences $\{x_n\}_{n \in \mathbb{N}}$, $\{y_n\}_{n \in \mathbb{N}} \subset X$ such that*

$$\theta \le x_n \le y_n \quad \text{and} \quad \|x_n\| > 2^n\|y_n\|, \quad n \in \mathbb{N}.$$

Set

$$z_n = \frac{x_n}{\|x_n\|}, \quad v_n = \frac{y_n}{2^n\|y_n\|}, \quad n \in \mathbb{N}.$$

Then

$$
\begin{aligned}
\theta \quad &< \quad z_n \\
&\le \quad \frac{x_n}{2^n\|y_n\|} \\
&\le \quad \frac{y_n}{2^n\|y_n\|} \\
&= \quad v_n
\end{aligned}
$$

and

$$\sum_{n=1}^{\infty} \|v_n\| = \sum_{n=1}^{\infty} \frac{1}{2^n}$$

$$= 1.$$

Thus, the series $\displaystyle\sum_{n=1}^{\infty} v_n$ *converges to some element* $v \in X$. *Define*

$$w_n = \begin{cases} v_1 + v_2 + \cdots + v_{2m}, & \text{if } n = 2m, \quad m \in \mathbb{N}, \\ \\ v_1 + v_2 + \cdots + v_{2m} + z_{2m+1}, & \text{if } n = 2m+1, \quad m \in \mathbb{N}. \end{cases}$$

We have

$$\theta < w_2 \le w_3 \le w_4 \le \ldots \le v, \quad \sup_{n\in\mathbb{N}} \|w_n\| \le 2.$$

On the other hand,

$$\|w_{2m+1} - w_{2m}\| = \|z_{2m+1}\|$$

$$= 1, \quad m \in \mathbb{N}.$$

Therefore, the sequence $\{w_n\}_{n\in\mathbb{N}}$ *is not convergent. This is a contradiction because the cone* $\mathscr{P}$ *is regular. Consequently the cone* $\mathscr{P}$ *is a normal cone. This completes the proof.*

Definition 1.14.21 *Let* $\mathscr{P}$ *be a cone in a Banach space* X. *If for any sequence* $\{x_n\}_{n\in\mathbb{N}} \subset X$ *that satisfies the condition*

$$x_1 \le x_2 \le \ldots \le x_n \le \ldots, \quad M = \sup_{n\in\mathbb{N}} \|x_n\| < \infty,$$

there exists an element $x \in X$ *so that* $\|x_n - x\| \to 0$, *as* $n \to \infty$, *then the cone* $\mathscr{P}$ *is said to be fully regular.*

Example 1.14.22 *The cone* $\mathscr{P}$ *in Example 4.38 is fully regular.*

Exercise 1.14.23 *Let* $X = \{x = \{x_n\}_{n \in \mathbb{N}} : x_n \to 0, \quad as \quad n \to \infty\}$. *In X define a norm*

$$\|x\| = \sup_{n \in \mathbb{N}} |x_n|.$$

Let

$$\mathscr{P} = \{x = \{x_n\}_{n \in \mathbb{N}} \in X : x_j \geq 0, \quad j \in \mathbb{N}\}.$$

Prove that $\mathscr{P}$ *is a regular cone and it is not a fully regular cone.*

Remark 1.14.24 *By the proof of Theorem 1.14.19, it follows that if* $\mathscr{P}$ *is a fully regular cone in a Banach space X, then* $\mathscr{P}$ *is a normal cone.*

Theorem 1.14.25 *Let* $\mathscr{P}$ *be a fully regular cone in a Banach space X. Then* $\mathscr{P}$ *is a regular cone.*

Proof 1.14.26 *By Remark 1.14.24, we have that* $\mathscr{P}$ *is a normal cone. Let N be the normal constant. Take a sequence* $\{x_n\}_{n \in \mathbb{N}} \subset X$ *such that*

$$x_1 \leq x_2 \leq \ldots \leq x_n \leq \ldots \leq y$$

for some $y \in X$. *Then*

$$y - x_n \leq y - x_1, \quad n \in \mathbb{N}.$$

Hence and the normality of the cone $\mathscr{P}$, *it follows that*

$$\|y - x_n\| \leq N\|y - x_1\|, \quad n \in \mathbb{N}.$$

Therefore $\sup_{n \in \mathbb{N}} \|x_n\| < \infty$. *Because* $\mathscr{P}$ *is a fully regular cone, there exists an element* $x \in X$ *such that*

$$\|x - x_n\| \to 0, \quad as \quad n \to \infty.$$

Thus, $\mathscr{P}$ *is a regular cone. This completes the proof.*

Definition 1.14.27 *Let* $\mathscr{P}$ *be a cone in a Banach space X.*

1. If $X = \mathscr{P} - \mathscr{P}$, *i.e., any element* $x \in X$ *can be expressed in the form* $x = y - z, y, z \in \mathscr{P}$, *then we say that the cone* $\mathscr{P}$ *is reproducing.*

2. If $X = \overline{\mathscr{P} - \mathscr{P}}$, then we say that the cone $\mathscr{P}$ is total.

Example 1.14.28 *The cone $\mathscr{P}$ in Example 1.14.2 is reproducing.*

Exercise 1.14.29 *Prove that the cone $\mathscr{P}$ in Example 1.14.3 is total.*

Theorem 1.14.30 *If $\mathscr{P}$ is a solid cone in a Banach space, then $\mathscr{P}$ is reproducing.*

Proof 1.14.31 *Take $x_0 \in \mathscr{P}^o$ arbitrarily. Since $\mathscr{P}^o$ is a solid cone, there is an $r > 0$ so that*

$$B(x_0, r) = \{x \in X : \|x - x_0\| \le r\} \subset \mathscr{P}.$$

Take $x \in X$, $x \ne \theta$. Note that

$$x_0 \pm \frac{r}{\|x\|} x \in B(x_0, r) \subset \mathscr{P}.$$

Set

$$y = \frac{\|x\|}{2r}\left(x_0 + \frac{r}{\|x\|}x\right),$$

$$z = \frac{\|x\|}{2r}\left(x_0 - \frac{r}{\|x\|}x\right).$$

Then $y, z \in \mathscr{P}$ and $y - z = x$. Thus, $\mathscr{P}$ is reproducing. This completes the proof.

Theorem 1.14.32 *Assume that $\mathscr{P}$ is a cone in a Banach space X and $B = \{x \in X : \|x\| \le 1\}$. Then the following assertions are equivalent.*

(i) *$\mathscr{P}$ is reproducing.*

(ii) *There exists a constant $r > 0$ such that any $x \in X$ can be expressed in the form $x = y - z$, where $y, z \in \mathscr{P}$ and $\|y\| \le r\|x\|$ and $\|z\| \le r\|x\|$.*

(iii) *There exists a constant $\gamma > 0$ such that*

$$\gamma B \subset B \cap \mathscr{P} - B \cap \mathscr{P}.$$

(iv) *There exists a constant* $\delta > 0$ *such that*

$$\delta B \subset \overline{B \cap \mathscr{P} - B \cap \mathscr{P}}.$$

Proof 1.14.33 $(i) \Longrightarrow (ii)$ *and* $(ii) \Longrightarrow (i)$ *Assume* (i). *We have that* $\mathscr{P}$ *is a reproducing cone in X. Let*

$$X_n = \left\{ x \in X : \exists y \in \mathscr{P} \quad such \quad that \quad x \leq y, \quad \|y\| \leq n\|x\| \right\},$$

$$n \in \mathbb{N}.$$

We have that $X = \bigcup_{n=1}^{\infty} X_n$. *Note that X is a set of the second category. Then there are* $n_1 \in \mathbb{N}$, $x_0 \in X$ *and* $R, r > 0$ *such that*

$$\overline{X_{n_1}} \supset T = \{x \in X : r < \|x - x_0\| < R\}.$$

Take $y_0, z_0 \in \mathscr{P}$ *and set* $-x_0 = y_0 - z_0$. *We choose* $n_2 \in \mathbb{N}$ *so that* $\|y_0\| \leq n_2\|x_0\|$. *Set*

$$T_0 = \{x \in X : r < \|x\| < R\}.$$

Now, take $n_3 \in \mathbb{N}$ *so that*

$$n_3 > n_1 + \frac{1}{r}(n_1 + n_2)\|x_0\|.$$

Let $x \in T_0$ *be arbitrarily chosen. Then* $y = x_0 + x \in T$. *Note that there exists a sequence* $\{x_j\}_{j \in \mathbb{N}} \subset T$ *such that* $x_j \to y$, *as* $j \to \infty$. *By the definition of* X_{n_1}, *it follows that there exist a sequence* $\{y_j\}_{j \in \mathbb{N}} \subset \mathscr{P}$ *so that* $x_j \leq y_j$, $j \in \mathbb{N}$, *and* $\|y_j\| \leq n_1\|x_j\|$, $j \in \mathbb{N}$. *Then*

$$x_j - x_0 \leq y_j + y_0 - z_0$$

$$\leq y_j + y_0 \in \mathscr{P}, \quad j \in \mathbb{N},$$

and

$$\|y_j + y_0\| \leq \|y_j\| + \|y_0\|$$

$$\leq n_1\|x_j\| + n_2\|x_0\|$$

$$\leq\ n_1\|x_j-x_0\|+n_1\|x_0\|+n_2\|x_0\|$$

$$=\ (n_1+n_2)\|x_0\|+n_1\|x_j-x_0\|$$

$$=\ (n_1+n_2)\frac{\|x_0\|}{r}+n_1\|x_j-x_0\|$$

$$\leq\ (n_1+n_2)\frac{\|x_0\|}{r}\|x_j-x_0\|+n_1\|x_j-x_0\|$$

$$=\ \left((n_1+n_2)\frac{\|x_0\|}{r}+n_1\right)\|x_j-x_0\|$$

$$\leq\ n_3\|x_j-x_0\|,\quad j\in\mathbb{N}.$$

*Therefore, $x_j-x_0\in X_{n_3}$, $j\in\mathbb{N}$, and $x_j-x_0\to y-x_0=x$, as $j\to\infty$.
Thus, $x\in\overline{X_{n_3}}$ and $\overline{X_{n_3}}\supset T_0$. Because $x\in\overline{X_{n_3}}$, we have that $\lambda x\in\overline{X_{n_3}}$
for any $\lambda\geq 0$. Therefore, $\overline{X_{n_3}}=X$. Now, we will prove that $X=
X_{n_3}$. For any $x\in X$, $x\neq\theta$, we choose $x_1\in X_{n_3}$ such that $\|x-x_1\|<
\dfrac{1}{3\cdot 2}\|x\|$. From $x_1\in X_{n_3}$, it follows that there exists an $y_1\in\mathscr{P}$ such
that $x_1\leq y_1$ and $\|y_1\|\leq n_3\|x_1\|$. As above, there exists an $x_2\in X_{n_3}$
and $y_2\in\mathscr{P}$ such that*

$$\|x-x_1-x_2\|<\frac{1}{3\cdot 2^2}\|x\|,\quad x_2\leq y_2,\quad \|y_2\|\leq n_3\|x_2\|.$$

*Continuing this process, we obtain a sequence $\{x_k\}_{k\in\mathbb{N}}\subset X_{n_3}$ and a
sequence $\{y_k\}_{k\in\mathbb{N}}\subset X$ such that*

$$\|x-x_1-\cdots-x_k\|<\frac{1}{3\cdot 2^k}\|x\|,\quad x_k\leq y_k,\quad \|y_k\|\leq n_3\|x_k\|,\quad k\in\mathbb{N}.$$

We have $x=\displaystyle\sum_{k=1}^{\infty}x_k$ and

$$\|x_k\|\ \leq\ \left\|x-\sum_{j=1}^{k-1}x_j\right\|+\left\|x-\sum_{j=1}^{k}x_j\right\|$$

$$\leq \frac{1}{3 \cdot 2^{k-1}} \|x\| + \frac{1}{3 \cdot 2^k} \|x\|$$

$$= \frac{1}{2^k} \|x\|, \quad k \in \mathbb{N}.$$

Therefore,

$$\sum_{k=1}^{\infty} \|y_k\| \leq n_3 \sum_{k=1}^{\infty} \|x_k\|$$

$$\leq n_3 \|x\| \sum_{k=1}^{\infty} \frac{1}{2^k}$$

$$= n_3 \|x\|$$

$$< \infty.$$

Thus, the series $\displaystyle\sum_{k=1}^{\infty} y_k$ *converges to some* $y \in \mathscr{P}$ *and* $x \leq y$, $\|y\| \leq n_3 \|x\|$. *This implies that* $x \in X_{n_3}$ *and* $X = X_{n_3}$. *Let now,* $x \in X$ *be arbitrarily chosen. Then* $x \in X_{n_3}$ *and there exists an* $y \in \mathscr{P}$ *so that* $\|y\| \leq n_3 \|x\|$. *Take* $z = y - x$. *Then* $z = y - x$ *and* $z \in \mathscr{P}$ *and*

$$\|y\| \leq n_3 \|x\|$$

$$\leq (n_3 + 1) \|x\|,$$

$$\|z\| = \|y - x\|$$

$$\leq \|y\| + \|x\|$$

$$\leq (n_3 + 1) \|x\|.$$

Thus, we get $(i) \Longrightarrow (ii)$. *If we assume* (ii), *then it follows* (i).

$(ii) \Longrightarrow (iii)$ *Let* $x \in B$ *be arbitrarily chosen. Then, by* (ii), *it follows* *that there exists an $r > 0$ such that*

$$x = y - z, \quad y, z \in \mathscr{P}, \quad \|y\|, \|z\| \le r\|x\| \le r.$$

Consider

$$y_1 = \frac{1}{2r} y, \quad z_1 = \frac{1}{2r} z.$$

Then $y_1, z_1 \in \mathscr{P}$ and $\|y_1\| \le \frac{1}{2}$, $\|z_1\| \le \frac{1}{2}$, i.e., $y_1, z_1 \in B$ and $y_1, z_1 \in B \cap \mathscr{P}$. Hence,

$$\frac{1}{2r} x = y_1 - z_1$$

and

$$\frac{1}{2r} B \subset B \cap \mathscr{P} - B \cap \mathscr{P}.$$

$(iii) \Longrightarrow (ii)$ *We have that there exists a constant $\gamma > 0$ such that*

$$\gamma B \subset B \cap \mathscr{P} - B \cap \mathscr{P}.$$

Take $x \in X$, $x \ne \theta$, arbitrarily. Set $x_1 = \gamma \dfrac{x}{\|x\|}$. Then $x_1 \in \gamma B$, $\|x_1\| = \gamma$ and

$$x_1 = y_1 - z_1, \quad y_1, z_1 \in B \cap \mathscr{P}.$$

Hence,

$$x = \frac{\|x\|}{\gamma} y_1 - \frac{\|x\|}{\gamma} z_1.$$

Set

$$y = \frac{\|x\|}{\gamma} y_1, \quad z = \frac{\|x\|}{\gamma} z_1.$$

Then $y, z \in \mathscr{P}$,

$$\|y\| \le \frac{1}{\gamma} \|x\|, \quad \|z\| \le \frac{1}{\gamma} \|x\|$$

and $x = y - z$.

$(iii) \Longrightarrow (iv)$ *We have that there exists a constant $\gamma > 0$ such that*

$$\gamma B \subset B \cap \mathscr{P} - B \cap \mathscr{P}.$$

Hence,

$$\gamma B \subset \overline{B \cap \mathscr{P} - B \cap \mathscr{P}}.$$

$((iv) \Longrightarrow (iii)$ *Let* $D = B \cap \mathscr{P} - B \cap \mathscr{P}$. *We have that there exists a* $\delta > 0$ *so that* $\delta B \subset \overline{D}$. *Take* $x \in \dfrac{\delta}{2}B$ *arbitrarily. Then* $2x \in \delta B$ *and there exists* $2x_1 \in D$ *such that*

$$\|2x - 2x_1\| < \frac{\delta}{2}.$$

Hence,
$$\|2^2 x - 2^2 x_1\| < \delta,$$

i.e., $2^2 x - 2^2 x_1 \in \delta B$. *Then there exists* $2^2 x_2 \in D$ *so that*

$$\|2^2 x - 2^2 x_1 - 2^2 x_2\| < \frac{\delta}{2}.$$

From here,
$$\|2^3 x - 2^3 x_1 - 2^3 x_2\| < \delta.$$

Continuing this process, we get a sequence $\{x_n\}_{n \in \mathbb{N}} \subset X$ *such that* $x_n \in \dfrac{1}{2^n} D$, $n \in \mathbb{N}$, *and*

$$\|x - x_1 - x_2 - \cdots - x_n\| < \frac{\delta}{2^{n+1}}, \quad n \in \mathbb{N}.$$

Therefore,
$$x = \sum_{n=1}^{\infty} x_n, \quad x_n = y_n - z_n, \quad y_n, z_n \in \mathscr{P}, \quad n \in \mathbb{N},$$

and
$$\|y_n\| \leq \frac{1}{2^n}, \quad \|z_n\| \leq \frac{1}{2^n}, \quad n \in \mathbb{N}.$$

This implies that
$$x = y - z, \quad y = \sum_{n=1}^{\infty} y_n, \quad z = \sum_{n=1}^{\infty} z_n, \quad y, z \in \mathscr{P},$$

and $\|y\|, \|z\| \leq 1$. *Consequently* $x \in D$ *and* $\dfrac{\delta}{2}B \subset D$. *This completes the proof.*

Definition 1.14.34 *Assume that $\mathscr{P}$ is a cone in a Banach space X. Then*

$$\mathscr{P}^* = \{f \in X^* : f(x) \geq 0, \quad x \in \mathscr{P}\}$$

is said to be a dual cone of $\mathscr{P}$.

We have that $\mathscr{P}^*$ is a nonempty closed convex set in X^* and $\lambda \mathscr{P}^* \subset \mathscr{P}^*$ for any $\lambda > 0$. Below we suppose that X is a Banach space.

Theorem 1.14.35 *$\mathscr{P}^*$ is a cone in X^* if and only if $\mathscr{P}$ is a total cone in X.*

Proof 1.14.36

> 1. Let $\mathscr{P}^*$ be a cone in X^*. Assume that $\mathscr{P}$ is not a total cone in X. Then there exists an $x_0 \in X \backslash \overline{(\mathscr{P} - \mathscr{P})}$. Note that $\overline{\mathscr{P} - \mathscr{P}}$ is a closed convex set. Hence and the first separation theorem of convex sets[2], it follows that there exist $f \in X^*$ and $c \in R$ such that
>
> $$f(x_0) < c < f(\overline{\mathscr{P} - \mathscr{P}})$$
>
> Because $f(\theta) = 0$, $\theta \in \mathscr{P} - \mathscr{P}$, we have that $c > 0$ and $f(x_0) < c$, $f \neq \theta$. On the other hand, for any $x \in \mathscr{P}$ and $\lambda > 0$, we have $\lambda x \in \mathscr{P} - \mathscr{P}$ and $-\lambda x \in \mathscr{P} - \mathscr{P}$. Thus,
>
> $$\lambda f(x) > c \quad and \quad -\lambda f(x) > c.$$
>
> Therefore,
>
> $$\frac{c}{\lambda} < f(x) < -\frac{c}{\lambda}$$
>
> and letting $\lambda \to \infty$, we get $f(x) = 0$ and then $f \in \mathscr{P}^* \cap (-\mathscr{P}^*)$. Thus, we get that $\mathscr{P}^* \cap (-\mathscr{P}^*) \neq \{\theta\}$. This is a contradiction because $\mathscr{P}^*$ is a cone in X^*. Consequently $\mathscr{P}$ is a total cone in X.

---

[2] **Theorem 1.14.37** *Suppose that A and B are two nonempty convex sets of a topological vector space E. If $A^0 \neq \emptyset$ and $A^0 \cap B = \emptyset$, where A^0 denotes the interior point set of A, then there exists a hyperplane H of E such that H sepoarates A and B, moreover, H separates strictly A^0 and B.*

2. Let $X = \overline{\mathscr{P} - \mathscr{P}}$, *i.e.,* $\mathscr{P}$ *is a total cone in* X. *Then, for any* $f \in \mathscr{P}^* \cap (-\mathscr{P}^*)$, $f(y) = 0$ *for all* $y \in \mathscr{P}$ *and for any* $x \in X$ *there exist sequences* $\{x_n\}_{n \in \mathbb{N}} \subset X$, $\{y_n\}_{n \in \mathbb{N}}, \{z_n\}_{n \in \mathbb{N}} \subset \mathscr{P}$ *such that*

$$x_n = y_n - z_n, \quad n \in \mathbb{N}, \quad and \quad \|x_n - x\| \to 0, \quad as \quad n \to \infty.$$

Then

$$\begin{aligned} 0 &= f(y_n) - f(z_n) \\[2mm] &= f(x_n) \\[2mm] &\to f(x), \quad as \quad n \to \infty. \end{aligned}$$

So, $f(x) = \theta$. *Therefore,* $f = \theta$ *and* $\mathscr{P}^* \cap (-\mathscr{P}^*) = \{\theta\}$. *Consequently* $\mathscr{P}^*$ *is a cone in* X^*. *This completes the proof.*

Theorem 1.14.38 *Let* $\mathscr{P}$ *be a cone in* X. *If* $\mathscr{P}$ *is reproducing, then* $\mathscr{P}^*$ *is normal.*

Proof 1.14.39 *Let* $f, g \in X^*$ *and* $\theta \le f \le g$. *Because* $\mathscr{P}$ *is reproducing, by Theorem 1.14.32,* $(i) \implies (ii)$, *it follows that there exists an* $r > 0$ *such that any* $x \in X$ *can be expressed in the form* $x = y - z$, $y, z \in \mathscr{P}$ *and*
$$\|y\| \le r\|x\|, \quad \|z\| \le r\|x\|.$$
Then

$$\begin{aligned} f(x) &= f(y) - f(z) \\[2mm] &\le f(y) \\[2mm] &\le g(y) \\[2mm] &\le \|g\|\|y\| \\[2mm] &\le \|g\|\|x\| \end{aligned}$$

and

$$-f(x) \quad = \quad f(z) - f(y)$$

$$\leq \quad f(z)$$

$$\leq \quad g(z)$$

$$\leq \quad \|g\| \|z\|$$

$$\leq \quad r \|g\| \|x\|$$

for $x \in \mathscr{P}$. Thus,

$$|f(x)| \leq r \|g\| \|x\|, \quad x \in \mathscr{P}.$$

This implies that

$$\|f\| \leq r \|g\|.$$

This completes the proof.

Theorem 1.14.40 *Assume that $\mathscr{P}$ is a cone in X. If $\mathscr{P}^*$ is reproducing in X^*, then $\mathscr{P}$ is normal in X.*

Proof 1.14.41 *We know that $\mathscr{P}^{**} = (\mathscr{P}^*)^*$. By Theorem 1.14.38, we have that $(\mathscr{P}^*)^*$ is normal. Because $\mathscr{P} \subset \mathscr{P}^{**}$, we conclude that $\mathscr{P}$ is normal. This completes the proof.*

2

Fixed-Point Index for Sums of Two Operators

The Leray-Schauder degree is an important tool in nonlinear analysis, allowing to establish the existence of fixed points for a mapping acting in a normed linear space. There are many interesting problems not set on the whole space, but instead the setting is a closed convex subset of a normed linear space, e.g., a cone. The fixed point index, a generalization of the Leray-Schauder degree, arises in a variety of mathematical problems. The computation of the fixed-point index may help in determining the fixed points for some nonlinear mappings defined on some closed convex subsets of Banach spaces. Since some applications involve noncompact maps, it is natural to try to extend the fixed point index to include as large a class of maps as possible.

Early in the 1970s, Amann [2,3] and Nussbaum [34,35] introduced the fixed-point index for strict set contractions and condensing mappings and have derived as results some fixed point theorems. As an extension, Li Guozhen [32] has defined the fixed point index of 1-set contractions and obtained some fixed-point theorems. More recently, Djebali and Mebarki [18] have developed a generalized fixed-point index theory for the sum of an h-expansive mapping and a k-set contraction when $0 \leq k < h - 1$ as well as in the limit case $k = h - 1$. Then some researchers have been interested in the extension of this index in various directions, we cite [10, 19, 27].

This chapter concerns the fixed-point index theory for the sum $T + F$ of two mappings. We present the definition of this index as well as some of its properties. Then several conditions allowing the computation of the fixed point index are shown. We will consider separately two cases: firstly the computation of fixed-point index in cones of Banach spaces is treated in Section 2.2. Then in Section 2.3, we will discuss the computation of fixed-point index in translates of cones. Some of the results of this chapter can be found in [18, 19, 23, 24, 27, 30, 44].

DOI: 10.1201/9781003381969-2

2.1   Auxiliary Results

In this section, we collect some auxiliary results needed in the sequel. Of particular importance for our purpose is the fixed-point index for strict set contractions and its properties as well as the Lipschitz invertibility of some classes of mappings.

Recall that a subset $D \neq \emptyset$ of a metric (more generally, topological) space Y is called a retract of Y if there exists a continuous map $r : Y \to D$, called a retraction, such that $r(x) = x$, $\forall x \in D$.

Example 2.1.1

> *1. Every closed ball $D = \overline{\mathscr{B}}(0,R) \subset Y$ is a retract of the normed space Y via the radial retraction $r : Y \to D$ defined by $r(x) = x$ if $x \in D$ and $r(x) = Rx/\|x\|$ otherwise.*
>
> *2. If X is a nonempty closed, convex subset of a Banach space E, then X is a retract of E (this is an easy consequence of Dugundji's extension theorem).*
> *In particular, every cone or translate of cone in E is a retract of E.*

Theorem 2.1.2 *(Dugundji's extension theorem). Let X and Y be two normed linear spaces, $A \subset X$ a closed subset and $f : A \to Y$ a continuous mapping. Then f has a continuous extension $\tilde{f} : A \to Y$ such that $\tilde{f}(X) \subset \text{Conv}(f(X))$, where $\text{Conv}(A)$ refers to the convex hull of X.*

The proofs of the results which will be presented in Sections 2.2 and 2.3 involve the fixed-point index for strict set contractions whose basic properties are collected in the following lemma.

Lemma 2.1.3 *Let X be a retract of a Banach space E. For every bounded open subset $U \subset X$ and every strict set contraction $f : \overline{U} \to X$ without fixed point on the boundary ∂U, there exists uniquely one integer $i(f,U,X)$ satisfying the following conditions:*
(a) (Normalization property). If $f : \overline{U} \to U$ is a constant map, then

$$i(f,U,X) = 1.$$

(b) (Additivity property). For any pair of disjoint open subsets U_1, U_2 in U such that f has no fixed point on $\overline{U} \setminus (U_1 \cup U_2)$, we have

$$i(f, U, X) = i(f, U_1, X) + i(f, U_2, X),$$

where $i(f, U_j, X) := i(f|_{\overline{U_j}}, U_j, X)$, $j = 1, 2$.

(c) (Homotopy Invariance property). The index $i(h(x, t), U, X)$ does not depend on the parameter $t \in [0, 1]$, where

(i) $h : [0, 1] \times \overline{U} \to X$ is continuous and $h(t, x)$ is uniformly continuous in t with respect to $x \in \overline{U}$,

(ii) $h(t, .) : \overline{U} \to X$ is a strict k-set contraction, where k does not depend on $t \in [0, 1]$,

(iii) $h(t, x) \neq x$, for every $t \in [0, 1]$ and $x \in \partial U$.

(d) (Preservation property). If Y is a retract of X and $f(\overline{U}) \subset Y$, then

$$i(f, U, X) = i(f, U \cap Y, Y),$$

where $i(f, U \cap Y, Y) := i(f|_{\overline{U \cap Y}}, U, Y)$.

(e) (Excision property). Let $V \subset U$ an open subset such that f has no fixed point in $\overline{U} \setminus V$. Then

$$i(f, U, X) = i(f, V, X).$$

(f) (Solvability property). If $i(f, U, X) \neq 0$, then f has a fixed point in U.

The proof of Lemma 2.1.3 can be found in [30, Theorem 1.3.5] or [16, 23, 31].

Proposition 2.1.4 *Assume that X is a closed convex set of a Banach space E, X_1 is a bounded closed convex subset of X, U is a nonempty open set of X with $U \subset X_1$. If $A : X_1 \to X$ is a strict set contraction, $A(X_1) \subset X_1$ and A has no fixed point in $X_1 \setminus U$, then*

$$i(A, U, X) = 1.$$

Proof 2.1.5 *Since X_1 is a closed subset of E, $\overline{U} \subset X_1$, by the preservation property of the fixed point index, it follows that*

$$i(A, U, X) = i(A, U, X_1). \tag{2.1}$$

Because A has no fixed points in $X_1 \backslash U$, by the excision property of the fixed point index, we get

$$i(A, U, X_1) = i(A, X_1, X_1). \tag{2.2}$$

Take $z_0 \in U \subset X_1$ and let

$$H(t, x) = t z_0 + (1-t)Ax, \quad t \in [0, 1], \quad x \in X_1.$$

We have $H : [0,1] \times X_1 \to X_1$ is continuous and bounded. Also, for any $t \in [0,1]$ and B a bounded set in X_1, we have

$$\alpha(H(t, B)) \leq (1-t)\alpha(A(B))$$

$$\leq (1-t)k\alpha(B).$$

So, $H(t, \cdot) : X_1 \to X_1$ is a strict set contraction for any $t \in [0,1]$. Hence, using the normality and the homotopy invariance of the fixed point index, we get

$$i(A, X_1, X_1) = i(H(0, \cdot), X_1, X_1)$$

$$= i(H(1, \cdot), X_1, X_1)$$

$$= i(z_0, X_1, X_1)$$

$$= 1.$$

From here and from (2.1), (2.2), we arrive at

$$i(A, U, X) = i(A, U, X_1)$$

$$= i(A, X_1, X_1)$$

$$= 1.$$

Corollary 2.1.6 *Assume that X is a closed convex set in E and U is a nonempty bounded open convex subset of X. If $A : \overline{U} \to X$ is a strict set contraction and $A(\overline{U}) \subset U$, then*

$$i(A, U, X) = 1.$$

Proof 2.1.7 *We apply Theorem 2.1.4 for* $X_1 = \overline{U}$. *Then*

$$i(A, U, X) = 1.$$

Given a real Banach space $(E, \|.\|)$, let $\mathscr{P} \neq \{0\}$ be a cone in E and $\mathscr{K} = \mathscr{P} + \theta$ $(\theta \in E)$ a θ-translate of $\mathscr{P}$. The following results are direct consequences of the properties of the index i in case of a translate of a cone, rather than in a cone.

Proposition 2.1.8 *Let* $U \subset \mathscr{K}$ *be a bounded open subset with* $\theta \in U$. *Assume that* $A : \overline{U} \to \mathscr{K}$ *is a strict set contraction that satisfies the so-called Leray-Schauder boundary condition type:*

$$Ax - \theta \neq \lambda(x - \theta), \quad \forall x \in \partial U, \forall \lambda \geq 1. \tag{2.3}$$

Then $i(A, U, \mathscr{K}) = 1$.

Proof 2.1.9 *Let*

$$H(t, x) = tAx + (1 - t)\theta, \quad t \in [0, 1], \quad x \in \overline{U}.$$

We have $H : [0, 1] \times \overline{U} \to \mathscr{K}$ *is continuous and* $H(t, \cdot) : \overline{U} \to \mathscr{K}$ *is a strict set contraction. Assume that there is* $(t_0, x_0) \in [0, 1] \times \partial U$ *such that*

$$H(t_0, x_0) = x_0.$$

Hence,

$$t_0 A x_0 + (1 - t_0)\theta = x_0.$$

If $t_0 = 0$, *then* $x_0 = \theta$. *This is a contradiction because* $\theta \in U$. *If* $t_0 \neq 0$, *then*

$$Ax_0 - \theta = \frac{1}{t_0}(x_0 - \theta),$$

which contradicts with (2.3). *Therefore,* $H(t, x) \neq x$ *for any* $(t, x) \in [0, 1] \times \partial U$. *Thus, by the homotopy invariance and the normality of the fixed point index, it follows*

$$i(A, U, X) = i(H(1, .), U, \mathscr{K})$$

$$= i(H(0, .), U, \mathscr{K})$$

$$= i(0, U, \mathscr{K})$$

$$= 1.$$

This completes the proof.

Corollary 2.1.10 *Let $U \subset \mathscr{K}$ be a bounded open subset with $\theta \in U$. Assume that $A : \overline{U} \to \mathscr{K}$ is a strict set contraction that satisfies the condition of type norm:*

$$\|Ax - \omega\| \leq \|x - \omega\| \quad and \quad Ax \neq x, \quad \forall x \in \partial U. \qquad (2.4)$$

Then $i(A, U, \mathscr{K}) = 1$.

Proof 2.1.11 *It is sufficient to prove that the condition (2.4) implies the condition (2.3). Indeed, assume by contradiction that some $x_0 \in \partial U$ and $\lambda_0 \geq 0$ exist and satisfy $Ax_0 - \theta \neq \lambda_0(x_0 - \theta)$. We consider two cases:*

(i) If $\lambda_0 = 1$, then $Ax_0 = x_0$, contradicting the hypothesis (2.4).

(ii) If $\lambda_0 > 1$, then $\|Ax_0 - \theta\| = \lambda_0\|x_0 - \theta\| > \|x_0 - \theta\|$, a contradiction of (2.4) is again reached.

Proposition 2.1.12 *Let U be a bounded open subset of $\mathscr{K}$. Assume that $A : \overline{U} \to \mathscr{K}$ is a strict set contraction and there is $v_0 \in \mathscr{P} \backslash \{0\}$ such that*

$$x - Ax \neq \lambda v_0, \ for \ all \ x \in \partial U, \lambda \geq 0. \qquad (2.5)$$

Then $i(A, U, \mathscr{K}) = 0$.

Proof 2.1.13 *Define the homotopy $H : [0, 1] \times \overline{U} \to \mathscr{K}$ by*

$$H(t, x) = Ax + t\lambda_0 v_0,$$

for some

$$\lambda_0 > \|v_0\|^{-1} \sup_{x \in \overline{U}}((\|x\| + \|Ax\|)). \qquad (2.6)$$

Such a choice is possible since U is a bounded subset and then so is $A(\overline{U})$. The operator H is continuous and uniformly continuous in t for each x, and the mapping $H(t,.)$ is strict set contraction for each $t \in [0,1]$. In addition, $H(t,.)$ has no fixed point on ∂U for each $t \in [0,1]$. On the contrary, there would exist some $x_0 \in \partial U$ and $t_0 \in [0,1]$ such that

$$x_0 = Ax_0 + t_0\lambda_0 v_0,$$

contradicting the hypothesis. By the homotopy invariance of the fixed point index, we have

$$i(A,U,\mathcal{K}) = i(H(0,\cdot),U,\mathcal{K})$$

$$= i(H(1,\cdot),U,\mathcal{K})$$

$$= 0.$$

Indeed, suppose that $i(H(1,.),U,\mathcal{K}) \neq 0$. Then there exists $x_1 \in U$ such that $Ax_1 + \lambda_0 v_0 = x_1$, which implies that $\lambda_0 \leq \|v_0\|^{-1}(\|x_1\| + \|Ax_1\|)$, contradicting (2.6).

Remark 2.1.14 *Letting $\theta = 0$, we obtain the computations of the index in case of the cone.*

One of the most common approaches used for the existence of positive solutions of various types of nonlinear problems relies on the use of the Krasnosel'skii's compression–expansion fixed point theorem.

Theorem 2.1.15 *(Krasnosel'skii's compression–expansion fixed point theorem). Let $\alpha, \beta > 0$, $\alpha \neq \beta$, $r := \min\{\alpha,\beta\}$ and $R := \max\{\alpha,\beta\}$. Assume that $A : \mathcal{P}_{r,R} \to \mathcal{P}$ is a strict set contraction and there exists $p \in \mathcal{P} \setminus \{0\}$ such that the following conditions are satisfied:*

$$\begin{array}{ll} Ax \neq \lambda x & \text{for } \|x\| = \alpha \text{ and } \lambda > 1, \\ Ax + \mu p \neq x & \text{for } \|x\| = \beta \text{ and } \mu > 0. \end{array} \quad (2.7)$$

Then A has a fixed point x in $\mathcal{P}$ with $r \leq \|x\| \leq R$.

Remark 2.1.16 *If $\beta < \alpha$, the conditions (2.7) represent a compression property of A upon the conical shell $\mathscr{P}_{r,R}$, while if $\alpha < \beta$, the conditions (2.7) express an expansion property of A upon $\mathscr{P}_{r,R}$.*

Proof 2.1.17 *We only give the proof in case of the cone expansion ($\alpha < \beta$). First, extend the mapping A to $\partial \mathscr{P}_R$ by Dugundji's extension theorem. Without loss of generality, assume that $Ax \neq x$ on $\partial \mathscr{P}_r$ and $Ax \neq x$ on $\partial \mathscr{P}_R$, otherwise we are finished. By Propositions 2.1.8 and 2.1.12, we have*

$$i(A, \mathscr{P}_r, \mathscr{P}) = 1 \text{ and } i(A, \mathscr{P}_R, \mathscr{P}) = 0.$$

The additivity property yields

$$i(A, \mathscr{P}_R \setminus \overline{\mathscr{P}_r}, \mathscr{P}) = 1.$$

By the existence property of the index, A has at least one fixed point in the closed set $(\overline{\mathscr{P}_R} \setminus \mathscr{P}_r)$.

Some Lipschitz Invertible Mappings

Let X be a linear normed space and I be the identity map of X.

   The case of expansive mapping is given in the following lemma.

Lemma 2.1.18 *Let $(X, \|.\|)$ be a normed linear space, $D \subset X$. If a mapping $T : D \to X$ is expansive with a constant $h > 1$, then the mapping $I - T : D \to (I - T)(D)$ is invertible and*

$$\|(I-T)^{-1}x - (I-T)^{-1}y\| \leq (h-1)^{-1}\|x-y\| \text{ for all } x, y \in (I-T)(D).$$

Proof 2.1.19 *For each $x, y \in D$, we have*

$$
\begin{aligned}
\|(I-T)x - (I-T)y\| &= \|(Tx - Ty) - (x-y)\| \\
&\geq (h-1)\|x-y\|,
\end{aligned}
\tag{2.8}
$$

which shows that $(I-T)^{-1} : (I-T)(D) \to D$ exists. Hence, for $x, y \in (I-T)(D)$, we have $(I-T)^{-1}x$, $(I-T)^{-1}y \in D$. Thus, using $(I-T)^{-1}x$, $(I-T)^{-1}y$ substitute for x, y in (2.8), respectively, we obtain

$$\|(I-T)^{-1}x - (I-T)^{-1}y\| \leq \frac{1}{h-1}\|x-y\|.$$

Some other examples of Lipschitz invertible mappings (see [45]) are presented below.

Let $lip(T) = \max\{h \geq 0 : d(Tx, Ty) \geq hd(x,y), \ \forall x, y \in E\}$ and $Lip(T)$ denotes the Lipschitz constant for T if T is a Lipschitz map.

1. Let $(E, \|.\|)$ be a Banach space and $T : E \to E$ be Lipschitzian map with constant $\beta > 0$. Assume that for each $z \in E$, the map $T_z : E \to E$ defined by $T_z x = Tx + z$ satisfies that T_z^p is expansive for some $p \in \mathbb{N}$ and is surjective. Then $(I - T)$ maps E onto E, the inverse of $I - T : E \to E$ exists, and

$$\|(I - T)^{-1}x - (I - T)^{-1}y\| \leq \gamma_p \|x - y\| \text{ for all } x, y \in E,$$

where

$$\gamma_p = \frac{\beta^p - 1}{(\beta - 1)(lip(T^p) - 1)}.$$

2. Let $(X, \|.\|)$ be a linear normed space, $M \subset X$. Assume that the mapping $T : M \to X$ is contractive with a constant $k < 1$, then the inverse of $I - T : M \to (I - T)(M)$ exist, and

$$\|(I - T)^{-1}x - (I - T)^{-1}y\| \leq (1 - k)^{-1}\|x - y\|$$

for all $x, y \in (I - T)(M)$.

3. Let $(E, \|.\|)$ be a Banach space and $T : E \to E$ be Lipschitzian map with constant $\beta \geq 0$. Assume that for each $z \in E$, the map $T_z : E \to E$ defined by $T_z x = Tx + z$ satisfies that T_z^p is contractive for some $p \in \mathbb{N}$. Then $(I - T)$ maps E onto E, the inverse of $I - T : E \to E$ exists, and

$$\|(I - T)^{-1}x - (I - T)^{-1}y\| \leq \rho_p \|x - y\| \text{ for all } x, y \in E,$$

where

$$\rho_p = \begin{cases} \dfrac{p}{1 - Lip(T^p)}, & \text{if } \beta = 1; \\[2ex] \dfrac{1}{1 - \beta}, & \text{if } \beta < 1; \\[2ex] \dfrac{\beta^p - 1}{(\beta - 1)(1 - Lip(T^p))}, & \text{if } \beta > 1. \end{cases}$$

2.2   Fixed Point Index on Cones

In this section, $\mathscr{P}$ will refer to a cone in a Banach space E, Ω is a subset of $\mathscr{P}$, and U is a bounded open subset of $\mathscr{P}$. For some constant $r > 0$, we will denote the conical shell by $\mathscr{P}_r = \mathscr{P} \cap \mathscr{B}_r$, where $\mathscr{B}_r = \{x \in E : \|x\| < r\}$ is the open ball centered at the origin with radius r. The results of the two first sub-sections can be found in [18].

2.2.1   The case where T is an h-expansive mapping and F is a k-set contraction with $0 \leq k < h - 1$

Assume that $T : \Omega \to E$ is an h-expansive mapping and $F : \overline{U} \to E$ is a k-set contraction. By Lemma 2.1.18, the operator $(I - T)^{-1}$ is $(h - 1)^{-1}$-Lipschitzian on $(I - T)(\Omega)$. Suppose that

$$F(\overline{U}) \subset (I - T)(\Omega) \tag{2.9}$$

and

$$x \neq Tx + Fx, \text{ for all } x \in \partial U \cap \Omega. \tag{2.10}$$

Then $x \neq (I - T)^{-1}Fx$, for all $x \in \partial U$ and the mapping $(I - T)^{-1}F : \overline{U} \to \mathscr{P}$ is a strict $k(h - 1)^{-1}$-set contraction. Indeed, $(I - T)^{-1}F$ is continuous and bounded; and for any bounded set B in U, we have

$$\alpha\big(((I - T)^{-1}F)(B)\big) \leq (h - 1)^{-1}\alpha(F(B)) \leq k(h - 1)^{-1}\alpha(B).$$

By Lemma 2.1.3, the fixed point index $i\big((I - T)^{-1}F, U, \mathscr{P}\big)$ is well defined. Thus, we put

$$i_*(T + F, U \cap \Omega, \mathscr{P}) = \begin{cases} i\big((I - T)^{-1}F, U, \mathscr{P}\big) & \text{if } U \cap \Omega \neq \emptyset \\ 0, & \text{if } U \cap \Omega = \emptyset. \end{cases} \tag{2.11}$$

This integer is called the generalized fixed point index of the sum $T + F$ on $U \cap \Omega$ with respect to the cone $\mathscr{P}$.

Theorem 2.2.1 *The fixed point index defined in (2.11) satisfies the following properties:*

(a) (Normalization property). If $U = \mathscr{P}_r$, $0 \in \Omega$, and $Fx = z_0 \in \mathscr{B}(-T0, (h-1)r) \cap \mathscr{P}$ for all $x \in \overline{\mathscr{P}_r}$, then

$$i_* (T + F, \mathscr{P}_r \cap \Omega, \mathscr{P}) = 1.$$

(b) (Additivity property). For any pair of disjoint open subsets U_1, U_2 in U such that $T + F$ has no fixed point on $(\overline{U} \setminus (U_1 \cup U_2)) \cap \Omega$, we have

$$i_* (T + F, U \cap \Omega, \mathscr{P}) = i_* (T + F, U_1 \cap \Omega, \mathscr{P}) + i_* (T + F, U_2 \cap \Omega, \mathscr{P}),$$

where $i_* (T + F, U_j \cap \Omega, X) := i_* (T + F|_{\overline{U_j}}, U_j \cap \Omega, \mathscr{P})$, $j = 1, 2$.

(c) (Homotopy Invariance property). The fixed point index $i_* (T + H(t, .), U \cap \Omega, \mathscr{P})$ does not depend on the parameter $t \in [0, 1]$ whenever

(i) $H : [0, 1] \times \overline{U} \to E$ is continuous and $H(t, x)$ is uniformly continuous in t with respect to $x \in \overline{U}$,

(ii) $H([0, 1] \times \overline{U}) \subset (I - T)(\Omega)$,

(iii) $H(t, .) : \overline{U} \to E$ is a l-set contraction with $0 \le l < h - 1$ and l does not depend on $t \in [0, 1]$,

(iv) $Tx + H(t, x) \ne x$, for all $t \in [0, 1]$ and $x \in \partial U \cap \Omega$.

(d) (Solvability property). If $i_* (T + F, U \cap \Omega, \mathscr{P}) \ne 0$, then $T + F$ has a fixed point in $U \cap \Omega$.

Proof 2.2.2 *Properties (b), (c), and (d) follow directly from (2.11) and the corresponding properties of the fixed point index for strict-set contractions (see Lemma 2.1.3). We only check that if $U = \mathscr{P}_r$, then*

$$i((I - T)^{-1} z_0, U, \mathscr{P}) = 1.$$

For this, we show that $y_0 := (I - T)^{-1} z_0 \in \mathscr{P}_r \cap \Omega$. We have $F(\overline{\mathscr{P}_r}) = \{z_0\} \subset (I - T)(\Omega)$, which gives $y_0 \in \Omega$ and since T is an expansive operator with $h > 1$ and $F(\overline{\mathscr{P}_r}) \subset \mathscr{B}(-T0, (h-1)r) \cap \mathscr{P}$, Lemma 2.1.18 guarantees that

$$\|(I - T)y_0 + T0\| = \|(I - T)y_0 - (I - T)0\|$$

$$\ge (h - 1)\|y_0\|.$$

Hence

$$(h-1)\|y_0\| \ \leq \ \|(I-T)y_0 + T0\|$$

$$= \ \|z_0 - (-T0)\|$$

$$< \ (h-1)r,$$

that is $y_0 = (I-T)^{-1}z_0 \in \mathscr{P}_r$. *By property (a) in Lemma 2.1.3, we deduce that*

$$i\left((I-T)^{-1}z_0, \mathscr{P}_r, \mathscr{P}\right) = 1.$$

Therefore $i_* (T + z_0, \mathscr{P}_r \cap \Omega, \mathscr{P}) = 1$, *which completes the proof.*

Remark 2.2.3 *Theorem 2.2.1 still holds if instead of the cone* $\mathscr{P}$, *we consider a retract* X *of* E. *In this case, the set* $\mathscr{P}_r$ *is replaced by* $X \cap \mathscr{B}_r$.

Other Properties of the Index i_*

Lemma 2.2.4 *Let* X *be a closed convex subset of a Banach space* E, U *a nonempty bounded open subset of* X *and* Ω *be a subset of* X. *Assume that* $T : \Omega \to E$ *is an h-expansive mapping and* $F : U \to E$ *is a k-set contraction with* $k < h - 1$ *such that* $F(\overline{U}) \subset (I-T)(\Omega)$, *and* $Tx + Fx \neq x$, *for all* $x \in U \cap \Omega$.
The index i_* *satisfying the following properties:*
(i) (Excision property). Let $V \subset U$ *be an open subset such that* $T + F$ *has no fixed point in* $(\overline{U} \backslash V) \cap \Omega$. *Then*

$$i_* (T+F, U \cap \Omega, X) = i_* (T+F, V \cap \Omega, X).$$

(ii) (Preservation property). If Y *is a retract of* X *and* $\Omega \subset Y$, *then*

$$i_* (T+F, U \cap \Omega, X) = i_* (T+F, U \cap \Omega, Y).$$

Proof 2.2.5 *Properties (i) and (ii) follow directly from the definition (2.11) and the corresponding properties of the fixed point index for strict set contractions in Lemma 2.1.3.*

Computation of the Index i_*

Lemma 2.2.6 *Let* X *be a closed convex subset of a Banach space* E, X_1 *a bounded closed convex subset of* X, Ω *be a subset of* X *and* U

*a nonempty bounded open convex subset of X with $U \subset X_1$. Assume
that $T : \Omega \to E$ is an h-expansive mapping and $F : X_1 \to E$ is a k-set
contraction with $k < h - 1$. If*

$$F(X_1) \subset (I - T)(X_1 \cap \Omega),$$

and

$$Tx + Fx \neq x, \text{ for all } x \in (X_1 \setminus U) \cap \Omega. \tag{2.12}$$

Then

$$i_* (T + F, U \cap \Omega, X) = 1.$$

*In particular, if X is a nonempty bounded convex closed subset of
E, $\Omega \subset X$, $F : X \to E$ is a k-set contraction with $k < h - 1$ and $F(X) \subset
(I - T)(\Omega)$, then*

$$i_* (T + F, X \cap \Omega, X) = 1.$$

Proof 2.2.7 *The mapping $(I - T)^{-1}F : X_1 \to X$ is a strict $k(h - 1)^{-1}$-
set contraction and it is readily seen that $(I - T)^{-1}F(X_1) \subset X_1$ such
that there is no fixed point of $(I - T)^{-1}F$ in $X_1 \setminus U$. Otherwise, there
would exist some $x_0 \in X_1 \setminus U$ such that $x_0 = (I - T)^{-1}Fx_0$. Hence,
if $x_0 \in \Omega$, we get a contradiction with the condition (2.12);
if not we get a contradiction with $(I - T)^{-1}Fx_0 \in \Omega$.
The result then follows from the definition (2.11) of the index i_* and
Proposition 2.1.4.*

Proposition 2.2.8 *Assume that $T : \Omega \subset \mathscr{P} \to E$ is an h-expansive
mapping $F : \overline{\mathscr{P}_r} \to E$ is a k-set contraction with $0 \leq k < h - 1$,
$F(\partial \mathscr{P}_r \cap \Omega) \subset \mathscr{P}$, and $tF(\overline{\mathscr{P}_r}) \subset (I - T)(\Omega)$ for all $t \in [0, 1]$.
If $0 \in \Omega$, $\|T0\| < (h - 1)r$, and*

$$Fx \not\geq x - Tx, \quad \forall x \in \partial \mathscr{P}_r \cap \Omega,$$

then $i_* (T + F, \mathscr{P}_r \cap \Omega, \mathscr{P}) = 1$.

Proof 2.2.9 *Consider the homotopic deformation $H : [0, 1] \times \overline{\mathscr{P}_r} \to E$
defined by*

$$H(t, x) = tFx.$$

*The operator H is continuous and uniformly continuous in t with re-
spect to x. Moreover, $H(t, .)$ is a k-set contraction for each t and the*

mapping $T + H(t, .)$ has no fixed point on $\partial \mathscr{P}_r \cap \Omega$ for each t. On the contrary, there would exist some $x_0 \in \partial \mathscr{P}_r \cap \Omega$ and $t_0 \in [0, 1]$ such that

$$x_0 = T x_0 + t_0 F x_0.$$

Consider two cases:
(i) If $t_0 = 0$, then $T x_0 = x_0$ and

$$
\begin{aligned}
(h-1)\|x_0\| &\leq \|(I-T)x_0 + T0\| \\
&= \|T0\| \qquad\qquad\qquad (2.13) \\
&< (h-1)r,
\end{aligned}
$$

which contradicts $x_0 \in \partial \mathscr{P}_r$.
(ii) If $t_0 \in (0, 1]$, then $F x_0 = \dfrac{1}{t_0}(x_0 - T x_0)$, where $\dfrac{1}{t_0} \geq 1$, which contradicts our assumption.
From the invariance under homotopy and the normalization properties of the generalized fixed point index in Theorem 2.2.1, we deduce that

$$
\begin{aligned}
i_* (T+F, \mathscr{P}_r \cap \Omega, \mathscr{P}) &= i_* (T+0, \mathscr{P}_r \cap \Omega, \mathscr{P}) \\
&= 1,
\end{aligned}
$$

which completes the proof.

Remark 2.2.10 *The results of Proposition 2.2.8 does not hold if $F(\partial \mathscr{P}_r \cap \Omega) \not\subset \mathscr{P}$. Indeed, let $\mathscr{P}$ be a cone in a Banach space E with non-empty interior and $r > 0$. We choose a point x_1 in the interior of $\mathscr{P}$ with $dist(x_1, E \backslash \mathscr{P}) > r$ and $\Omega = \mathscr{P}_{2\|x_1\|}$. With $T = 2I$ and $F : \overline{\mathscr{P}}_r \ni x \to -x_1 \in -\mathscr{P} \subset E$, all assumptions of the latter proposition are satisfied. However, $0 \in -\mathscr{P} \subset E \backslash \mathscr{P}$ implies that $i_* (T+F, \mathscr{P}_r \cap \Omega, \mathscr{P}) = 0$.*

Corollary 2.2.11 *Assume that $T : \Omega \subset \mathscr{P} \to E$ is an h-expansive mapping, $F : \overline{\mathscr{P}}_r \to E$ is a k-set contraction with $0 \leq k < h-1$, $F(\partial \mathscr{P}_r \cap \Omega) \subset \mathscr{P}$, and $tF(\overline{\mathscr{P}}_r) \subset (I-T)(\Omega)$, for all $t \in [0, 1]$. If $0 \in \Omega$, $\|T0\| < (h-1)r$, and*

$$Tx + Fx < x, \quad \forall x \in \partial \mathscr{P}_r \cap \Omega.$$

Then $T + F$ has a fixed point in $\mathscr{P}_r \cap \Omega$.

Proof 2.2.12 *Since F and T satisfy the assumptions of Proposition 2.2.8,*

$$i_* (T+F, \mathscr{P}_r \cap \Omega, \mathscr{P}) = 1.$$

This Corollary 2.2.11 then follows from the solvability property of the index i_*.

Proposition 2.2.13 *Assume that $T : \Omega \subset \mathscr{P} \to E$ is an h-expansive mapping, $F : \overline{\mathscr{P}_r} \to E$ is a k-set contraction with $0 \le k < h - 1$, and $tF(\overline{\mathscr{P}_r}) \subset (I-T)(\Omega)$, for all $t \in [0,1]$. If $0 \in \Omega$, $\|T0\| < (h-1)r$, and*

$$Fx \ne \lambda(x - Tx), \quad \text{for all } x \in \partial \mathscr{P}_r \cap \Omega \text{ and } \lambda \ge 1. \tag{2.14}$$

Then $i_* (T+F, \mathscr{P}_r \cap \Omega, \mathscr{P}) = 1$.

Proof 2.2.14 *Define the homotopic deformation $H : [0,1] \times \overline{\mathscr{P}_r} \to E$ by*

$$H(t,x) = tFx.$$

The operator H is continuous and uniformly continuous in t for each x. Moreover, $H(t,.)$ is a k-set contraction for each t and the mapping $T + H(t,.)$ has no fixed point on $\partial \mathscr{P}_r \cap \Omega$. Otherwise, there would exist some $x_0 \in \partial \mathscr{P}_r \cap \Omega$ and $t_0 \in [0,1]$ such that

$$x_0 = Tx_0 + t_0 Fx_0.$$

We may distinguish between two cases:
(i) If $t_0 = 0$, then $Tx_0 = x_0$, as in (2.13), which implies that $\|x_0\| < r$, contradiction.
(ii) If $t_0 \in (0,1]$, then $Fx_0 = \dfrac{1}{t_0}(x_0 - Tx_0)$, where $\dfrac{1}{t_0} \ge 1$, leading again to a contradiction with Hypothesis (2.14).
By properties (a) and (d) of the fixed point index in Theorem 2.2.1, we deduce that

$$\begin{aligned} i_* (T+F, \mathscr{P}_r \cap \Omega, \mathscr{P}) &= i_* (T+0, \mathscr{P}_r \cap \Omega, \mathscr{P}) \\ &= 1. \end{aligned}$$

The following results are immediate consequences of Proposition 2.2.13.

Corollary 2.2.15 *Assume that $T : \Omega \subset \mathscr{P} \to E$ is an h-expansive mapping, $F : \overline{\mathscr{P}_r} \to E$ is a k-set contraction with $0 \le k < h - 1$, and $tF(\overline{\mathscr{P}_r}) \subset (I - T)(\Omega)$, for all $t \in [0,1]$. If $0 \in \Omega$, $\|T0\| < (h-1)r$, and*

$$\|Fx\| \le \|x - Tx\| \ \text{and} \ Tx + Fx \ne x \ \text{for all} \ x \in \partial \mathscr{P}_r \cap \Omega, \quad (2.15)$$

then $i_* (T + F, \mathscr{P}_r \cap \Omega, \mathscr{P}) = 1$.

Proof 2.2.16 *It is sufficient to prove that the condition (2.15) implies the condition (2.14) in Proposition 2.2.13. For this, assume by contradiction that some $x_0 \in \partial \mathscr{P}_r \cap \Omega$ and $\lambda_0 \ge 0$ exist and satisfy $Fx_0 = \lambda_0(x_0 - Tx_0)$. Then two cases are discussed separately:*
(i) If $\lambda_0 = 1$, then $Tx_0 + Fx_0 = x_0$ and a contradiction is reached.
(ii) If $\lambda_0 > 1$, then $\|Fx_0\| = \lambda_0 \|x_0 - Tx_0\| > \|x_0 - Tx_0\|$, whence a contradiction.

Corollary 2.2.17 *Assume that $T : \Omega \subset \mathscr{P} \to E$ is an h-expansive mapping, $F : \overline{\mathscr{P}_r} \to E$ is a k-set contraction with $0 \le k < h - 1$, and $tF(\overline{\mathscr{P}_r}) \subset (I - T)(\Omega)$, for all $t \in [0,1]$. If $0 \in \Omega$, $\|T0\| < (h-1)r$, and*

$$\|Fx\| + \|Tx\| < \|x\|, \ \text{for all} \ x \in \partial \mathscr{P}_r \cap \Omega, \quad (2.16)$$

then $T + F$ has a fixed point in $\mathscr{P}_r \cap \Omega$.

Proof 2.2.18 *Since the condition (2.16) implies the analogue one (2.14) in Proposition 2.2.13, then*

$$i_* (T + F, \mathscr{P}_r \cap \Omega, \mathscr{P}) = 1.$$

Corollary 2.2.17 then follows from the solvability property of the index $i*$.

Proposition 2.2.19 *Let U be a bounded open subset of $\mathscr{P}$ with $0 \in U$. Assume that $T : \Omega \subset \mathscr{P} \to E$ is an h-expansive mapping, $F : \overline{U} \to E$ is a k-set contraction with $0 \le k < h - 1$, and $F(\overline{U}) \subset (I - T)(\Omega)$. If*

$$Fx \ne (I - T)(\lambda x), \ \text{for all} \ x \in \partial U, \lambda \ge 1 \ \text{and} \ \lambda x \in \Omega, \quad (2.17)$$

then $i_* (T + F, U \cap \Omega, \mathscr{P}) = 1$.

Proof 2.2.20 *The mapping* $(I-T)^{-1}F : \overline{U} \to \mathscr{P}$ *is a strict* $k(h-1)^{-1}$- *set contraction and it is readily seen that the following condition of Leray-Schauder type is satisfied*

$$(I-T)^{-1}Fx \neq \lambda x, \text{ for all } x \in \partial U \text{ and } \lambda \geq 1.$$

In fact, if there exist $x_0 \in \partial U$ *and* $\lambda_0 \geq 1$ *such that* $(I-T)^{-1}Fx_0 = \lambda_0 x_0$. *Then* $Fx_0 = (I-T)(\lambda_0 x_0)$, *which contradicts our assumption. Our claim then follows from (2.11) and Proposition 2.1.8 for* $\theta = 0$.

Proposition 2.2.21 *Let* U *be a bounded open subset of* $\mathscr{P}$ *with* $0 \in U \cap \Omega$. *Assume that* $T : \Omega \subset \mathscr{P} \to E$ *is an h-expansive mapping,* $F : \overline{U} \to E$ *is a k-set contraction with* $0 \leq k < h-1$, *and* $F(\overline{U}) \subset (I-T)(\Omega)$. *If*

$$\|Fx+T0\| \leq (h-1)\|x\| \text{ and } Tx+Fx \neq x, \text{ for all } x \in \partial U \cap \Omega, \tag{2.18}$$

then $i_*(T+F, U \cap \Omega, \mathscr{P}) = 1$.

Proof 2.2.22 *According to Lemma 2.1.18, we can see that* $(I-T)^{-1}F : \overline{U} \to \mathscr{P}$ *is a strict* $k(h-1)^{-1}$-*set contraction. From the inclusion* $F(\overline{U}) \subset (I-T)(\Omega)$, *for each* $x \in \overline{U}$, *we can find some* $y \in \Omega$ *such that* $Fx = y - Ty$. *In what follows, we check that the condition (2.4) in Corollary 2.1.10 is satisfied for* $\theta = 0$. *For each* $x \in \overline{U}$, $(I-T)^{-1}Fx \in \Omega$ *and*

$$T((I-T)^{-1}Fx) + Fx = (I-T)^{-1}Fx,$$

which implies that

$$\|T((I-T)^{-1}Fx) - T0\| \leq \|(I-T)^{-1}Fx\| + \|Fx+T0\|.$$

T being expansive with constant h, we have

$$\|T((I-T)^{-1}Fx) - T0\| \geq h\|(I-T)^{-1}Fx\|.$$

Therefore,

$$\|(I-T)^{-1}Fx\| \leq (h-1)^{-1}\|Fx+T0\|. \tag{2.19}$$

From (2.19) and the assumption (2.18), we conclude that for all $x \in \partial U$,

$$\|(I-T)^{-1}Fx\| \leq (h-1)^{-1}\|Fx+T0\|$$

$$\leq \|x\|.$$

Our claim then follows from (2.11) and Corollary 2.1.10 for $\theta = 0$.

The following result follows immediately from Corollary 2.1.6.

Proposition 2.2.23 *Assume that $T : \overline{\Omega} \subset \mathscr{P} \to E$ is an h-expansive mapping, $F : \overline{\mathscr{P}_r} \to E$ is a k-set contraction with $0 \leq k < h-1$, and $F(\overline{\mathscr{P}_r}) \subset (I-T)(\Omega)$. If further $(I-T)^{-1}F(\overline{\mathscr{P}_r}) \subset \mathscr{P}_r$, then $i_*(T + F, \mathscr{P}_r \cap \Omega, \mathscr{P}) = 1$.*

As a particular case, we get

Corollary 2.2.24 *Assume that $T : \overline{\Omega} \subset \mathscr{P} \to E$ is an h-expansive mapping, $F : \overline{\mathscr{P}_r} \to E$ is a k-set contraction with $0 \leq k < h-1$, and $F(\overline{\mathscr{P}_r}) \subset (I-T)(\Omega)$. If $0 \in \Omega$ and*

$$\|Fx+T0\| < (h-1)r, \text{ for all } x \in \overline{\mathscr{P}_r}, \tag{2.20}$$

then $i_*(T+F, \mathscr{P}_r \cap \Omega, \mathscr{P}) = 1$.

Proof 2.2.25 *From (2.19) and the assumption (2.20), for any $x \in \overline{\mathscr{P}_r}$, we conclude that*

$$\|(I-T)^{-1}Fx\| \leq \frac{1}{h-1}\|Fx+T0\| < r,$$

which implies that $(I-T)^{-1}F(\overline{\mathscr{P}_r}) \subset \mathscr{P}_r$.

Proposition 2.2.26 *Assume that $T : \overline{\Omega} \subset \mathscr{P} \to E$ is an h-expansive mapping, $F : \overline{U} \to E$ is a k-set contraction with $0 \leq k < h-1$, and $F(\overline{U}) \subset (I-T)(\Omega)$. If there exists $u_0 \in \mathscr{P} \setminus \{0\}$ such that*

$$Fx \neq (I-T)(x - \lambda u_0), \text{ for all } \lambda \geq 0 \text{ and } x \in \partial U \cap (\Omega + \lambda u_0), \tag{2.21}$$

then $i_*(T+F, U \cap \Omega, \mathscr{P}) = 0$.

Proof 2.2.27 *The mapping* $(I-T)^{-1}F : \overline{U} \to \mathscr{P}$ *is a strict* $k(h-1)^{-1}$-
set contraction and for some $u_0 \in \mathscr{P}\backslash\{0\}$ *this operator satisfies*

$$x - (I-T)^{-1}Fx \neq \lambda u_0, \ \forall x \in \partial U, \forall \lambda \geq 0.$$

By (2.11) and Proposition 2.1.12 for $\theta = 0$, *we deduce that*

$$i_* (T+F, U \cap \Omega, \mathscr{P}) = i((I-T)^{-1}F, U, \mathscr{P}) = 0,$$

proving our claim.

Proposition 2.2.28 *Assume that* $T : \Omega \subset \mathscr{P} \to E$ *is an h-expansive
mapping,* $F : \overline{U} \to E$ *a k-set contraction with* $0 \leq k < h - 1$, *and*
$F(\overline{U}) \subset (I-T)(\Omega)$. *Suppose further that there exists* $u_0 \in \mathscr{P}\backslash\{0\}$ *such
that* $T(x - \lambda u_0) \in \mathscr{P}$, *for all* $\lambda \geq 0$ *and* $x \in \partial U \cap (\Omega + \lambda u_0)$, *and one
of the following conditions holds:*
(a) $Fx \nleq x - \lambda u_0, \ \forall x \in \partial U, \forall \lambda \geq 0.$
(b) $Fx \in \mathscr{P}$, $\|Fx\| > N\|x - \lambda u_0\|, \ \forall x \in \partial U, \forall \lambda \geq 0$, *and the cone* $\mathscr{P}$
is normal with constant N.
Then
$$i_* (T+F, U \cap \Omega, \mathscr{P}) = 0.$$

Proof 2.2.29 *We show that conditions (a) or (b) imply that*

$$Fx \neq (I-T)(x - \lambda u_0), \ \text{for all } \lambda \geq 0 \text{ and } x \in \partial U \cap (\Omega + \lambda u_0).$$

On the contrary, assume the existence of $\lambda_0 \geq 0$ *and* $x_0 \in \partial U \cap (\Omega + \lambda_0 u_0)$ *such that*
$$Fx_0 = (I-T)(x_0 - \lambda_0 u_0).$$
Then $x_0 - \lambda_0 u_0 - Fx_0 = T(x_0 - \lambda_0 u_0) \in \mathscr{P}$. *If condition (a) holds, then
a contradiction is achieved. Otherwise, we deduce that*

$$Fx_0 \leq x_0 - \lambda_0 u_0.$$

Since $\mathscr{P}$ *is normal with constant N, we deduce that*

$$\|Fx_0\| \leq N\|x_0 - \lambda_0 u_0\|,$$

contradicting condition (b) and ending the proof of Proposition 2.2.28.

2.2.2 The case where T is an h-expansive mapping and F is an $(h-1)$-set contraction

Let $\mathscr{P}$ be a cone in a Banach space E, Ω a subset of $\mathscr{P}$, and U a bounded open subset of $\mathscr{P}$.

Suppose that $T : \Omega \to E$ is an h-expansive mapping and $F : \overline{U} \to E$ is an $(h-1)$-set contraction. Since $(I-T)^{-1}$ is $(h-1)^{-1}$-Lipschitzian, then $(I-T)^{-1}F : \overline{U} \to \mathscr{P}$ is a 1-set contraction. Assume further that

$$tF(\overline{U}) \subset (I-T)(\Omega), \ \forall t \in [0,1] \tag{2.22}$$

and

$$0 \notin \overline{(I-T-F)(\partial U \cap \Omega)}. \tag{2.23}$$

Then there exists $\gamma > 0$ such that

$$\inf_{x \in \partial U \cap \Omega} \|x - Tx - Fx\| \geq \gamma.$$

Thus

$$0 \notin (I-T-kF)(\partial U \cap \Omega), \forall k \in (1 - \frac{\gamma}{M}, 1),$$

where $M = \sup_{x \in \overline{U}} \|Fx\| + \gamma$. Indeed, for all $x \in \partial U \cap \Omega$, we have

$$\|0 - (x - Tx - kFx)\| \geq \|x - Tx - Fx\| - (1-k)\|Fx\|$$

$$\geq \gamma - (1-k)M > 0.$$

In other words, $x \neq (I-T)^{-1}kFx$, for all $x \in \partial U$ and $k \in (1 - \frac{\gamma}{M}, 1)$. Clearly, $(I-T)^{-1}kF$ is a strict k-set contraction mapping. As a consequence, by (2.11) and Lemma 2.1.3, the fixed point index $i_*(T + kF, U \cap \Omega, \mathscr{P})$ is well defined. Thus we set

$$i_*(T + F, U \cap \Omega, \mathscr{P}) = i_*(T + kF, U \cap \Omega, \mathscr{P})$$

$$= i((I-T)^{-1}kF, U, \mathscr{P}), \tag{2.24}$$

$$k \in (1 - \frac{\gamma}{M}, 1).$$

However, we must check that $i_*(T+F, U \cap \Omega, \mathcal{P})$ does not depend on the parameter $k \in (1 - \dfrac{\gamma}{M}, 1)$. For this, let $G_j = k_j F : \overline{U} \to E$ be $k_j(h-1)$-set contractions with $k_j \in (1 - \dfrac{\gamma}{M}, 1)$ $(j = 1, 2)$. Then $\|G_j x - Fx\| = (1 - k_j)\|Fx\| \leq (1 - k_j)M < \gamma$, $\forall x \in \partial U$. Define the convex deformation $H : [0, 1] \times \overline{U} \to E$:

$$H(t, x) = tG_1 x + (1 - t)G_2 x.$$

The operator H is continuous, uniformly continuous in t for each x, and $H([0, 1] \times \overline{U}) \subset (I - T)(\Omega)$. In addition, $H(t, .)$ is a $\tilde{k}(h-1)$-set contraction for each t, where $\tilde{k} = \max(k_1, k_2)$ and $T + H(t, .)$ has no fixed point on $\partial U \cap \Omega$ for each t. In fact, for all $x \in \partial U \cap \Omega$, we have

$$\|x - Tx - H(t, x)\| = \|x - Tx - tG_1 x - (1 - t)G_2 x\|$$

$$\geq \|x - Tx - Fx\| - t\|Ax - G_1 x\|$$

$$- (1 - t)\|Fx - G_2 x\|$$

$$> \gamma - t\gamma - (1 - t)\gamma = 0.$$

From the invariance by homotopy property of the index in Theorem 2.2.1, we deduce that

$$i_*(T + G_1, U \cap \Omega, \mathcal{P}) = i_*(T + G_2, U \cap \Omega, \mathcal{P}),$$

which shows that the index $i_*(T + F, U \cap \Omega, \mathcal{P})$ does not depend on k.

The generalized fixed point index defined in (2.24) satisfies some properties grouped in the following:

Theorem 2.2.30

(a) (Normalization property). If $U = \mathcal{P}_r = \mathcal{P} \cap \mathcal{B}_r$ *and* $Fx = z_0 \in \mathcal{B}(-T0, (h-1)r) \cap \mathcal{P}$, *for all* $x \in \overline{\mathcal{P}_r}$, *then* $i_*(T + F, \mathcal{P}_r \cap \Omega, \mathcal{P}) = 1$.

(b) (Additivity property). For any pair of disjoint open subsets U_1, U_2 *in*
U *such that*
$0 \notin (I - T - F)((\overline{U} \setminus (U_1 \cup U_2)) \cap \Omega)$, *we have*

$$i_*(T + F, U \cap \Omega, \mathcal{P}) = i_*(T + F, U_1 \cap \Omega, \mathcal{P}) + i_*(T + F, U_2 \cap \Omega, \mathcal{P}),$$

where $i_*(T + F, U_j \cap \Omega, X) := i_*(f|_{\overline{U_j}}, U_j \cap \Omega, \mathcal{P})$, $j = 1, 2$.

(c) (Homotopy Invariance property). The fixed point index $i_*(T + H(t,.), U \cap \Omega, \mathscr{P})$ does not depend on the parameter $t \in [0,1]$, where

(i) $H : [0,1] \times \overline{U} \to E$ is continuous and $H(t,x)$ is uniformly continuous in t with respect to $x \in \overline{U}$,

(ii) $H(t,.) : \overline{U} \to E$ is an $(h-1)$-set contraction,

(iii) $tH([0,1] \times \overline{U}) \subset (I-T)(\Omega)$, for all $t \in [0,1]$,

(iv) $0 \notin \overline{(I-T-H(t,.))(\partial U \cap \Omega)}$, for all $t \in [0,1]$,

(d) (Solvability property). If $i_*(T + F, U \cap \Omega, \mathscr{P}) \neq 0$, then $0 \in \overline{(I-T-F)(U \cap \Omega)}$.

Proof 2.2.31

(a) *Since F is a constant mapping, it is a 0-set contraction, which implies that $(I-T)^{-1}F$ is a 0-set contraction. As in the proof of Theorem 2.2.1, part (a), $y_0 = (I-T)^{-1}z_0 \in \mathscr{P}_r$. By the normalization property in Lemma 2.1.3, we deduce that*

$$i((I-T)^{-1}z_0, \mathscr{P}_r, \mathscr{P}) = 1.$$

Therefore $i_*(T + z_0, \mathscr{P}_r \cap \Omega, \mathscr{P}) = 1$, proving our claim.

(b) *Let*

$$\gamma = \inf_{(\overline{U} \setminus (U_1 \cup U_2)) \cap \Omega} \|x - Tx - Fx\| > 0.$$

Suppose that $G = kF : \overline{U} \to E$ is a $k(h-1)$-set contraction and

$$\|Gx - Fx\| < \gamma, \forall x \in \overline{U} \setminus (U_1 \cup U_2) \cap \Omega. \tag{2.25}$$

From (2.24), we have

$$i_*(T + F, U \cap \Omega, \mathscr{P}) = i_*(T + G, U \cap \Omega, \mathscr{P})$$

and

$$i_*(T + F, U_j \cap \Omega, \mathscr{P}) = i_*(T + G, U_j \cap \Omega, \mathscr{P}), \quad j = 1,2.$$

Hence, $T + G$ has no fixed point in $\overline{U} \setminus (U_1 \cup U_2) \cap \Omega$. In fact, if the exists $x_0 \in \overline{U} \setminus (U_1 \cup U_2) \cap \Omega$ such that $x_0 = Tx_0 + Gx_0$, then

$$\gamma \leq \|x_0 - Tx_0 - Fx_0\| = \|x_0 - Tx_0 - Gx_0 + Gx_0 - Fx_0\|$$
$$= \|Gx_0 - Fx_0\|,$$

*which contradicts (4.33). The claim follows from (2.24) and property
(b) of the fixed point index in Theorem 2.2.1.*

(c) *By the property of the function H, there exist $\gamma > 0$ and $N > 0$ such
that*

$$\|x - Tx - H(t,x)\| \geq \gamma, \text{ for all } x \in \partial U \cap \Omega \text{ and } t \in [0,1],$$

*as well as $\|H(t,x)\| \leq N$, for all $x \in \overline{U}$ and $t \in [0,1]$. Let $K(t,x) = kH(t,x)$, where $k \in (1 - \frac{\gamma}{2N}, 1)$. Then for all $x \in \partial U \cap \Omega$ and $t \in [0,1]$,
we have*

$$\|x - Tx - K(t,x)\| = \|x - Tx - H(t,x)\| + \|H(t,x) - K(t,x)\|$$

$$\geq \gamma - (k-1)N > \gamma - \frac{\gamma}{2} > 0.$$

*Obviously, $K(t,.) : \overline{U} \to E$ is a $k(h-1)$-set contraction, where k does
not depend on $t \in [0,1]$ and $K([0,1] \times \overline{U}) \subset (I-T)(\Omega)$.*

*Then our claim follows from (2.24) and property (c) of the fixed
point index in Theorem 2.2.1.*

(d) *Consider a sequence $(k_n)_n \subset (0,1)$ such that $k_n \to 1$, as $n \to \infty$ and
define the function $G_n = k_n F$ $n \in \mathbb{N}$. Then $G_n : \overline{U} \to E$ is a $k(h-1)$-set
contraction. Since $\|Fx\| < \infty$, $\forall x \in \overline{U}$, we obtain that*

$$\|Fx - G_n x\| = \|Fx - k_n Fx\|$$

$$= (1-k_n)\|F_n x\| \to 0, \text{ as } n \to +\infty.$$

Hence, there exists $n_0 > 0$ such that for every $n \geq n_0$

$$\|Fx - G_n x\| < \gamma, \text{ where } 0 < \gamma < \inf_{x \in \partial U \cap \Omega} \|x - Tx - Fx\|.$$

From (2.24),

$$i_*(T + G_n, U \cap \Omega, \mathscr{P}) = i_*(T + F, U \cap \Omega, \cap P) \neq 0.$$

*Thus, property (d) in Theorem 2.2.1 guaranties that for all $n \in \mathbb{N}$, the
mapping $T + G_n$ has a fixed point x_n in $U \cap \Omega$. Consequently,*

$$\|x_n - Tx_n - Fx_n\| = \|x_n - Tx_n - G_n x_n + G_n x_n - Fx_n\|$$

$$= \|G_n x_n - Fx_n\| \to 0, \text{ as } n \to +\infty.$$

Then $x_n - Tx_n - Fx_n \to 0$, as $n \to +\infty$, that is $0 \in \overline{(I-T-F)(U \cap \Omega)}$.

Remark 2.2.32 *Regarding the additivity property in Theorem 2.2.30, we cannot replace the condition* $0 \notin (I-T-F)((\overline{U}\setminus(U_1\cup U_2))\cap \Omega)$ *by the weaker one that* $T+F$ *has no fixed point on* $(\overline{U}\setminus(U_1\cup U_2))\cap \Omega$. *In fact, let us consider the Banach space* c_0 *of real sequences converging to zero with the sup-norm and the cone* $\mathscr{P}$ *of sequences* $(a_n)_n$ *with only positive entries* a_n. *Let* $r : \overline{\mathscr{P}}_5 \to \overline{\mathscr{P}}_1$ *be the radial retraction,* $s : \overline{\mathscr{P}}_1 \ni (a_1, a_2, \ldots) \to (1, a_1, a_2, \ldots) \in \overline{\mathscr{P}}_1$ *the well-known shift map, and let* $\hat{F} := s \circ r$. *For* $T = 2I, F = -\hat{F}$, *and* $U = \Omega = \mathscr{P}_5, U_1 = \mathscr{P}_3 \setminus \overline{\mathscr{P}}_2, U_2 = \mathscr{P}_5 \setminus \overline{\mathscr{P}}_4$, *we get*

$$i_*(T+F, \mathscr{P}_5, \mathscr{P}) \;=\; 1 \neq 0 + 0$$

$$=\; i_*(T+F, U_1, \mathscr{P}) + i_*(T+F, U_2, \mathscr{P}).$$

Remark 2.2.33 *Notice that a sufficient condition for (3.11) holds is*

$$\exists \delta > 0, \; \forall x \in \partial U \cup \Omega, \; \|x - Tx - Fx\| \geq \delta.$$

According to Theorem 2.2.1 and in a way similar to the one used to show Propositions 2.2.13, 2.2.15, 2.2.19, and 2.2.21, we can show the following results. These results have been established in [18].

Proposition 2.2.34 *Assume that* $T : \Omega \subset \mathscr{P} \to E$ *is an h-expansive mapping and* $F : \overline{\mathscr{P}}_r \to E$ *is a* $(h-1)$*-set contraction with* $tF(\overline{\mathscr{P}}_r) \subset (I-T)(\Omega)$, *for all* $t \in [0,1]$ *and* $0 \notin (I-T-F)(\partial \mathscr{P}_r \cap \Omega)$. *If* $0 \in \Omega$, $\|T0\| < (h-1)r$, *and*

$$Fx \neq \lambda(x-Tx) \text{ for all } x \in \partial \mathscr{P}_r \cap \Omega \text{ and } \lambda > 1,$$

then $i_*(T+F, \mathscr{P}_r \cap \Omega, \mathscr{P}) = 1$.

Proposition 2.2.35 *Assume that* $T : \Omega \subset \mathscr{P} \to E$ *is an h-expansive mapping and* $F : \overline{\mathscr{P}}_r \to E$ *is an* $(h-1)$*-set contraction with* $tF(\overline{\mathscr{P}}_r) \subset (I-T)(\Omega)$, *for all* $t \in [0,1]$ *and* $0 \notin (I-T-F)(\partial \mathscr{P}_r \cap \Omega)$. *If* $0 \in \Omega$, $\|T0\| < (h-1)r$, *and*

$$\|Fx\| \leq \|x - Tx\| \text{ for all } x \in \partial \mathscr{P}_r \cap \Omega,$$

then $i_*(T+F, \mathscr{P}_r \cap \Omega, \mathscr{P}) = 1$.

Proposition 2.2.36 *Let U be a bounded open subset of $\mathscr{P}$ such that $0 \in U$. Assume that $T : \Omega \subset \mathscr{P} \to E$ is an h-expansive mapping and $F : \overline{U} \to E$ is an $(h-1)$-set contraction with $F(\overline{U}) \subset (I-T)(\Omega)$ and $0 \notin \overline{(I-T-F)(\partial U \cap \Omega)}$. If*

$$Fx \neq (I-T)(\lambda x), \quad \text{for all } x \in \partial U, \lambda > 1 \text{ and } \lambda x \in \Omega,$$

then $i_* (T+F, U \cap \Omega, \mathscr{P}) = 1$.

Proposition 2.2.37 *Let U be a bounded open subset of $\mathscr{P}$ such that $0 \in U \cap \Omega$. Assume that $T : \Omega \subset \mathscr{P} \to E$ is an h-expansive mapping and $F : \overline{U} \to E$ is an $(h-1)$-set contraction with $F(\overline{U}) \subset (I-T)(\Omega)$ and $0 \notin \overline{(I-T-F)(\partial U \cap \Omega)}$. If*

$$\|Fx + T0\| \leq (h-1)\|x\| \quad \text{for all } x \in \partial U \cap \Omega, \qquad (2.26)$$

then $i_* (T+F, U \cap \Omega, \mathscr{P}) = 1$.

Proposition 2.2.38 *Assume that $T : \Omega \subset \mathscr{P} \to E$ is an h-expansive mapping such that $F : \overline{U} \to E$ is an $(h-1)$-set contraction with $tF(\overline{U}) \subset (I-T)(\Omega)$, for all $t \in [0,1]$ and $0 \notin \overline{(I-T-F)(\partial U \cap \Omega)}$. If there exists $u_0 \in \mathscr{P} \backslash \{0\}$ such that*

$$\gamma Fx \neq (I-T)(x - \lambda u_0), \text{ for all } \lambda \geq 0, x \in \partial U \cap (\Omega + \lambda u_0), \quad (2.27)$$

and $\gamma \in (0,1)$, then $i_* (T+F, U \cap \Omega, \mathscr{P}) = 0$.

Proof 2.2.39 *The mapping $(I-T)^{-1}F : \overline{U} \to \mathscr{P}$ is a 1-set contraction. Suppose that $i_* (T+F, U \cap \Omega, \mathscr{P}) \neq 0$. Then, from (2.24), for $k \in (k_0, 1)$ with some $0 < k_0 < 1$, we can see that*

$$i((I-T)^{-1}kF, U, \mathscr{P}) \neq 0.$$

For each $k \in (k_0, 1)$ and $r > 0$, define the homotopy:

$$H(t,x) = (I-T)^{-1}kFx + tru_0, \text{ for } x \in \overline{U} \text{ and } t \in [0,1].$$

The operator H is continuous and uniformly continuous in t for each x. Moreover, $H(t,.)$ is a strict k-set contraction for each t and

$$H([0,1] \times \overline{U}) = (I-T)^{-1}kF(U) + tru_0 \subset \mathscr{P}.$$

We check that $H(t,x) \neq x$, for all $(t,x) \in [0,1] \times \partial U$. If $H(t_0,x_0) = x_0$ for some $(t_0,x_0) \in [0,1] \times \partial U$, then

$$x_0 - t_0 r u_0 = (I-T)^{-1} k F x_0,$$

and so $x_0 - t_0 r u_0 \in \Omega$. Hence

$$(I-T)(x_0 - t_0 r u_0) = k F x_0,$$

for $x_0 \in \partial U \cap (\Omega + t_0 r u_0)$, contradicting the assumption (2.27). By the property (c) of the index in Lemma 2.1.3, for $k \in (k_0, 1)$, we deduce that

$$i((I-T)^{-1} k F + r u_0, U \cap \Omega, \mathscr{P}) = i((I-T)^{-1} k F, U, \mathscr{P}) \neq 0.$$

By property (f) of the index in Lemma 2.1.3, for each $k \in (k_0, 1)$ and $r > 0$, there exists $x_r \in U$ such that

$$x_r - (I-T)^{-1} k F x_r = r u_0. \tag{2.28}$$

Letting $r \to +\infty$ in (2.28), the left-hand side of (2.28) is bounded while the right-hand side is not, which is a contradiction. Therefore,

$$i_*(T+F, U \cap \Omega, \mathscr{P}) = 0,$$

which completes the proof.

Proposition 2.2.40 *Assume that $T : \Omega \subset \mathscr{P} \to E$ is an expansive mapping with constant $h > 1$ such that $F : \overline{U} \to E$ is an $(h-1)$-set contraction with $tF(\overline{U}) \subset (I-T)(\Omega)$, for all $t \in [0,1]$ and $0 \notin (I-T-F)(\partial U \cap \Omega)$. Suppose that there exists $u_0 \in \mathscr{P} \setminus \{0\}$ such that $T(x - \lambda u_0) \in \mathscr{P}$, for all $\lambda \geq 0$ and $x \in \partial U \cap (\Omega + \lambda u_0)$, and one of the following conditions is satisfied:*
(a) $\gamma F x \not\leq x - \lambda u_0$, for all $x \in \partial U$, $\lambda \geq 0$, and $\gamma \in (0,1)$.
(b) $F x \in \mathscr{P}$, $\gamma \|F x\| > N \|x - \lambda u_0\|$, for all $x \in \partial U$, $\lambda \geq 0$, $\gamma \in (0,1)$, and the cone $\mathscr{P}$ is normal with constant N.
Then $i_*(T+F, U \cap \Omega, \mathscr{P}) = 0$.

Proof 2.2.41 *The proof is similar to that of Proposition 2.2.28.*

Proposition 2.2.42 *Assume that $T : \Omega \subset \mathscr{P} \to E$ is an expansive mapping with constant $h > 1$ and $F : \overline{U} \to E$ is an $(h-1)$-set contraction with $F(\partial U) \subset \mathscr{P}$ and $tF(\overline{U}) \subset (I-T)(\Omega), \forall t \in [0,1]$ and $0 \notin (I-T-F)(\partial U \cap \Omega)$. Suppose further that there exists $u_0 \in \mathscr{P}\setminus\{0\}$ such that*

$$c_0 Fx \nleq x - T(x - \lambda u_0), \text{ for all } \lambda \geq 0, x \in \partial U \cap (\Omega + \lambda u_0) \quad (2.29)$$

and $c_0 \in (0,1)$. Then $i_* (T+F, U \cap \Omega, \mathscr{P}) = 0$.

Proof 2.2.43 *The mapping $(I-T)^{-1}F : \overline{U} \to \mathscr{P}$ is a 1-set contraction. By contradiction, suppose that $i_* (T+F, U \cap \Omega, \mathscr{P}) \neq 0$. From the definition 2.24, for $k \in (k_0, 1)$ with $c_0 \leq k_0 < 1$, we have*

$$i((I-T)^{-1}kF, U, \mathscr{P}) \neq 0.$$

For each $k \in (k_0, 1)$ and $r > 0$, consider the homotopic deformation

$$H(t,x) = (I-T)^{-1}kF + tru_0, \text{ for } x \in \overline{U} \text{ and } t \in [0,1].$$

The operator H is continuous and uniformly continuous in t, for each x. Moreover, $H(t,.)$ is a strict k-set contraction, for each t and $H([0,1] \times \overline{U}) \subset \mathscr{P}$. We prove that $H(t,x) \neq x$ for all $(t,x) \in [0,1] \times \partial U$. If $H(t_0, x_0) = x_0$ for some $(t_0, x_0) \in [0,1] \times \partial U$, then

$$x_0 - t_0 ru_0 = (I-T)^{-1}kFx_0,$$

and so $x_0 - t_0 ru_0 \in \Omega$. Hence

$$x_0 - t_0 ru_0 - T(x_0 - t_0 ru_0) = kFx_0,$$

for $x_0 \in \partial U \cap (\Omega + t_0 ru_0)$, which implies that

$$x_0 - T(x_0 - t_0 ru_0) \geq kFx_0 \geq c_0 Fx_0,$$

contradicting the assumption (2.29). As a consequence, by property (c) of the index in Lemma 2.1.3, for $k \in (k_0, 1)$, we get

$$i((I-T)^{-1}kF + ru_0, U \cap \Omega, \mathscr{P}) = i((I-T)^{-1}kF, U, \mathscr{P}) \neq 0.$$

By the existence property of the index in Lemma 2.1.3, for each $k \in (k_0, 1)$ and $r > 0$, there exists $x_r \in U$ such that

$$x_r - (I - T)^{-1} k F x_r = r u_0. \tag{2.30}$$

Letting $r \to +\infty$ in (2.30), the left side of (2.30) is bounded, but the right side is not, leading to a contradiction. Therefore

$$i_*(T + F, U \cap \Omega, \mathscr{P}) = 0.$$

The proof is complete.

2.2.3 The case where T is a nonlinear expansive mapping and F is an k-set contraction

Let (X, d) be a metric space. Following [46], we put

Definition 2.2.44 *The mapping $T : X \to X$ is said to be nonlinear expansive, if there exists a function $\phi : [0, +\infty) \to [0, +\infty)$ such that*

$$d(Tx, Ty) \geq \phi(d(x, y)), \ \forall x, y \in X,$$

with $\phi(t) > t$, $\forall t > 0$ and

$$\exists r > 0, \quad \omega = \inf_{t \in (0, 2r]} \frac{\phi(t) - t}{t} > 0, \tag{2.31}$$

that is T is $(\omega + 1)$-expansive. We will denote by $D = \mathscr{B}_r = \mathscr{B}(0, r)$.

Lemma 2.2.45 *Let $(X, \|.\|)$ be a linear normed space and $T : D \to X$ a nonlinear expansive mapping. Then the inverse of $A := I - T : D \to (I - T)(D)$ exists and is ω^{-1}-Lipschitzian.*

Proof 2.2.46 *For each $x, y \in D$, $x \neq y$, and $0 < \|x - y\| \leq 2r$, we have*

$$\begin{aligned}
\|Ax - Ay\| &= \|(Tx - Ty) - (x - y)\| \\
&\geq \phi(\|x - y\|) - \|x - y\| \tag{2.32} \\
&\geq \omega \|x - y\|,
\end{aligned}$$

showing that A is injective. Thus $A^{-1} : D \to A(D)$ exists. Taking $x, y \in A(D)$ and using (2.32), we get

$$\|A^{-1}x - A^{-1}y\| \leq \omega^{-1} \|x - y\|, \ \text{for all } x, y \in A(D).$$

In what follows, $\mathscr{P}$ will refer to a cone in a Banach space E. Let $\Omega = \mathscr{P} \cap \mathscr{B}_r$ and U be a bounded open subset of $\mathscr{P}$ such that $U \cap \Omega \neq \emptyset$.

Assume that $F : \overline{U} \to E$ is a k-set contraction and $T : \Omega \to X$ is a nonlinear expansive mapping with $0 \leq k < \omega$. By Lemma 2.2.45, the operator $(I - T)^{-1}$ is ω^{-1}-Lipschitzian on $(I - T)(\Omega)$.

Suppose further that $F(\overline{U}) \subset (I - T)(\Omega)$, and $x \neq Tx + Fx$, for all $x \in \partial U \cap \Omega$. Then $x \neq (I - T)^{-1} Fx$, for all $x \in \partial U$ and the mapping $(I - T)^{-1} F : \overline{U} \to \mathscr{P}$ is a strict $k\omega^{-1}$-set contraction. From Lemma 2.1.3, the fixed point index $i((I - T)^{-1} F, U, \mathscr{P})$ is well defined. Thus we put

$$i_* (T + F, U \cap \Omega, \mathscr{P}) = i((I - T)^{-1} F, U, \mathscr{P}). \qquad (2.33)$$

Notice that, for $U \cap \Omega = \emptyset$, the index $i_* (T + F, U \cap \Omega, \mathscr{P}) = 0$.

Next, we compute the fixed point index for the class of mappings under consideration by appealing to Lemma 2.1.3.

Proposition 2.2.47 *Assume that $T : \Omega \to E$ is a nonlinear expansive mapping. Let $\rho > 0$ and $F : \overline{\mathscr{P}_\rho} \to E$ be a k-set contraction with $0 \leq k < \omega$, $tF(\overline{\mathscr{P}_\rho}) \subset (I - T)(\Omega)$, for all $t \in [0, 1]$ and $F(\partial \mathscr{P}_\rho \cap \Omega) \subset \mathscr{P}$. If $0 \in \Omega$, $\|T0\| < \omega\rho$ and*

$$Fx \ngeq (I - T)x, \quad \text{for all } x \in \partial \mathscr{P}_\rho \cap \Omega,$$

then $i_* (T + F, \mathscr{P}_\rho \cap \Omega, \mathscr{P}) = 1$.

Proof 2.2.48 *First, we show that $y_0 := (I - T)^{-1} 0 \in \mathscr{P}_\rho \cap \Omega$. If $y_0 = 0$ we are finished, otherwise from the inclusion $tF(\overline{\mathscr{P}_\rho}) \subset (I - T)(\Omega)$ for all $t \in [0, 1]$, for each $x \in \overline{\mathscr{P}_\rho}$, we can find some $y \in \Omega$ such that $0Fx = y - Ty$, that is $(I - T)^{-1} 0 \in \Omega$, which implies that $y_0 \in \Omega$. Further, Lemma 2.2.45 guarantees that*

$$\|(I - T)y_0 + T0\| = \|(I - T)y_0 - (I - T)0\|$$

$$\geq \omega\|y_0\|.$$

Thus

$$\omega\|y_0\| \leq \|(I - T)y_0 + T0\|$$

$$= \|0 + T0\| < \omega\rho.$$

Define the homotopic deformation $H : [0,1] \times \overline{\mathscr{P}_\rho} \to E$ by

$$H(t,x) = (I - T)^{-1} tFx.$$

The operator H is continuous in x and uniformly continuous in t for each x. Moreover, $H(.,t)$ is a strict $k\omega^{-1}$-set contraction for each t and $H(t,.)$ has no fixed point on $\partial \mathscr{P}_\rho$ for each t. Otherwise, there exist $x_0 \in \partial \mathscr{P}_\rho$ and $t_0 \in [0,1]$ such that $x_0 = (I - T)^{-1} t_0 Fx_0$. Hence $t_0 Fx_0 = x_0 - Tx_0$. We may distinguish between two cases. If $t_0 = 0$, then $Tx_0 = x_0$ and

$$\begin{aligned} \omega\|x_0\| &\leq \|(I-T)x_0 + T0\| \\ &= \|T0\| < \omega\rho, \end{aligned}$$

contradicting $x_0 \in \partial \mathscr{P}_\rho$. If $t_0 \in (0,1]$, then $Fx_0 \geq t_0 Fx_0 = x_0 - Tx_0$, leading again to a contradiction with the hypothesis. By the homotopy invariance and normalization properties of the fixed point index in Lemma 2.1.3, we deduce that

$$\begin{aligned} i_*(T + F, \mathscr{P}_\rho \cap \Omega, \mathscr{P}) &= i((I - T)^{-1} F, \mathscr{P}_\rho, \mathscr{P}) \\ &= i((I - T)^{-1} 0, \mathscr{P}_\rho, \mathscr{P}) = 1. \end{aligned}$$

Proposition 2.2.49 *Let U be an bounded open subset of $\mathscr{P}$ with $0 \in U \cap \Omega$. Assume that $T : \Omega \to E$ is a nonlinear expansive mapping, $F : \overline{U} \to E$ is a k-set contraction with $0 \leq k < \omega$, and $F(\overline{U}) \subset (I - T)(\Omega)$. If*

$$\|Fx + T0\| \leq (\omega - 1)\|x\| \quad \text{and} \quad Tx + Fx \neq x, \quad \text{for all } x \in \partial U \cap \Omega, \tag{2.34}$$

then $i_*(T + F, U \cap \Omega, \mathscr{P}) = 1$.

Proof 2.2.50 *According to Lemma 2.2.45, the mapping $(I - T)^{-1} F : \overline{U} \to \mathscr{P}$ is a strict $k\omega^{-1}$-set contraction. From the inclusion $F(\overline{U}) \subset (I - T)(\Omega)$, for each $x \in \overline{U}$, we can find some $y \in \Omega$ such that $Fx = y - Ty$. Then $(I - T)^{-1} Fx \in \Omega$ and for any $x \in \overline{U}$, we have*

$$T((I - T)^{-1} Fx) + Fx = (I - T)^{-1} Fx,$$

which implies that

$$\|T((I-T)^{-1}Fx)-T0\| \le \|(I-T)^{-1}Fx\| + \|Fx+T0\|.$$

Since $(I-T)^{-1}Fx \in \Omega$ *with* $(I-T)^{-1}Fx \ne 0$, *then*

$$\|T((I-T)^{-1}Fx)-T0\| \ge \phi(\|(I-T)^{-1}Fx\|) - \|(I-T)^{-1}Fx\|$$

$$\ge \omega\|(I-T)^{-1}Fx\|.$$

Therefore,

$$\|(I-T)^{-1}Fx\| \le (\omega-1)^{-1}\|Fx+T0\|. \tag{2.35}$$

From (2.35) *and the assumption* (2.34), *we conclude that*

$$\|(I-T)^{-1}Fx\| \le (\omega-1)^{-1}\|Fx+T0\|$$
$$\le \|x\|, \forall x \in \partial U. \tag{2.36}$$

The claim of the proposition then follows from (2.33) *and Corollary 2.1.10 for* $\theta = 0$.

Corollary 2.2.51 *Assume that* $T : \Omega \to E$ *is a nonlinear expansive mapping,* $F : \overline{\mathscr{P}_\rho} \to E$ *is a k-set contraction with* $0 \le k < \omega$, *and* $F(\overline{\mathscr{P}_\rho}) \subset (I-T)(\Omega)$. *If* $0 \in \Omega$ *and*

$$\|Fx+T0\| < (\omega-1)\rho, \text{ for all } x \in \overline{\mathscr{P}_\rho}, \tag{2.37}$$

then $i_*(T+F, \mathscr{P}_\rho \cap \Omega, \mathscr{P}) = 1$.

Proof 2.2.52 *From* (2.35) *and Assumption* (2.37), *for any* $x \in \overline{\mathscr{P}_\rho}$, *we have*

$$\|(I-T)^{-1}Fx\| \le (\omega-1)^{-1}\|Fx+T0\| < \rho,$$

which implies that $(I-T)^{-1}F(\overline{\mathscr{P}_\rho}) \subset \mathscr{P}_\rho$. *Our claim follows from* (2.33) *and Corollary 2.1.6.*

Proposition 2.2.53 *Assume that* $T : \Omega \to E$ *is a nonlinear expansive mapping,* $F : \overline{U} \to E$ *is a k-set contraction with* $0 \le k < \omega$ *and* $F(\overline{U}) \subset (I-T)(\Omega)$. *If there exists* $u_0 \in \mathscr{P} \setminus \{0\}$ *such that*

$$T(x-\lambda u_0) \in \mathscr{P} \text{ and } Fx \nleq x - \lambda u_0 \text{ for all } (x,\lambda) \in \partial U \times [0,+\infty),$$

then $i_*(T+F, U \cap \Omega, \mathscr{P}) = 0$.

Proof 2.2.54 *Given $u_0 \in \mathscr{P} \setminus \{0\}$, we show that*

$$x - (I-T)^{-1} Fx \neq \lambda u_0, \ \forall x \in \partial U, \forall \lambda \geq 0.$$

On the contrary, assume the existence of $x_0 \in \partial U$ and $\lambda_0 \geq 0$ such that

$$x_0 - (I-T)^{-1} Fx_0 = \lambda_0 u_0$$

Then

$$x_0 - \lambda_0 u_0 = (I-T)^{-1} Fx_0 \in \Omega.$$

Hence $x_0 - \lambda_0 u_0 - Fx_0 = T(x_0 - \lambda_0 u_0) \in \mathscr{P}$, that is $Fx_0 \leq x_0 - \lambda_0 u_0$, which contradicts our assumption.

Consequently, by (2.33) and Proposition 2.1.12, we get

$$i_* (T+F, U \cap \Omega, \mathscr{P}) = i((I-T)^{-1} F, U, \mathscr{P}) = 0,$$

which completes the proof.

2.2.4 The case where $(I-T)$ is a Lipschitz invertible mapping and F is a k-set contraction

Let $\mathscr{P}$ be a cone in a Banach space E, Ω a subset of $\mathscr{P}$, and U a bounded open subset of $\mathscr{P}$.

Assume that $T : \Omega \to E$ is a mapping such that $(I-T)$ is Lipschitz invertible with constant $\gamma > 0$ and $F : \overline{U} \to E$ is a k-set contraction.

Suppose that

$$0 \leq k < \gamma^{-1},$$
$$F(\overline{U}) \subset (I-T)(\Omega),$$

and

$$x \neq Tx + Fx, \ \text{for all } x \in \partial U \cap \Omega.$$

Then $x \neq (I-T)^{-1} Fx$, for all $x \in \partial U$ and the mapping $(I-T)^{-1} F : \overline{U} \to \mathscr{P}$ is a strict γk-set contraction. Indeed, $(I-T)^{-1} F$ is continuous and bounded and for any bounded set B in U, we have

$$\alpha(((I-T)^{-1} F)(B)) \leq \gamma \alpha(F(B)) \leq \gamma k \alpha(B).$$

By Lemma 2.1.3, the fixed point index $i((I-T)^{-1} F, U, \mathscr{P})$ is well defined. Thus we put

$$i_* (T+F, U \cap \Omega, \mathscr{P}) = i((I-T)^{-1} F, U, \mathscr{P}). \qquad (2.38)$$

Proposition 2.2.55 *Assume that the mapping $T : \Omega \subset \mathscr{P} \to E$ is such that $(I - T)$ is Lipschitz invertible with constant $\gamma > 0$, $F : \overline{U} \to E$ is a k-set contraction with $0 \leq k < \gamma^{-1}$, and $tF(\overline{U}) \subset (I - T)(\Omega)$ for all $t \in [0, 1]$. If $(I - T)^{-1}0 \in U$, and*

$$(I - T)x \neq \lambda F x \text{ for all } x \in \partial U \cap \Omega \text{ and } 0 \leq \lambda \leq 1, \qquad (2.39)$$

then $i_* (T + F, U \cap \Omega, \mathscr{P}) = 1$.

Proof 2.2.56 *Consider the homotopic deformation $H : [0, 1] \times \overline{U} \to \mathscr{P}$ given by*

$$H(t, x) = (I - T)^{-1} t F x.$$

The operator H is continuous and uniformly continuous in t for each x. Moreover, $H(t, .)$ is a strict set contraction for each t and the mapping $H(t, .)$ has no fixed point on ∂U for each t. Otherwise, there would exist some $x_0 \in \partial U \cap \Omega$ and $t_0 \in [0, 1]$ such that

$$x_0 - T x_0 = t_0 F x_0,$$

which contradicts our assumption.
From the invariance under homotopy and the normalization property of the index fixed point, we deduce that

$$i_* ((I - T)^{-1} F, U, \mathscr{P}) = i_* ((I - T)^{-1} 0, U, \mathscr{P}) = 1.$$

Consequently, from (2.38), we deduce that

$$i_* (T + F, U \cap \Omega, \mathscr{P}) = 1,$$

which completes the proof.

As a consequence of Proposition 2.2.55, we have the following two results.

Corollary 2.2.57 *Assume that the mapping $T : \Omega \subset \mathscr{P} \to E$ be such that $(I - T)$ is Lipschitz invertible with constant $\gamma > 0$, $F : \overline{U} \to E$ is a k-set contraction with $0 \leq k < \gamma^{-1}$, and $tF(\overline{U}) \subset (I - T)(\Omega)$ for all $t \in [0, 1]$. If $(I - T)^{-1}0 \in U$, and*

$$\|Fx\| \leq \|x - Tx\| \text{ and } Tx + Fx \neq x \text{ for all } x \in \partial U \cap \Omega,$$

then $i_* (T + F, U \cap \Omega, \mathscr{P}) = 1$.

Proof 2.2.58 *It is sufficient to prove that the assumption (2.39) is satisfied. For this, assume by contradiction that some $x_0 \in \partial U \cap \Omega$ and $0 \le \lambda_0 \le 1$ exist and satisfy $x_0 - Tx_0 = \lambda_0 Fx_0$. Then two cases are discussed separately:*
(i) If $\lambda_0 = 1$, then $Tx_0 + Fx_0 = x_0$ and a contradiction is reached.
(ii) If $0 \le \lambda_0 < 1$, then $\|x_0 - Tx_0\| = \lambda_0 \|Fx_0\| < \|Fx_0\|$, whence a contradiction.

Corollary 2.2.59 *Assume that the mapping $T : \Omega \subset \mathscr{P} \to E$ is such that $(I - T)$ is Lipschitz invertible with constant $\gamma > 0$, $F : \overline{U} \to E$ is a k-set contraction with $0 \le k < \gamma^{-1}$, and $tF(\overline{U}) \subset (I-T)(\Omega)$ for all $t \in [0,1]$. If $(I-T)^{-1}0 \in U$,*

$$Fx \in \mathscr{P} \ \text{for all} \ x \in \partial U \cap \Omega,$$

and

$$Fx \not\geq x - Tx \ \text{for all} \ \ x \in \partial U \cap \Omega,$$

then $i_*(T + F, \cap \Omega, \mathscr{P}) = 1$.

Proof 2.2.60 *It is easy to see that Assumption (2.39) is satisfied. Otherwise, there exist some $x_0 \in \partial \mathscr{P}_r \cap \Omega$ and $0 < \lambda_0 \le 1$ such that*

$$x_0 = \lambda_0 Fx_0 + Tx_0.$$

Then $x_0 - Tx_0 = \lambda_0 Fx_0 \le Fx_0$, which contradicts our assumption.

Proposition 2.2.61 *Let U be a bounded open subset of $\mathscr{P}$ with $0 \in U$. Assume that the mapping $T : \Omega \subset \mathscr{P} \to E$ is such that $(I - T)$ is Lipschitz invertible with constant $\gamma > 0$, $F : \overline{U} \to E$ is a k-set contraction with $0 \le k < \gamma^{-1}$, and $F(\overline{U}) \subset (I-T)(\Omega)$. If*

$$Fx \ne (I-T)(\lambda x) \ \text{for all} \ x \in \partial U, \lambda \ge 1 \ \text{and} \ \lambda x \in \Omega,$$

then $i_*(T + F, U \cap \Omega, \mathscr{P}) = 1$.

Proof 2.2.62 *The mapping $(I-T)^{-1}F : \overline{U} \to \mathscr{P}$ is a strict γk-set contraction and it is readily seen that the following condition of Leray-Schauder type is satisfied*

$$(I-T)^{-1}Fx \ne \lambda x, \ \text{for all} \ x \in \partial U \ \text{and} \ \lambda \ge 1.$$

In fact, if there exist $x_0 \in \partial U$ and $\lambda_0 \geq 1$ such that $(I-T)^{-1}Fx_0 = \lambda_0 x_0$, then $Fx_0 = (I-T)(\lambda_0 x_0)$, contradicting our assumption. The claim then follows from (2.38) and Proposition 2.1.8 with $\theta = 0$.

Proposition 2.2.63 *Let U be a bounded open subset of $\mathscr{P}$ with $0 \in U \cap \Omega$. Assume that the mapping $T : \Omega \subset \mathscr{P} \to E$ is such that $(I-T)$ is Lipschitz invertible with constant $\gamma > 0$, $F : \overline{U} \to E$ is a k-set contraction with $0 \leq k < \gamma^{-1}$, and $F(\overline{U}) \subset (I-T)(\Omega)$. If further*

$$\gamma\|Fx+T0\| \leq \|x\| \quad \text{and} \quad Tx+Fx \neq x \quad \text{for all } x \in \partial U \cap \Omega, \quad (2.40)$$

then $i_*(T+F, U \cap \Omega, \mathscr{P}) = 1$.

Proof 2.2.64 *The mapping $(I-T)^{-1}F : \overline{U} \to \mathscr{P}$ is a strict γk-set contraction. $(I-T)$ is Lipschitz invertible with constant $\gamma > 0$, for each $x \in \overline{U}$*

$$
\begin{aligned}
\|(I-T)^{-1}Fx\| &= \|(I-T)^{-1}Fx - (I-T)^{-1}(I-T)0\| \\
&\leq \gamma\|Fx+T0\|.
\end{aligned}
\quad (2.41)
$$

Therefore, from (2.41) and Assumption (2.40), we conclude that for all $x \in \partial U$,

$$\|(I-T)^{-1}Fx\| \leq \gamma\|Fx+T0\| \leq \|x\|.$$

Our claim then follows from (2.38) and Corollary 2.1.10 for $\theta = 0$.

The following result is an immediate consequence of Proposition 2.1.6.

Proposition 2.2.65 *Assume that the mapping $T : \Omega \subset \mathscr{P} \to E$ is such that $(I-T)$ is Lipschitz invertible with constant $\gamma > 0$, $F : \overline{U} \to E$ is a k-set contraction with $0 \leq k < \gamma^{-1}$, and $F(\overline{U}) \subset (I-T)(\Omega)$. If further*

$$(I-T)^{-1}F(\overline{U}) \subset U,$$

then $i_*(T+F, U \cap \Omega, \mathscr{P}) = 1$.

In particular, we get

Corollary 2.2.66 *Assume that the mapping $T : \Omega \subset \mathscr{P} \to E$ is such that $(I - T)$ is Lipschitz invertible with constant $\gamma > 0$, $F : \overline{\mathscr{P}_r} \to E$ is a k-set contraction with $0 \leq k < \gamma^{-1}$, and $F(\overline{\mathscr{P}_r}) \subset (I - T)(\Omega)$. If $0 \in \Omega$ and*

$$\gamma\|Fx + T0\| < r, \text{ for all } x \in \overline{\mathscr{P}_r}, \tag{2.42}$$

then $i_*(T + F, \mathscr{P}_r \cap \Omega, \mathscr{P}) = 1$.

Proof 2.2.67 *From (2.41) and Assumption (2.42), for any $x \in \overline{\mathscr{P}_r}$, we have*

$$\|(I - T)^{-1}Fx\| \leq \gamma\|Fx + T0\| < r,$$

which implies that $(I - T)^{-1}F(\overline{\mathscr{P}_r}) \subset \mathscr{P}_r$.

Proposition 2.2.68 *Assume that the mapping $T : \Omega \subset \mathscr{P} \to E$ is such that $(I - T)$ is Lipschitz invertible with constant $\gamma > 0$, $F : \overline{U} \to E$ is a k-set contraction with $0 \leq k < \gamma^{-1}$, and $F(\overline{U}) \subset (I - T)(\Omega)$. If there exists $u_0 \in \mathscr{P} \setminus \{0\}$ such that*

$$Fx \neq (I - T)(x - \lambda u_0), \text{ for all } \lambda \geq 0 \text{ and } x \in \partial U \cap (\Omega + \lambda u_0), \tag{2.43}$$

then the fixed point index $i_*(T + F, U \cap \Omega, \mathscr{P}) = 0$.

Proof 2.2.69 *The mapping $(I - T)^{-1}F : \overline{U} \to \mathscr{P}$ is a strict γk-set contraction and for some $u_0 \in \mathscr{P} \setminus \{0\}$ this operator satisfies*

$$x - (I - T)^{-1}Fx \neq \lambda u_0, \ \forall x \in \partial U, \forall \lambda \geq 0.$$

By (2.38) and Proposition 2.1.12, we deduce that

$$i_*(T + F, U \cap \Omega, \mathscr{P}) = i((I - T)^{-1}F, U, \mathscr{P}) = 0.$$

Proposition 2.2.70 *Assume that the mapping $T : \Omega \subset \mathscr{P} \to E$ is such that $(I - T)$ is Lipschitz invertible with constant $\gamma > 0$, $F : \overline{U} \to E$ is a k-set contraction with $0 \leq k < \gamma^{-1}$, and $F(\overline{U}) \subset (I - T)(\Omega)$. Suppose further that there exists $u_0 \in \mathscr{P} \setminus \{0\}$ such that $T(x - \lambda u_0) \in \mathscr{P}$, for all $\lambda \geq 0$ and $x \in \partial U \cap (\Omega + \lambda u_0)$.*

If one of the following conditions holds
(a) $Fx \not\leq x, \ \forall x \in \partial U$.
(b) $Fx \in \mathscr{P}$, $\|Fx\| > N\|x\|$, $\forall x \in \partial U$, and the cone $\mathscr{P}$ is normal with constant N,
then $i_*(T + F, U \cap \Omega, \mathscr{P}) = 0$.

Proof 2.2.71 *We show that conditions (a) or (b) imply that*

$$Fx \neq (I - T)(x - \lambda u_0), \text{ for all } \lambda \geq 0 \text{ and } x \in \partial U \cap (\Omega + \lambda u_0).$$

On the contrary, assume the existence of $\lambda_0 \geq 0$ and $x_0 \in \partial U \cap (\Omega + \lambda_0 u_0)$ such that

$$Fx_0 = (I - T)(x_0 - \lambda_0 u_0).$$

Then $x_0 - Fx_0 = T(x_0 - \lambda_0 u_0) + \lambda_0 u_0 \in \mathscr{P}$. If condition (a) holds, then a contradiction is achieved. Otherwise, we deduce that

$$Fx_0 \leq x_0.$$

Since $\mathscr{P}$ is normal, we deduce that

$$\|Fx_0\| \leq N\|x_0\|,$$

contradicting condition (b) and ending the proof of this Proposition.

2.3 Fixed-Point Index on Translates of Cones

Let E be a Banach space, $\mathscr{P} \subset E$ ($\mathscr{P} \neq \{0\}$) be a cone. Given $\theta \in E$ ($\theta \neq 0$), consider the translate of $\mathscr{P}$, namely

$$\mathscr{K} = \mathscr{P} + \theta = \{x + \theta, \ x \in \mathscr{P}\}.$$

Then $\mathscr{K}$ is a closed convex of E, hence a retract of E.

Let Ω be any subset of $\mathscr{K}$ and U a bounded open of $\mathscr{K}$ such that $U \cap \Omega \neq \emptyset$. We denote by $\overline{U}$ and ∂U the closure and the boundary of U relative to $\mathscr{K}$.

For $0 < r < R$, define

$$
\begin{aligned}
\mathscr{K}_r &= \mathscr{K} \cap \{x \in E : \|x - \theta\| < r\}, \\
\partial \mathscr{K}_r &= \mathscr{K} \cap \{x \in E : \|x - \theta\| = r\},
\end{aligned}
$$

The fixed point index $i_*(T + F, U \cap \Omega, \mathscr{K})$ defined by

$$i_*(T + F, U \cap \Omega, \mathscr{K}) = \begin{cases} i((I - T)^{-1}F, U, \mathscr{K}), & \text{if } U \cap \Omega \neq \emptyset \\ 0 & \text{if } U \cap \Omega = \emptyset, \end{cases} \tag{2.44}$$

is well defined whenever $T : \Omega \to E$ is a mapping such that $(I - T)$ is Lipschitz invertible with constant $\gamma > 0$ and $F : \overline{U} \to E$ is a k-set contraction with $0 \le k < \gamma^{-1}$ and $F(\overline{U}) \subset (I - T)(\Omega)$.

Proposition 2.3.1 *Let U be a bounded open subset of $\mathscr{K}$ with $\theta \in U$. Assume that the mapping $T : \Omega \subset \mathscr{K} \to E$ is such that $(I - T)$ is Lipschitz invertible with constant $\gamma > 0$, $F : \overline{U} \to E$ is a k-set contraction with $0 \le k < \gamma^{-1}$, and $F(\overline{U}) \subset (I - T)(\Omega)$. If*

$$Fx \ne (I - T)(\lambda x + (1 - \lambda)\theta) \ \textit{for all } x \in \partial U,$$

$$\lambda \ge 1 \textit{ and } \lambda x + (1 - \lambda)\theta \in \Omega, \tag{2.45}$$

then $i_*(T + F, U \cap \Omega, \mathscr{P}) = 1$.

Proof 2.3.2 *Define the homotopic deformation $H : [0, 1] \times \overline{U} \to \mathscr{K}$ by*

$$H(t, x) = t(I - T)^{-1}Fx + (1 - t)\theta.$$

Then, the operator H is continuous and uniformly continuous in t for each x, and the mapping $H(t, .)$ is a strict γk-set contraction for each t. Moreover, $H(t, .)$ has no fixed point on ∂U for each t. Otherwise, there would exist some $x_0 \in \partial U$ and $t_0 \in [0, 1]$ such that $\frac{1}{t_0}x_0 + (1 - \frac{1}{t_0})\theta \in \Omega$ for $t_0 \ne 0$, and

$$t_0(I - T)^{-1}Fx_0 + (1 - t_0)\theta = x_0.$$

We may distinguish between two cases:
(i) If $t_0 = 0$, then $x_0 = \theta$, which is a contradiction.
(ii) If $t_0 \in (0, 1]$, then $Fx_0 = (I - T)(\frac{1}{t_0}x_0 + (1 - \frac{1}{t_0})\theta)$, which contradicts our assumption.
The properties of invariance by homotopy and normalization of the fixed point index guarantee that

$$i((I - T)^{-1}F, U, \mathscr{K}) = i(\theta, U, \mathscr{K}).$$

Consequently, by (2.44), we deduce that $i_*(T + F, U \cap \Omega, \mathscr{K}) = 1$.

Proposition 2.3.3 *Let U be a bounded open subset of $\mathscr{K}$ with $\theta \in U$. Assume that the mapping $T : \Omega \subset \mathscr{K} \to E$ is such that $(I - T)$ is Lipschitz invertible with constant $\gamma > 0$, $F : \overline{U} \to E$ is a k-set contraction with $0 \leq k < \gamma^{-1}$ and $F(\overline{U}) \subset (I - T)(\Omega)$. If*

$$\|Fx - T\theta - \theta\| \leq \|x - \theta\| \ \text{ and } \ Tx + Fx \neq x, \ \text{for all } x \in \partial U \cap \Omega,$$
$$(2.46)$$

then $i_*(T + F, U \cap \Omega, \mathscr{P}) = 1$.

Proof 2.3.4 *The mapping $(I - T)^{-1}F : \overline{U} \to \mathscr{P}$ is a strict γk-set contraction. Since $(I - T)$ is Lipschitz invertible with constant $\gamma > 0$, for each $x \in \overline{U}$*

$$\|(I - T)^{-1}Fx - \theta\| = \|(I - T)^{-1}Fx - (I - T)^{-1}(I - T)\theta\|$$

$$\leq \gamma\|Fx + T\theta - \theta\|.$$
$$(2.47)$$

Therefore, for all $x \in \partial U$, we get

$$\|(I - T)^{-1}Fx - \theta\| \leq \gamma\|Fx + T\theta - \theta\|$$

$$\leq \|x - \theta\|.$$

The claim then follows from (2.44) and Corollary 2.1.10, which completes the proof.

Proposition 2.3.5 *Assume that the mapping $T : \Omega \subset \mathscr{K} \to E$ is such that $(I - T)$ is Lipschitz invertible with constant $\gamma > 0$, $F : \overline{U} \to E$ is a k-set contraction with $0 \leq k < \gamma^{-1}$, and $(tF(\overline{U}) + (1 - t)\theta) \subset (I - T)(\Omega)$ for all $t \in [0, 1]$. If $(I - T)^{-1}\theta \in U$, and*

$$(I - T)x \neq \lambda Fx + (1 - \lambda)\theta \ \text{ for all } x \in \partial U \cap \Omega \text{ and } 0 \leq \lambda \leq 1, \quad (2.48)$$

then $i_*(T + F, U \cap \Omega, \mathscr{K}) = 1$.

Proof 2.3.6 *Define the homotopic deformation $H : [0, 1] \times \overline{U} \to E$ by*

$$H(t, x) = tFx + (1 - t)\theta.$$

Then, the operator H is continuous and uniformly continuous in t for each x, and the mapping $H(t,.)$ is a k-set contraction for each t. Moreover, $T + H(t,.)$ has no fixed point on $\partial U \cap \Omega$ for each t. Otherwise, there would exist some $x_0 \in \partial U \cap \Omega$ and $t_0 \in [0,1]$ such that

$$Tx_0 + t_0 Fx_0 + (1-t_0)\theta = x_0,$$

then $x_0 - Tx_0 = t_0 Fx_0 + (1-t_0)\theta$, leading to a contradiction with the hypothesis. By (2.44), the invariance by homotopy property of the generalized fixed point index i_* and the normalization property of the fixed point index i, we conclude that

$$i_*(T+F, U \cap \Omega, \mathscr{K}) = i_*(T+\theta, U \cap \Omega, \mathscr{K})$$

$$= i((I-T)^{-1}\theta, U \cap \Omega, \mathscr{K}) = 1.$$

Corollary 2.3.7 *Assume that the mapping $T : \Omega \subset \mathscr{K} \to E$ is such that $(I-T)$ is Lipschitz invertible with constant $\gamma > 0$, $F : \overline{U} \to E$ is a k-set contraction with $0 \le k < \gamma^{-1}$, and $\left(tF(\overline{U}) + (1-t)\theta\right) \subset (I-T)(\Omega)$ for all $t \in [0,1]$. If $(I-T)^{-1}\theta \in U$,*

$$Fx \in \mathscr{K} \text{ for all } x \in \Omega \cap \partial U,$$

and

$$Fx \not\ge (I-T)x \text{ for all } x \in \partial U \cap \Omega, \tag{2.49}$$

then $i_*(T+F, U \cap \Omega, \mathscr{K}) = 1$.

Proof 2.3.8 *It is easy to see that Assumption (2.48) is satisfied. Otherwise, there exist some $x_0 \in \partial U \cap \Omega$ and $0 \le \lambda_0 \le 1$ such that $x_0 - Tx_0 = \lambda_0 Fx_0 + (1-\lambda_0)\theta$. Then*

$$Fx_0 - x_0 + Tx_0 = (1-\lambda_0)(Fx_0 - \theta) \in \mathscr{P},$$

which leads us to a contradiction with (2.49).

Proposition 2.3.9 *Assume that the mapping $T : \Omega \subset \mathscr{K} \to E$ is such that $(I-T)$ is Lipschitz invertible with constant $\gamma > 0$, $F : \overline{U} \to E$ is a k-set contraction with $0 \le k < \gamma^{-1}$, and $F(\overline{U}) \subset (I-T)(\Omega)$. If there exists $u_0 \in \mathscr{P} \setminus \{0\}$ such that*

$$Fx \ne (I-T)(x - \lambda u_0), \text{ for all } \lambda \ge 0 \text{ and } x \in \partial U \cap (\Omega + \lambda u_0), \tag{2.50}$$

then $i_*(T+F, U \cap \Omega, \mathscr{K}) = 0$.

Proof 2.3.10 *The mapping* $(I-T)^{-1}F:\overline{U}\to\mathscr{K}$ *is a strict* γk-set con-traction. *Suppose that* $i_*(T+F,U\cap\Omega,\mathscr{K})\neq 0$. *Then,*

$$i((I-T)^{-1}F,U,\mathscr{P})\neq 0.$$

For each $r>0$, *define the homotopy:*

$$H(t,x)=(I-T)^{-1}Fx+tru_0,\ \text{for}\ x\in\overline{U}\ \text{and}\ t\in[0,1].$$

The operator H *is continuous and uniformly continuous in* t *for each* x. *Moreover,* $H(t,.)$ *is a strict* γk-set contraction for each t and

$$H([0,1]\times\overline{U})=(I-T)^{-1}F(U)+tru_0\subset\mathscr{K}.$$

We check that $H(t,x)\neq x$, *for all* $(t,x)\in[0,1]\times\partial U$. *If* $H(t_0,x_0)=x_0$ *for some* $(t_0,x_0)\in[0,1]\times\partial U$, *then*

$$x_0-t_0ru_0=(I-T)^{-1}Fx_0,$$

and so $x_0-t_0ru_0\in\Omega$. *Hence*

$$(I-T)(x_0-t_0ru_0)=Fx_0,$$

for $x_0\in\partial U\cap(\Omega+t_0ru_0)$, *contradicting the assumption* (2.50).
By the homotopy invariance property of the fixed point index, we de-duce that

$$i((I-T)^{-1}F+ru_0,U\cap\Omega,\mathscr{P})=i((I-T)^{-1}F,U,\mathscr{P})\neq 0.$$

Thus the solvability property of the fixed point index, for each $r>0$, *there exists* $x_r\in U$ *such that*

$$x_r-(I-T)^{-1}Fx_r=ru_0. \tag{2.51}$$

Letting $r\to+\infty$ *in* (2.51), *the left-hand side of* (2.51) *is bounded while the right-hand side is not, which is a contradiction. Therefore,*

$$i_*(T+F,U\cap\Omega,\mathscr{P})=0.$$

As a consequence, we get

Proposition 2.3.11 *Assume that the mapping $T : \Omega \subset \mathcal{K} \to E$ is such that $(I - T)$ is Lipschitz invertible with constant $\gamma > 0$, $F : \overline{U} \to E$ is a k-set contraction with $0 \le k < \gamma^{-1}$, and $F(\overline{U}) \subset (I - T)(\Omega)$. Suppose that there exist $u_0 \in \mathscr{P} \backslash \{0\}$ such that $T(x - \lambda u_0) \in \mathscr{P}$, for all $x \in \partial U \cap (\Omega + \lambda u_0)$, and $\lambda \ge 0$.*

If one of the following conditions holds:
(a) $Fx \nleq x$, $\forall x \in \partial U$,
(b) $Fx \in \mathcal{K}$, $\|Fx - \theta\| > N\|x - \theta\|$, $\forall x \in \partial U$, and the cone $\mathscr{P}$ is normal with constant N,
then $i_*(T + F, U \cap \Omega, \mathscr{P}) = 0$.

The proof of this proposition similar to the proof of Proposition 2.2.70 and is omitted.

In case where T is an h-expansive mapping, i.e., $\gamma = (h - 1)^{-1}$, Propositions 2.3.1, 2.3.3, and 2.3.9 take the form of Propositions 2.3.12, 2.3.13, and 2.3.14, respectively.

Proposition 2.3.12 *Assume that $T : \Omega \subset \mathcal{K} \to E$ is an h-expansive mapping with a constant $h > 1$, $F : \overline{U} \to E$ is a k-set contraction with $0 \le k < h - 1$, and $F(\overline{U}) \subset (I - T)(\Omega)$. If $\theta \in U$ and*

$$Fx \ne (I - T)(\lambda x + (1 - \lambda)\theta), \forall x \in \partial U, \lambda \ge 1 \text{ and } \lambda x + (1 - \lambda)\theta \in \Omega,$$

then $i_*(T + F, U \cap \Omega, \mathcal{K}) = 1$.

Proposition 2.3.13 *Assume that $T : \Omega \subset \mathcal{K} \to E$ is an h-expansive mapping, $F : \overline{U} \to E$ is a k-set contraction with $0 \le k < h - 1$, and $F(\overline{U}) \subset (I - T)(\Omega)$. If $\theta \in U \cap \Omega$ and*

$$\|Fx - T\theta - \theta\| \le (h - 1)\|x - \theta\| \text{ and } Tx + Fx \ne x, \forall x \in \partial U \cap \Omega,$$

then $i_*(T + F, U \cap \Omega, \mathcal{K}) = 1$.

Proposition 2.3.14 *Assume that $T : \Omega \subset \mathcal{K} \to E$ is an h-expansive mapping, $F : \overline{U} \to E$ is a k-set contraction with $0 \le k < h - 1$, and $F(\overline{U}) \subset (I - T)(\Omega)$. If there exists $u_0 \in \mathscr{P} \backslash \{0\}$ such that*

$$Fx \ne (I - T)(x - \lambda u_0), \forall x \in \partial U \cap (\Omega + \lambda u_0), \lambda \ge 0, \qquad (2.52)$$

then $i_*(T + F, U \cap \Omega, \mathcal{K}) = 0$.

3

Positive Fixed Points for Sums of Two Operators

This chapter is devoted to the study of the existence, positivity, and localization of solutions for abstract equations of the form

$$Tx + Fx = x, \quad x \in D, \qquad (3.1)$$

where D is a closed convex subset of a Banach space.

After computing the generalized fixed point index i_*, several fixed point theorems are obtained. Some extensions of Krasnosel'skii type expansion-compression fixed point theorems as well as of Leggett-Williams type fixed point theorems on cones are obtained for sums of two operators. Some of the results of this chapter can be found in [13, 14, 18, 19].

3.1 Krasnosel'skii's Compression–Expansion Fixed Point Theorems Type

3.1.1 Fixed point in conical annulus

In this section, four fixed point theorems of cone compression and expansion for an expansive operator perturbed by a k-set contraction are established.

In what follows, $\mathscr{P}$ will refer to a cone in a Banach space $(E, \|.\|)$ and Ω be a subset of $\mathscr{P}$. For $0 < r < R$, we define the conical annulus:

$$\begin{aligned} \mathscr{P}_{r,R} &= \mathscr{P} \cap \{x \in E : r < \|x\| < R\}, \\ \overline{\mathscr{P}}_{r,R} &= \mathscr{P} \cap \{x \in E : r \le \|x\| \le R\}. \end{aligned}$$

In the sequel, if $\alpha < \beta$, then the conditions represent a compressive form while if $\beta < \alpha$, these conditions express an expansive form.

DOI: 10.1201/9781003381969-3

Theorem 3.1.1 *(Order version).* *Let $0 \in \Omega$; $\alpha, \beta > 0$, $\alpha \neq \beta$, $r = \min(\alpha, \beta)$ and $R = \max(\alpha, \beta)$ such that $\mathscr{P}_{r,R} \cap \Omega \neq \emptyset$. Assume that $T : \Omega \subset \mathscr{P} \to E$ is an h-expansive mapping such that $\|T0\| < (h-1)\beta$, and $F : \overline{\mathscr{P}_R} \to E$ a k-set contraction with $0 \leq k < h - 1$. Suppose that the following conditions hold:*

$$F(\partial \mathscr{P}_\beta \cap \Omega) \subset \mathscr{P};$$

$$tF(\overline{\mathscr{P}_R}) \subset (I - T)(\Omega) \text{ for every } t \in [0,1];$$

there is $u_0 \in \mathscr{P} \backslash \{0\}$ with $T(x - \lambda u_0) \in \mathscr{P}$ for all $\lambda \geq 0$, $x \in \partial \mathscr{P}_\alpha \cap (\Omega + \lambda u_0)$. If

$$Fx \nleq x - \lambda u_0 \text{ for } x \in \partial \mathscr{P}_\alpha, \ \lambda \geq 0,$$
$$Fx \ngeq x - Tx \text{ for } x \in \partial \mathscr{P}_\beta \cap \Omega,$$

then $T + F$ has at least one fixed point $x \in \mathscr{P}_{r,R} \cap \Omega$.

Proof 3.1.2 *We only present the proof of the compressive form. It is analogous for the expansive form. By Propositions 2.2.8 and 2.2.28, we have*

$$i_*(T + F, \mathscr{P}_R \cap \Omega, \mathscr{P}) = 1 \text{ and } i_*(T + F, \mathscr{P}_r \cap \Omega, \mathscr{P}) = 0.$$

The additivity property of the index i_* in Theorem 2.2.1 yields

$$i_*(T + F, \mathscr{P}_{r,R} \cap \Omega, \mathscr{P}) = 1.$$

By the solvability property of the index i_*, $T + F$ has at least one fixed point in the conical annulus $\mathscr{P}_{r,R} \cap \Omega$, proving our claim.

Theorem 3.1.3 *(Norm version).* *Let $\mathscr{P} \subset E$ a normal cone with constant N, $0 \in \Omega$, $\alpha, \beta > 0$, $\alpha \neq \beta$, $r = \min(\alpha, \beta)$ and $R = \max\{\alpha, \beta\}$ such that $\mathscr{P}_{r,R} \cap \Omega \neq \emptyset$. Assume that $T : \Omega \subset \mathscr{P} \to E$ is an h-expansive mapping and $F : \overline{\mathscr{P}_R} \to E$ a k-set contraction with $0 \leq k < h - 1$, and $F(\overline{\mathscr{P}_R}) \subset (I - T)(\Omega)$. Suppose that there exists $u_0 \in \mathscr{P} \backslash \{0\}$ such that $T(x - \lambda u_0) \in \mathscr{P}$, for all $\lambda \geq 0$ and $x \in \partial \mathscr{P}_\beta \cap (\Omega + \lambda u_0)$ and the following conditions are satisfied:*

$$\|Fx + T0\| \leq (h - 1)\|x\|, \forall x \in \partial \mathscr{P}_\alpha,$$

$$Fx \in \mathscr{P}, \ \|Fx\| > N\|x - \lambda u_0\|, \forall x \in \partial \mathscr{P}_\beta.$$

Then $T + F$ has at least one fixed point $x \in \overline{\mathscr{P}}_{r,R} \cap \Omega$.

Proof 3.1.4 *We only give the proof in the expansion case. Without loss of generality, assume that $Tx + Fx \neq x$ on $\partial \mathscr{P}_r \cap \Omega$ and $Tx + Fx \neq x$ on $\partial \mathscr{P}_R \cap \Omega$, otherwise we are finished. By Propositions 2.2.21 and 2.2.28, we have*

$$i_* (T + F, \mathscr{P}_r \cap \Omega, \mathscr{P}) = 1 \ \text{ and } \ i_* (T + F, \mathscr{P}_R \cap \Omega, \mathscr{P}) = 0.$$

The additivity property of the index i_* yields

$$i_* (T + F, \mathscr{P}_{r,R} \cap \Omega, \mathscr{P}) = -1.$$

By the solvability property of the index i_*, the sum $T + F$ has at least one fixed point in the conical annulus $\overline{\mathscr{P}}_{r,R} \cap \Omega$.

Theorem 3.1.5 *(Homotopy version 1). Let $0 \in \Omega$; $\alpha, \beta > 0$, $\alpha \neq \beta$, $r = \min (\alpha, \beta)$ and $R = \max (\alpha, \beta)$ such that $\mathscr{P}_{r,R} \cap \Omega \neq \emptyset$. Assume that $T : \Omega \subset \mathscr{P} \to E$ is an h-expansive mapping and $F : \overline{\mathscr{P}}_R \to E$ a k-set contraction with $0 \leq k < h - 1$. Suppose that $\|T0\| < (h-1)\beta$,*

$$tF(\overline{\mathscr{P}}_R) \subset (I - T)(\Omega) \ \text{ for all } t \in [0,1], \tag{3.2}$$

and that there exists $u_0 \in \mathscr{P} \backslash \{0\}$ such that the following conditions are satisfied:

$$\begin{aligned} Fx \ &\neq \ (I-T)(x - \lambda u_0) \text{ for all } \lambda \geq 0, \ x \in \partial \mathscr{P}_\alpha \cap (\Omega + \lambda u_0), \\ Fx \ &\neq \ \lambda(x - Tx) \text{ for all } \lambda \geq 1, x \in \partial \mathscr{P}_\beta. \end{aligned}$$

Then $T + F$ has a fixed point $x \in \mathscr{P}_{r,R} \cap \Omega$.

Proof 3.1.6 *Suppose that $\beta < \alpha$. By Propositions 2.2.13 and 2.2.26, we have*

$$i_* (T + F, \mathscr{P}_r \cap \Omega, \mathscr{P}) = 1 \text{ and } i_* (T + F, \mathscr{P}_R \cap \Omega, \mathscr{P}) = 0.$$

The additivity property of the index i_* yields

$$i_* (T + F, \mathscr{P}_{r,R} \cap \Omega, \mathscr{P}) = -1.$$

By the solvability property of the index i_*, the sum $T + F$ has at least one fixed point in the conical annulus $\mathscr{P}_{r,R} \cap \Omega$.

Theorem 3.1.7 (*Homotopy version 2*). *Let* $\alpha, \beta > 0$, $\alpha \neq \beta$, $r = \min(\alpha, \beta)$ *and* $R = \max(\alpha, \beta)$ *such that* $\mathscr{P}_{r,R} \cap \Omega \neq \emptyset$. *Assume that* $T : \Omega \subset \mathscr{P} \to E$ *is an h-expansive mapping and* $F : \overline{\mathscr{P}_R} \to E$ *a k-set contraction with* $0 \leq k < h - 1$. *Suppose that* $F(\overline{\mathscr{P}_R}) \subset (I - T)(\Omega)$, *and there exists* $u_0 \in \mathscr{P} \backslash \{0\}$ *such that the following conditions hold:*

$$Fx \;\neq\; (I - T)(x - \lambda u_0) \;\text{ for all } \lambda > 0, \; x \in \partial \mathscr{P}_\alpha \cap (\Omega + \lambda u_0),$$
$$Fx \;\neq\; (I - T)(\lambda x) \;\text{ for all } \lambda > 1, \, x \in \partial \mathscr{P}_\beta, \text{ and } \lambda x \in Omega.$$

Then $T + F$ *has a fixed point* $x \in \overline{\mathscr{P}_{r,R}} \cap \Omega$.

Proof 3.1.8 *Suppose that* $\alpha < \beta$. *Without loss of generality, assume that* $Tx + Fx \neq x$ *on* $\partial \mathscr{P}_r \cap \Omega$ *and* $Tx + Fx \neq x$ *on* $\partial \mathscr{P}_R \cap \Omega$, *otherwise the result is obvious. By Propositions 2.2.19 and 2.2.26, we have*

$$i_* (T + F, \mathscr{P}_r \cap \Omega, \mathscr{P}) = 0 \text{ and } i_* (T + F, \mathscr{P}_R \cap \Omega, \mathscr{P}) = 1.$$

The additivity property of the index i_* *yields*

$$i_* (T + F, \mathscr{P}_{r,R} \cap \Omega, \mathscr{P}) = 1.$$

By the solvability property of the index i_*, *the sum* $T + F$ *has at least one fixed point in the conical annulus* $\overline{\mathscr{P}_{r,R}} \cap \Omega$.

3.1.2   Vector version

Theorems 3.1.1 and 3.1.5 are adapted for the treatment of systems of nonlinear equations. The compression–expansion conditions are given componentwise which allows the nonlinear term of a system to have different behaviors both in components and in variables.

In all what follows, we shall consider two Banach spaces $(E_1, \|.\|_1), (E_2, \|.\|_2)$; two cones $\mathscr{P}_1 \subset E_1, \mathscr{P}_2 \subset E_2$, the product space $E := E_1 \times E_2$, the corresponding cone $\mathscr{P} := \mathscr{P}_1 \times \mathscr{P}_2$ of E, and $\mathscr{P}^* := \mathscr{P}_1^* \times \mathscr{P}_2^*$, where $\mathscr{P}_i^* = \mathscr{P}_i \backslash \{0\}$, $i = 1, 2$. We shall use the same symbol $\leq$ to denote the partial order relations induced by $\mathscr{P}$ in E, and by $\mathscr{P}_i$ in $E_i (i = 1, 2)$. For $\alpha_i, \beta_i > 0$ with $\alpha_i \neq \beta_i$, we let $\alpha = (\alpha_1, \alpha_2)$, $\beta = (\beta_1, \beta_2)$, $r_i = \min(\alpha_i, \beta_i)$, $R_i = \max(\alpha_i, \beta_i)$ for $i = 1, 2$, and $r = (r_1, r_2)$, $R = (R_1, R_2)$. We also consider two subsets

Ω_i of $\mathscr{P}_i$ with $0 \in \Omega_i$ $(i = 1, 2)$, we denote $\Omega = \Omega_1 \times \Omega_2$ and we use the notations:

$$\mathscr{P}_r = (\mathscr{P}_1)_{r_1} \times (\mathscr{P}_2)_{r_2},$$

$$\partial \mathscr{P}_r = [\partial(\mathscr{P}_1)_{r_1} \times (\mathscr{P}_2)_{r_2}] \cup [(\mathscr{P}_1)_{r_1} \times \partial(\mathscr{P}_2)_{r_2}]$$

$$\mathscr{P}_{r,R} = \{x = (u, v) \in \mathscr{P} : r_1 < \|u\|_1 < R_1, \ r_2 < \|v\|_2 < R_2\};$$

$$U_1 = \{u \in \Omega_1 : \|u\|_1 < \alpha_1\};$$

$$U_2 = \{u \in \Omega_1 : \|u\|_1 < \beta_1\};$$

$$V_1 = \{v \in \Omega_2 : \|v\|_2 < \alpha_2\};$$

$$V_2 = \{v \in \Omega_2 : \|v\|_2 < \beta_2\}.$$

Now, we combine the result of Propositions 2.2.8 and 2.2.28 to establish the vector version of Theorem 3.1.1.

Theorem 3.1.9 *For each $i \in \{1, 2\}$, let $T_i : \Omega_i \subset \mathscr{P}_i \to E_i$ be an h_i-expansive mapping such that $\|T_i 0\|_i < (h_i - 1)\beta_i$, and $F_i : (\overline{\mathscr{P}_1})_{R_1} \times (\overline{\mathscr{P}_2})_{R_2} \to E_i$ be a k_i-set contraction with $0 \leq k_i < h_i - 1$. Assume that $\mathscr{P}_{r,R} \cap \Omega \neq \emptyset$ and the following conditions hold:*

$$F_i(\partial \mathscr{P}_R \cap \Omega) \subset \mathscr{P}_i;$$

$$t F_i(\overline{\mathscr{P}_R}) \subset (I_i - T_i)(\Omega_i) \quad \text{for all } t \in [0, 1];$$

there is $w^0 \in \mathscr{P}^*$, $w^0 = (w_1^0, w_2^0)$ with $T_i(y - \lambda w_i^0) \in \mathscr{P}_i$ for all $\lambda \geq 0$ and $y \in \partial(\mathscr{P}_i)_{\alpha_i} \cap (\Omega_i + \lambda w_i^0)$;

If

$$F_i x \not\leq x_i - \lambda w_i^0 \quad \text{for } x \in \partial \mathscr{P}_\alpha, \ \lambda \geq 0,$$

$$\tag{3.3}$$

$$F_i x \not\geq x_i - T_i x_i \quad \text{for } x \in \partial \mathscr{P}_\beta \cap \Omega,$$

then the operator $T + F = (T_1 + F_1, T_2 + F_2)$ has at least one fixed point $x = (u, v) \in \mathscr{P}_{r,R} \cap \Omega$.

Proof 3.1.10 *First, we prove the following results:*

(i) $i_*(T+F,U_1 \times V_1, \mathscr{P}) = 0$,

(ii) $i_*(T+F,U_1 \times V_2, \mathscr{P}) = 0$,

(iii) $i_*(T+F,U_2 \times V_1, \mathscr{P}) = 0$,

(iv) $i_*(T+F,U_2 \times V_2, \mathscr{P}) = 1$.

Claim (i). Define the homotopic deformation $H : [0,1] \times \overline{U}_1 \times \overline{V}_1 \to \mathscr{P}$ *by*

$$H(t,u,v) = (I-T)^{-1}F(u,v) + (1-t)\lambda w^0,$$

more exactly, $H(t,u,v) = (H_1(t,u,v), H_2(t,u,v))$, *where for* $i = 1,2$,

$$H_i(t,u,v) = (I_i - T_i)^{-1}F_i(u,v) + (1-t)\lambda w_i^0.$$

Here the number $\lambda \geq 0$ *will be chosen later. For* $i = 1,2$, *the operator* H_i *is continuous and uniformly continuous in t for each* $x \in \overline{U}_1 \times \overline{V}_1$. *Moreover,* $H_i(t,.)$ *is a strict set contraction for each t, and the operator* $H(t,.)$ *has no fixed points on* $\partial(U_1 \times V_1)$ *for each t. Otherwise, there exist* $(u_0, v_0) \in \partial(U_1 \times V_1)$ *and* $t_0 \in [0,1]$ *such that*

$$H(t_0, u_0, v_0) = (u_0, v_0).$$

We have $\partial(U_1 \times V_1) = (\partial U_1 \times V_1) \cup (U_1 \times \partial V_1)$. *Assume that* $(u_0, v_0) \in \partial U_1 \times V_1$. *Then from*

$$(I_1 - T_1)^{-1}F_1(u_0, v_0) + (1-t_0)\lambda w_1^0 = u_0,$$

we have $u_0 - (1-t_0)\lambda w_1^0 \in \Omega_1$ *and using (3.3)*

$$u_0 - (1-t_0)\lambda w_1^0 - F_1(u_0, v_0) = T_1(u_0 - (1-t_0)\lambda w_1^0) \in \mathscr{P}_1,$$

whence $F_1(u_0, v_0) \leq u_0 - (1-t_0)\lambda w_1^0$, *which is in contradiction with (3.3).*

Similarly, we obtain a contradiction in the case $(u_0, v_0) \in U_1 \times \partial V_1$. Hence $H(t,.)$ has no fixed points on $\partial(U_1 \times V_1)$ for each t. From the homotopy invariance property of the index i_*, we have

$$i(H(1,.), U_1 \times V_1, \mathscr{P}) = i(H(0,.), U_1 \times V_1, \mathscr{P}).$$

One has $H(1,.) = (I - T)^{-1} F$, and

$$H(0,.) = (I - T)^{-1} F(u, v) + \lambda w^0.$$

Since the operator $(I - T)^{-1}F$ is bounded, the set $(I - T)^{-1}F(\overline{U}_1 \times \overline{V}_1)$ is bounded. Hence, if $i(H(0,.), U_1 \times V_1, \mathscr{P}) \neq 0$, there exists $x = (x_1, x_2) \in U_1 \times V_1$ such that

$$(I - T)^{-1}F(x) = x + \lambda w^0, \tag{3.4}$$

which is not possible if $\lambda > 0$ is large enough. Consequently, $i(H(0,.), U_1 \times V_1, \mathscr{P}) = 0$ and thus

$$i_*(T + F, U_1 \times V_1, \mathscr{P}) = 0.$$

Claim (ii). *Define the homotopic deformation $H : [0,1] \times \overline{U}_1 \times \overline{V}_2 \to \mathscr{P}$ by*

$$H(t, u, v) = ((I_1 - T_1)^{-1}F_1(u, v) + (1 - t)\lambda w_1^0, \ (I_2 - T_2)^{-1}tF_2(u, v)),$$

where $\lambda \geq 0$. For $i = 1, 2$, the operator H_i is continuous and uniformly continuous in t for each x. Moreover, $H_i(t,.)$ is a strict set contraction for each t, and $H(t,.)$ has no fixed point on $\partial(U_1 \times V_2)$ for each t. Otherwise, there exist $(u_0, v_0) \in (U_1 \times \partial V_2) \cup (\partial U_1 \times V_2)$ and $t_0 \in [0,1]$ such that

$$H(t_0, u_0, v_0) = (u_0, v_0).$$

Assume that $(u_0, v_0) \in U_1 \times \partial V_2$. Then

$$(I_2 - T_2)^{-1}tF_2(u_0, v_0) = v_0.$$

It means that

$$F_2(u_0, v_0) \geq tF_2(u_0, v_0) = v_0 - T_2 v_0,$$

that is $F_2(u_0,v_0) \geq v_0 - T_2 v_0$, which contradicts the assumption (3.3). The case $(u_0,v_0) \in \partial U_1 \times V_2$ is similar to the proof of claim (i). From the homotopy invariance property of the index i_*, we deduce that

$$i(H(1,.),U_1 \times V_2, \mathscr{P}) = i(H(0,.),U_1 \times V_2, \mathscr{P}),$$

where $H(1,.) = (I-T)^{-1}F$ and

$$H(0,.) = \left((I_1-T_1)^{-1}F_1(u,v) + \lambda w_1^0,\ 0\right).$$

Since the operator $(I_1-T_1)^{-1}F_1$ is bounded, the set $(I_1-T_1)^{-1}F_1(\overline{U}_1 \times \overline{V}_2)$ is bounded. Hence, if $i(H(0,.),U_1 \times V_2, \mathscr{P}) \neq 0$, there exists $x = (x_1,x_2) \in U_1 \times V_2$ such that

$$(I_1-T_1)^{-1}F_1(x) = x_1 - \lambda w_1^0, \quad x_2 = 0.$$

For $\lambda > 0$ sufficiently large, the first equality does not hold. Thus $i(H(0,.),U_1 \times V_2, \mathscr{P}) = 0$, that is

$$i_*(T+F,U_1 \times V_2, \mathscr{P}) = 0.$$

Claim (iii). *This case is similar to the previous one. Now the homotopy is*

$$H(t,u,v) = \left((I_1-T_1)^{-1}tF_1(u,v),\ (I_2-T_2)^{-1}F_2(u,v) + (1-t)\lambda w_2^0\right).$$

For $\lambda \geq 0$ sufficiently large, we arrive at the conclusion

$$i_*(T+F,U_2 \times V_1, \mathscr{P}) = 0.$$

Claim (iv). *Define the homotopic deformation $H : [0,1] \times \overline{U}_2 \times \overline{V}_2 \to \mathscr{P}$,*

$$H(t,u,v) = (H_1(t,u,v),H_2(t,u,v)),$$

by

$$H_i(t,u,v) = (I_i-T_i)^{-1}tF_i(u,v) \quad for \ i=1,2.$$

Both operators H_1 and H_2 are continuous, uniformly continuous in t with respect to x, and strict set contractions for each t. Also $H(t,.)$ has no fixed points on $\partial(U_2 \times V_2)$. Otherwise, there exist

$(u_0, v_0) \in (\partial U_2 \times V_2) \cup (U_2 \times \partial V_2)$ *and* $t_0 \in [0,1]$ *such that* $H(t_0, u_0, v_0) = (u_0, v_0)$. *If* $(u_0, v_0) \in \partial U_2 \times V_2$, *then*

$$(I_1 - T_1)^{-1} t F_1(u_0, v_0) = u_0,$$

which implies

$$F_1(u_0, v_0) \geq t F_1(u_0, v_0) = u_0 - T_1 u_0.$$

This gives $F_1(u_0, v_0) \geq u_0 - T_1 u_0$, *which contradicts (3.3).*
Similarly, we arrive at a contradiction if $(u_0, v_0) \in U_2 \times \partial V_2$. *Therefore,*

$$i(H(1,.), U_2 \times V_2, \mathscr{P}) = i(H(0,.), U_2 \times V_2, \mathscr{P}).$$

One has $H(1,.) = (I - T)^{-1} F$ *and*

$$H(0,.) = ((I_1 - T_1)^{-1} 0, \ (I_2 - T_2)^{-1} 0).$$

For $i = 1,2$, *from the expansivity of* T_i *and the condition* $\|T_i 0\|_i < (h_i - 1)\beta_i$, *we have* $(I_i - T_i)^{-1} 0 \in (\mathscr{P}_i)_{\beta_i}$. *Then the definition of the index* i_* *and the normalization property of the fixed point index yield*

$$i_*(T + F, U_2 \times V_2, \mathscr{P}) = i((I - T)^{-1} 0, U_2 \times V_2, \mathscr{P}) = 1.$$

Thus all the claims (i)–(iv) are proved.
    Next, four possible cases for $\mathscr{P}_{r,R}$ *are discussed.*
    Case 1: $\alpha_1 < \beta_1$ *and* $\alpha_2 < \beta_2$. *In this case*

$$\begin{aligned}
\mathscr{P}_{r,R} \cap \Omega &= \{(u,v) \in \mathscr{P} : \alpha_1 < \|u\|_1 < \beta_1, \ \alpha_2 \leq \|v\|_2 < \beta_2\} \cap \Omega \\[2mm]
&= U_2 \times V_2 \setminus \{(U_1 \times V_1) \cup ((U_1 \times V_2) \setminus (U_1 \times V_1)) \\[2mm]
&\quad \cup ((U_2 \times V_1) \setminus (U_1 \times V_1))\}.
\end{aligned}$$

Using the claims (i)–(iv) and the additivity property of the index i_* *yields*

$$i_*(T + F, \mathscr{P}_{r,R} \cap \Omega, \mathscr{P}) = 1.$$

Consequently, by the solvability property of the index, $T + F$ *has at least one fixed point in the vector conical shell* $\mathscr{P}_{r,R} \cap \Omega$.

Case 2: $\alpha_1 < \beta_1$ *and* $\alpha_2 > \beta_2$. *In this case*

$$\mathscr{P}_{r,R} \cap \Omega = \{(u,v) \in \mathscr{P} : \alpha_1 < \|u\|_1 < \beta_1, \beta_2 \leq \|v\|_2 < \alpha_2\} \cap \Omega$$

$$= U_2 \times V_1 \setminus \{(U_1 \times V_2) \cup ((U_1 \times V_1) \setminus (U_1 \setminus V_2))$$

$$\cup ((U_2 \times V_2) \setminus (U_1 \setminus V_2))\}.$$

It follows that
$$i_* (T + F, \mathscr{P}_{r,R} \cap \Omega, \mathscr{P}) = -1,$$

which again guarantees the existence of at least one fixed point of $T +$
F *in* $\mathscr{P}_{r,R} \cap \Omega$.

   Case 3: $\alpha_1 > \beta_1$ *and* $\alpha_2 < \beta_2$. *One has*

$$\mathscr{P}_{r,R} \cap \Omega = \{(u,v) \in \mathscr{P} : \beta_1 < \|u\|_1 < \alpha_1, \alpha_2 \leq \|v\|_2 < \beta_2\} \cap \Omega$$

$$= U_1 \times V_2 \setminus \{(U_2 \times V_1) \cup ((U_2 \times V_2) \setminus (U_2 \times V_1))$$

$$\cup ((U_1 \times V_1) \setminus (U_2 \times V_1))\}.$$

Then
$$i_* (T + F, \mathscr{P}_{r,R} \cap \Omega, \mathscr{P}) = -1,$$

whence the conclusion.

   Case 4: $\alpha_1 > \beta_1$ *and* $\alpha_2 > \beta_2$. *Then*

$$\mathscr{P}_{r,R} \cap \Omega = \{(u,v) \in \mathscr{P} : \beta_1 < \|u\|_1 < \alpha_1, \beta_2 \leq \|v\|_2 < \alpha_2\} \cap \Omega$$

$$= U_1 \times V_1 \setminus \{(U_2 \times V_1) \cup (U_1 \times V_2)\}$$

$$= U_1 \times V_1 \setminus \{(U_2 \times V_2) \cup ((U_2 \times V_1) \setminus (U_2 \times V_2))$$

$$\cup ((U_1 \times V_2) \setminus (U_2 \times V_2))\}$$

In this case
$$i_* (T + F, \mathscr{P}_{r,R} \cap \Omega, \mathscr{P}) = 1,$$

which again implies the existence of at least one fixed point of $T + F$ in $\mathscr{P}_{r,R} \cap \Omega$.

Theorem 3.1.11 leads to a vector version of Theorem 3.1.5.

Theorem 3.1.11 *For each $i \in \{1,2\}$, let $T_i : \Omega_i \subset \mathscr{P}_i \to E_i$ be an h_i-expansive mapping such that $\|T_i 0\|_i < (h_i - 1)\beta_i$, and $F_i : (\overline{\mathscr{P}_1})_{R_1} \times (\overline{\mathscr{P}_2})_{R_2} \to E_i$ be a k_i-set contraction with $0 \leq k_i < h_i - 1$. Assume that $\mathscr{P}_{r,R} \cap \Omega \neq \emptyset$, $tF_i(\overline{\mathscr{P}_R}) \subset (I_i - T_i)(\Omega_i)$ for every $t \in [0,1]$, and that there exists $w^0 = (w_1^0, w_2^0) \in \mathscr{P}^*$ such that the following conditions are satisfied:*

$$F_i(x) \neq (I_i - T_i)(x_i - \lambda w_i^0), \ \forall x \in \partial \mathscr{P}_\alpha, \lambda \geq 0 \text{ and } x_i \in (\Omega_i + \lambda w_i^0),$$
$$(3.5)$$
$$F_i(x) \neq \lambda (x_i - T_i x_i), \ \forall x \in \partial \mathscr{P}_\beta \cap \Omega \text{ and } \lambda \geq 1. \qquad (3.6)$$

Then the operator $T + F = (T_1 + F_1, T_2 + F_2)$ has at least one fixed point $x = (u,v) \in \mathscr{P}_{r,R} \cap \Omega$.

Proof 3.1.12 *According to Propositions 2.2.13, 2.2.26 and in a way similar to the one used to show Theorem 3.1.9, we can show Theorem 3.1.11.*

The following example illustrates the last theoretical result.

Example 3.1.13 *Consider the integral system for $t \in [a,b]$,*

$$x_1(t) = (x_1(t))^3 + p_1(t)x_1(t) - \int_a^b K_1(t,s,x_1(s),x_2(s)) \, ds$$
$$(3.7)$$
$$x_2(t) = (x_2(t))^3 + p_2(t)x_2(t) - \int_a^b K_2(t,s,x_1(s),x_2(s)) \, ds,$$

where the functions $p_i : [a,b] \to \mathbb{R}_+$ are continuous and $K_i : [a,b] \times [a,b] \times \mathbb{R}_+^2 \to \mathbb{R}_+$ are continuous together with their first partial derivatives.

We make the following assumptions:

$(\mathscr{H}_0)$ $\ 1 < p_i^{(1)} := \min_{a \leq t \leq b} p_i(t) \leq p_i^{(2)} := \max_{a \leq t \leq b} p_i(t) \text{ for } i = 1,2.$

($\mathscr{H}_1$) *There exist* $R_1, R_2 > 0$ *such that*

$$\int_a^b K_i(t,s,u)\,ds < R_i^3 + (p_i(t) - 1)R_i,$$

for all $(t,u) \in [a,b] \times [0,R_1] \times [0,R_2]$ *and* $i = 1,2$.

($\mathscr{H}_2$) *There exists* $t_0 \in [a,b]$ *and* r_1, r_2 *such that* $0 < r_i < R_i$ *and*

$$\int_a^b K_i(t_0,s,u)\,ds > r_i^3 + (p_i(t_0) - 1)r_i,$$

for all $u \in [0,r_1] \times [0,r_2]$ *and* $i = 1,2$.

The main existence result on system (3.7) is

Theorem 3.1.14 *Under Assumptions* ($\mathscr{H}_0$)-($\mathscr{H}_2$), *the integral system (3.7) has at least one solution* $x = (x_1, x_2) \in \mathscr{C}\left([a,b], \mathbb{R}_+^2\right)$ *with*

$$r_i \le \|x_i\|_\infty \le R_i, \ i = 1,2.$$

Proof 3.1.15 *We apply Theorem 3.1.11. Here* $E_1 = E_2 = \mathscr{C}[a,b]$ *with norm* $\|u\|_\infty = \max_{t \in [a,b]} |u(t)|$ *and*

$$\mathscr{P}_1 = \mathscr{P}_2 = \{u \in \mathscr{C}[a,b] : u(t) \ge 0 \text{ for all } t \in [a,b]\}.$$

For $i = 1,2$, *we define the operators* $T_i : (\overline{\mathscr{P}_i})_{R_i} \to E_i$ *and* $F_i : \overline{\mathscr{P}}_R = (\overline{\mathscr{P}_1})_{R_1} \times (\overline{\mathscr{P}_2})_{R_2} \to E_i$ *by*

$$(T_i x_i)(t) = (x_i(t))^3 + p_i(t)x_i(t)$$

and

$$F_i(x_1, x_2)(t) = -\int_a^b K_i(t,s,x_1(s),x_2(s))\,ds.$$

Then the integral system (3.7) is equivalent to the operator equation

$$x = (T_1 x_1 + F_1(x_1,x_2), T_2 x_2 + F_2(x_1,x_2)),$$

where $x = (x_1, x_2)$. *We check that all assumptions of Theorem 3.1.11 are satisfied, with* $\Omega_i = (\overline{\mathscr{P}_i})_{R_i}$.

(a) For i = 1, 2,

$$\|T_i y - T_i z\|_\infty \ge p_i^{(1)} \|y - z\|_\infty \quad \text{for all} \quad y, z \in (\overline{\mathscr{P}_i})_{R_i},$$

that is, in view of $(\mathscr{H}_0)$, the mapping $T_i : (\overline{\mathscr{P}_i})_{R_i} \to E_i$ is expansive with constant $h_i = p_i^{(1)} > 1$.

(b) If $x = (x_1, x_2) \in \overline{\mathscr{P}}_R = (\overline{\mathscr{P}_1})_{R_1} \times (\overline{\mathscr{P}_2})_{R_2}$, then $\|x_i\|_\infty \le R_i$ for $i = 1, 2$, and $(\mathscr{H}_1)$ guarantees that for each $i \in \{1, 2\}$,

$$\|F_i x\|_\infty \le R_i^3 + (p_i^{(2)} - 1) R_i, \tag{3.8}$$

which shows that $F_i(\overline{\mathscr{P}}_R)$ is uniformly bounded. For $x \in \overline{\mathscr{P}}_R$, differentiating $(F_i x)(t)$ with respect to t yields

$$(F_i x)'(t) = -\int_a^b \frac{\partial K_i}{\partial t}(t, s, x(s)) \, ds.$$

Hence, there are a constants N_i such that

$$\|(F_i x)'\|_\infty \le N_i \tag{3.9}$$

for all $x \in \overline{\mathscr{P}}_R$. The estimates (3.8) and (3.9) show that $F_i(\overline{\mathscr{P}}_R)$ is a uniformly bounded equicontinuous subset of E_i. Thus, from the Ascoli-Arzelà compactness criterion, F_i maps bounded sets of $\mathscr{P}$ into relatively compact sets. Also the continuity of K_i, implies that F_i are continuous. Therefore, the mappings $F_i : \overline{\mathscr{P}}_R \to E_i$, $i = 1, 2$, are completely continuous, i.e., they are 0-set contractions.

(c) Check of (3.5). Let $w^0 = (w_1^0, w_2^0) \in \mathscr{P}^*$ and assume that there exist $i \in \{1, 2\}$, $v = (v_1, v_2) \in \partial \mathscr{P}_r$ with $v_i - \lambda_1 w_i^0 \in (\overline{\mathscr{P}_i})_{R_i}$, and $\lambda_1 \ge 0$ such that

$$F_i v = (I_i - T_i)(v_i - \lambda_1 w_i^0).$$

Then

$$\int_a^b K_i(t_0, s, v(s)(s)) \, ds = -(F_i v)(t_0)$$

$$= -(I_i - T_i)(v_i - \lambda_1 w_i^0)(t_0)$$

$$= (v_i(t_0) - \lambda_1 w_i^0(t_0))^3$$

$$+ (p_i(t_0) - 1)(v_i(t_0) - \lambda_1 w_i^0(t_0))$$

$$\leq (v_i(t_0))^3 + (p_i(t_0) - 1)v_i(t_0)$$

$$\leq r_i^3 + (p_i(t_0) - 1)r_i,$$

which contradicts $(\mathcal{H}_2)$. *Hence (3.5) holds.*

(d) Check of (3.6). Assume that there exist $i \in \{1,2\}$, $u = (u_1, u_2) \in \partial \mathcal{P}_R$, *and* $\lambda_0 \geq 1$ *such that*

$$F_i u = \lambda_0 (u_i - T_i u_i).$$

Let $t_i \in [a,b]$ *with* $u_i(t_i) = R_i$. *From* $(\mathcal{H}_1)$, *we have*

$$- (F_i u)(t_i) = \int_a^b K_i(t_i, s, u(s)) \, ds < R_i^3 + (p_i(t_i) - 1) R_i.$$

On the other hand,

$$- (F_i u)(t_i) = -\lambda_0 (u_i(t_i) - (T_i u_i)(t_i))$$

$$= \lambda_0 (R_i^3 + (p_i(t_i) - 1) R_i)$$

$$\geq R_i^3 + (p_i(t_i) - 1) R_i,$$

which yields a contradiction. Hence (3.6) holds.

(e) It remains to check that $\mu F_i(\overline{\mathcal{P}_R}) \subset (I_i - T_i)((\overline{\mathcal{P}_i})_{R_i})$ *for all* $\mu \in [0,1]$ *and* $i = 1,2$. *For this, let* $z \in \overline{\mathcal{P}_R}$ *and* $\mu \in [0,1]$ *be arbitrarily fixed. We have to prove that there exists* $x = (x_1, x_2) \in \overline{\mathcal{P}_R}$ *such that for* $i = 1,2$, $\mu F_i z = (I_i - T_i) x_i$, *equivalently*

$$T_i x_i - x_i = -\mu F_i z,$$

or explicitly

$$(x_i(t))^3 + (p_i(t) - 1) x_i(t) = -\mu (F_i z)(t), \quad t \in [a,b].$$

To this aim, we can easily show that for each function $y \in C[a,b]$ with

$$0 \le y(t) \le R_i^3 + (p_i(t) - 1)R_i \text{ for all } t \in [a,b],$$

there exists a unique function $x_i \in (\overline{\mathscr{P}_i})_{R_i}$ which solves the equation

$$(x_i(t))^3 + (p_i(t) - 1)x_i(t) = y(t), t \in [a,b].$$

In virtue of $(\mathscr{H}_1)$, such a function y is $y = -\mu F_i z$.
Therefore Theorem 3.1.11 applies and gives the conclusion.

3.1.3 Extensions

In the more general case, where the mapping $(I - T)$ is Lipschitz invertible and the conical shell $\mathscr{P}_r$ is replaced by an any bounded open subsets of $\mathscr{P}$, Theorems 3.1.1, 3.1.3, 3.1.5, and 3.1.7 become Theorems 3.1.16, 3.1.17, 3.1.18, and 3.1.19, respectively.

In what follows, the assumption (i) represents a compressive form while (ii) expresses an expansive form. The proofs of the following theorems are based on the results presented in Section 2.2.4; so we omit the details.

Theorem 3.1.16 *(Order version). Let E be a Banach space, $\mathscr{P} \subset E$ a normal cone with constant N, and U_1 and U_2 two bounded open subsets of $\mathscr{P}$ such that $\overline{U_1} \subset U_2$ and $0 \in U_1 \cap \Omega$, where $\Omega \subset \mathscr{P}$ be such that $(U_2 \setminus U_1) \cap \Omega \ne \emptyset$. Assume that the mapping $T : \Omega \to E$ be such that $(I - T)$ is Lipschitz invertible with constant $\gamma > 0$, $(I - T)^{-1}0 \in U_1$, $F : \overline{U} \to E$ is a k-set contraction with $0 \le k < \gamma^{-1}$, $tF(\overline{U_2}) \subset (I - T)(\Omega)$ for all $t \in [0,1]$.*
Let $u_0 \in \mathscr{P}^*$ be such that $T(x - \lambda u_0) \in \mathscr{P}$, for all $\lambda \ge 0$ and $x \in (\partial U_1 \cup \partial U_2) \cap (\Omega + \lambda u_0)$, and suppose that one of the following conditions is satisfied:
(i) $Fx \not\le x, \forall x \in \partial U_1$ and $Fx \in \mathscr{P}$, $Fx \not\ge x - Tx, \forall x \in \partial U_2 \cap \Omega$.
(ii) $Fx \in \mathscr{P}$, $Fx \not\ge x - Tx, \forall x \in \partial U_1 \cap \Omega$ and $Fx \not\le x, \forall x \in \partial U_2$.
Then $T + F$ has a fixed point $(\overline{U_2} \setminus U_1) \cap \Omega$.

Theorem 3.1.17 *(Norm version). Let E be a Banach space, $\mathscr{P} \subset E$ a cone and U_1 and U_2 two bounded open subsets of $\mathscr{P}$ such that $\overline{U_1} \subset U_2$ and $0 \in U_1 \cap \Omega$, where $\Omega \subset \mathscr{P}$ be such that $(U_2 \setminus U_1) \cap \Omega \ne \emptyset$. Assume* *that the mapping $T : \Omega \to E$ be such that $(I - T)$ is Lipschitz invertible with constant $\gamma > 0$, $F : \overline{U} \to E$ is a k-set contraction with $0 \leq k < \gamma^{-1}$, and $F(\overline{U_2}) \subset (I - T)(\Omega)$. Let $u_0 \in \mathscr{P} \backslash \{0\}$ be such that $T(x - \lambda u_0) \in \mathscr{P}$, for all $\lambda \geq 0$ and $x \in (\partial U_1 \cup \partial U_2) \cap (\Omega + \lambda u_0)$, and suppose that one of the following conditions is satisfied:*

(i) $Fx \in \mathscr{P}$, $\|Fx\| > N\|x\|, \forall x \in \partial U_1$, and $\gamma \|Fx + T0\| \leq \|x\|, \forall x \in \partial U_2$.

(ii) $\gamma \|Fx + T0\| \leq \|x\|, \forall x \in \partial U_1$ and $Fx \in \mathscr{P}$, $\|Fx\| > N\|x\|, \forall x \in \partial U_2$.

Then $T + F$ has at least one fixed point in $(\overline{U_2} \backslash U_1) \cap \Omega$.

Theorem 3.1.18 *(Homotopy version 1). Let E be a Banach space, $\mathscr{P} \subset E$ a normal cone with constant N, and U_1 and U_2 two bounded open subsets of $\mathscr{P}$ such that $\overline{U_1} \subset U_2$ and $0 \in U_1 \cap \Omega$, where $\Omega \subset \mathscr{P}$ be such that $(U_2 \backslash U_1) \cap \Omega \neq \emptyset$. Assume that the mapping $T : \Omega \to E$ be such that $(I - T)$ is Lipschitz invertible with constant $\gamma > 0$, $(I - T)^{-1} 0 \in U_1$, $F : \overline{U} \to E$ is a k-set contraction with $0 \leq k < \gamma^{-1}$, $tF(\overline{U_2}) \subset (I - T)(\Omega)$ for all $t \in [0, 1]$. Suppose that there exists $u_0 \in \mathscr{P} \backslash \{0\}$ such that either one of the following conditions holds:*

(i) $Fx \neq (I - T)(x - \lambda u_0), \forall \lambda \geq 0, x \in \partial U_1 \cap (\Omega + \lambda u_0)$ and $(I - T)x \neq \lambda Fx, \forall x \in \partial U_2 \cap \Omega$ and $0 \leq \lambda \leq 1$,

(ii) $(I - T)x \neq \lambda Fx, \forall x \in \partial U_1 \cap \Omega$ and $0 \leq \lambda \leq 1$ and $Fx \neq (I - T)(x - \lambda u_0), \forall \lambda \geq 0, x \in \partial U_2 \cap (\Omega + \lambda u_0)$.

Then $T + F$ has a fixed point in $(U_2 \backslash U_1) \cap \Omega$.

Theorem 3.1.19 *(Homotopy version 2). Let E be a Banach space, $\mathscr{P} \subset E$ a cone, and U_1 and U_2 two bounded open subsets of $\mathscr{P}$ such that $\overline{U_1} \subset U_2$ and $0 \in U_1 \cap \Omega$, where $\Omega \subset \mathscr{P}$ be such $(U_2 \backslash U_1) \cap \Omega \neq \emptyset$. Assume that the mapping $T : \Omega \to E$ be such that $(I - T)$ is Lipschitz invertible with constant $\gamma > 0$, $F : \overline{U_2} \to E$ is a k-set contraction with $0 \leq k < \gamma^{-1}$, and $F(\overline{U_2}) \subset (I - T)(\Omega)$. Suppose that there exists $u_0 \in \mathscr{P} \backslash \{0\}$ such that either one of the following conditions holds:*

(i) $Fx \neq (I - T)(x - \lambda u_0), \forall \lambda \geq 0, x \in \partial U_1 \cap (\Omega + \lambda u_0)$ and $Fx \neq (I - T)(\lambda x), \forall x \in \partial U_2 \cap \Omega, \lambda \geq 1$ and $\lambda x \in \Omega$.

(ii) $Fx \neq (I - T)(\lambda x), \forall x \in \partial U_1 \cap \Omega, \lambda \geq 1$ and $\lambda x \in \Omega$, and $Fx \neq (I - T)(x - \lambda u_0), \forall \lambda \geq 0, x \in \partial U_2 \cap (\Omega + \lambda u_0)$.

Then $T + F$ has a fixed point in $(\overline{U_2} \backslash U_1) \cap \Omega$.

3.1.4   Case of translates of cones

In this section, let $\mathscr{P} \subset E$ ($\mathscr{P} \neq \{0\}$) be a cone. Given $\theta \in E$, we consider the translate of $\mathscr{P}$, namely

$$\mathscr{K} = \mathscr{P} + \theta = \{x + \theta, \ x \in \mathscr{P}\}.$$

For $0 < r < R$, we define

$$\begin{aligned}
\mathscr{K}_r &= \mathscr{K} \cap \{x \in E : \|x - \theta\| < r\}, \\
\partial \mathscr{K}_r &= \mathscr{K} \cap \{x \in E : \|x - \theta\| = r\}, \\
\mathscr{K}_{r,R} &= \mathscr{K} \cap \{x \in E : r < \|x - \theta\| < R\}.
\end{aligned}$$

As a consequence of the results obtained in section 2.3, we deduce the following fixed point theorems of compression–expansion type on translates of cones.

Theorem 3.1.20 *(Order version). Let E be a Banach space, $\mathscr{P} \subset E$ a cone, $\mathscr{K} = \mathscr{P} + \theta$ ($\theta \in E$) a translate of $\mathscr{P}$, and U_1 and U_2 two bounded open subsets of $\mathscr{P}$ such that $\overline{U_1} \subset U_2$. Let Ω be a subset of $\mathscr{K}$ such that $(U_2 \setminus U_1) \cap \Omega \neq \emptyset$. Assume that the mapping $T : \Omega \to E$ is such that $(I - T)$ is Lipschitz invertible with constant $\gamma > 0$, $(I - T)^{-1}\theta \in U_1$, $F : \overline{U_2} \to E$ is a k-set contraction with $0 \leq k < \gamma^{-1}$, $tF(\overline{U_2}) + (1 - t)\theta \subset (I - T)(\Omega)$ for all $t \in [0,1]$, and let $u_0 \in \mathscr{P} \setminus \{0\}$ be such that $T(x - \lambda u_0) \in \mathscr{P}$, for all $\lambda \geq 0$ and $x \in (\partial U_1 \cup \partial U_2) \cap (\Omega + \lambda u_0)$, and suppose that one of the following conditions is satisfied:*
(i) $Fx \not\leq x, \forall x \in \partial U_1$ and $Fx \in \mathscr{K}$, $Fx \not\geq x - Tx, \forall x \in \partial U_2 \cap \Omega$.
(ii) $Fx \in \mathscr{K}$, $Fx \not\geq x - Tx, \forall x \in \partial U_1 \cap \Omega$ and $Fx \not\leq x, \forall x \in \partial U_2$.
Then $T + F$ has a fixed point in $(\overline{U_2} \setminus U_1) \cap \Omega$.

Theorem 3.1.21 *(Norm version). Let E be a Banach space, $\mathscr{P} \subset E$ a normal cone with constant N, $\mathscr{K} = \mathscr{P} + \theta$ ($\theta \in E$) a translate of $\mathscr{P}$, and U_1 and U_2 two bounded open subsets of $\mathscr{K}$ such that $\overline{U_1} \subset U_2$ and $\theta \in U_1 \cap \Omega$, where $\Omega \subset \mathscr{K}$ be such that $(U_2 \setminus U_1) \cap \Omega \neq \emptyset$. Assume that the mapping $T : \Omega \to E$ is such that $(I - T)$ is Lipschitz invertible with constant $\gamma > 0$, $F : \overline{U_2} \to E$ is a k-set contraction with $0 \leq k < \gamma^{-1}$, and $F(\overline{U_2}) \subset (I - T)(\Omega)$. Let $u_0 \in \mathscr{P} \setminus \{0\}$ be such that $T(x - \lambda u_0) \in \mathscr{P}$, for all $\lambda \geq 0$ and $x \in (\partial U_1 \cup \partial U_2) \cap (\Omega + \lambda u_0)$, and suppose that one of the following conditions is satisfied:*

(i) $Fx \in \mathcal{K}$, $\|Fx - \theta\| > N\|x - \theta\|$, *for all* $x \in \partial U_1$,
and $\gamma\|Fx + T\theta - \theta\| \leq \|x - \theta\|$, *for all* $x \in \partial U_2$.
(ii) $\gamma\|Fx + T\theta - \theta\| \leq \|x - \theta\|$, *for all* $x \in \partial U_1$
and $Fx \in \mathcal{K}$, $\|Fx - \theta\| > N\|x - \theta\|, \forall x \in \partial U_2$.
Then $T + F$ *has at least one fixed point in* $(\overline{U_2} \setminus U_1) \cap \Omega$.

Theorem 3.1.22 *(Homotopy version). Let E be a Banach space, $\mathcal{P} \subset E$ a cone, and $\mathcal{K} = \mathcal{P} + \theta$ a translate of $\mathcal{P}$. Let U_1 and U_2 be two open bounded subsets of $\mathcal{K}$ such that $\theta \in \overline{U_1} \subset U_2$. Let $\Omega \subset \mathcal{K}$ with $\theta \in \Omega$ and $(U_2 \setminus U_1) \cap \Omega \neq \emptyset$. Assume that the mapping $T : \Omega \to E$ is such that $(I - T)$ is Lipschitz invertible with constant $\gamma > 0$, $F : \overline{U_2} \to E$ is a k-set contraction with $0 \leq k < \gamma^{-1}$, and $F(\overline{U_2}) \subset (I - T)(\Omega)$. Suppose that there exists $u_0 \in \mathcal{P} \setminus \{0\}$ such that either one of the following conditions holds:*
(i) $Fx \neq (I - T)(x - \lambda u_0), \forall \lambda \geq 0, x \in \partial U_1 \cap (\Omega + \lambda u_0)$
and $Fx \neq (I - T)(\lambda x + (1 - \lambda)\theta)$ *for all* $x \in \partial U$, $\lambda \geq 1$ *and* $\lambda x + (1 - \lambda)\theta \in \Omega$,
(ii) $Fx \neq (I - T)(\lambda x + (1 - \lambda)\theta)$ *for all* $x \in \partial U$, $\lambda \geq 1$ *and* $\lambda x + (1 - \lambda)\theta \in \Omega$,
and $Fx \neq (I - T)(x - \lambda u_0), \forall \lambda \geq 0, x \in \partial U_2 \cap (\Omega + \lambda u_0)$.
Then $T + F$ *has a fixed point* $x \in (U_2 \setminus U_1) \cap \Omega$.

3.1.5 Further extensions

In this section, we establish some new extensions of the Krasnoselskii's compression–expansion fixed point theorem for sums of two mappings: one is an expansive operator and the other one is completely continuous.

Let X be a normed linear space with norm $\|.\|$ and let $\mathcal{P} \subset X$ be a wedge, i.e., a closed convex subset of X, with $\lambda \mathcal{P} \subset \mathcal{P} \neq \{0\}$ for every $\lambda \in \mathbb{R}_+$. If in addition $\mathcal{P} \cap (-\mathcal{P}) = \{0\}$, then $\mathcal{P}$ is a cone, and we say that $x < y$ if and only if $y - x \in \mathcal{P} \setminus \{0\}$. For two numbers $0 < r < R$, define the conical shell $\mathcal{P}_{r,R}$ by

$$\mathcal{P}_{r,R} := \{x \in \mathcal{P} : r \leq \|x\| \leq R\}.$$

Let E be a Banach space, K a subset of E, and $\mathcal{P} \subset E$ a wedge such that $\mathcal{P} \subset T(K)$. New fixed point theorem has been obtained in the case where $T : K \to E$ is an expansive mapping and $F : K \to E$ is a mapping such that $I - F$ is completely continuous one.

Theorem 3.1.23 *Let K be a subset of a Banach space X and $\mathscr{P} \subset X$ a wedge. Assume that $T : K \to X$ is an h-expansive mapping and $F : K \to X$ is a mapping such that $I - F : K \to \mathscr{P}$ is completely continuous one with $\mathscr{P} \subset T(K)$. Let $\alpha, \beta > 0$, $\alpha \neq \beta$, $p \in \mathscr{P} \setminus \{0\}$, $r := \min\{\alpha, \beta\}$ and $R := \max\{\alpha, \beta\}$.*

Suppose that the following conditions are satisfied:

$$x \neq \lambda Tx + Fx \ \text{for } x \in T^{-1}(\mathscr{P}), \|Tx\| = \alpha \ \text{and } \lambda > 1. \quad (3.10)$$

$$x \neq Tx + Fx - \mu p \ \text{for } x \in T^{-1}(\mathscr{P}), \|Tx\| = \beta \ \text{and } \mu > 0. \quad (3.11)$$

Then $T + F$ has a fixed point x in $T^{-1}(\mathscr{P})$ such that $r \leq \|Tx\| \leq R$.

Proof 3.1.24 *By Lemma 2.1.18, the operator $T^{-1} : T(K) \to K$ is a $\dfrac{1}{h}$-contraction. Then the operator N defined by*

$$
\begin{aligned}
N : \mathscr{P} \ &\to \ \mathscr{P} \\
y \ &\to \ Ny = T^{-1}y - FT^{-1}y
\end{aligned}
$$

is well defined and it is completely continuous.

Claim (1): We show that Condition (3.30) implies

$$Ny \neq \lambda y \ \text{for } \|y\| = \alpha \ \text{and } \lambda > 1.$$

On the contrary, assume the existence of $\lambda_0 > 1$ and $y_1 \in \mathscr{P}$ with $\|y_1\| = \alpha$ such that

$$Ny_1 = \lambda_0 y_1.$$

Let $x_1 := T^{-1}y_1$. Then

$$x_1 - Fx_1 = \lambda_0 Tx_1.$$

The hypotheses $y_1 \in \mathscr{P}, \|y_1\| = \alpha$ imply that $x_1 \in T^{-1}(\mathscr{P})$ and

$\|Tx_1\| = \alpha$, *which lead to a contradiction with Condition (3.30).*

Claim (2): We show that Condition (3.11) entails

$$Ny + \mu p \neq y \text{ for } \|y\| = \beta \text{ and } \mu > 0.$$

On the contrary, assume the existence of $\mu_0 > 1$ and $y_2 \in \mathcal{P}$ with $\|y_2\| = \beta$ such that

$$y_2 - Ny_2 = \mu_0 p.$$

Let $x_2 := T^{-1}y_2$. Then

$$x_2 = Tx_2 + Fx_2 - \mu_0 p.$$

The hypotheses $y_2 \in \mathcal{P}$, $\|y_2\| = \beta$ imply that $x_2 \in T^{-1}(\mathcal{P})$ and $\|Tx_2\| = \beta$, which lead to a contradiction with Condition (3.11).

Consequently, by Theorem 2.1.15, the operator N has a fixed point $y \in \mathcal{P}$ such that $r \leq \|y\| \leq R$, that is

$$T^{-1}y - FT^{-1}y = y.$$

Let $x := T^{-1}y$. Then $x \in T^{-1}(\mathcal{P})$, it is a fixed point of $T + F$, and

$$r \leq \|Tx\| \leq R.$$

If in addition $\mathcal{P}$ is a cone, then as a consequence of Theorem 3.1.23, we derive some cone compression and expansion fixed point theorems. The first one is formulated in terms of the partial order relation induced by $\mathcal{P}$ while the second one is of norm type.

Corollary 3.1.25 *Let K be a subset of a Banach space X and $\mathcal{P} \subset X$ a cone. Assume that $T : K \to X$ is an h-expansive mapping and $F : K \to X$ is a mapping such that $I - F : K \to \mathcal{P}$ is completely continuous with $\mathcal{P} \subset T(K)$. Let $\alpha, \beta > 0$, $\alpha \neq \beta$, $r := \min\{\alpha, \beta\}$ and $R := \max\{\alpha, \beta\}$. Suppose that the following conditions are satisfied:*

$$x \not\geq Tx + Fx \text{ for } x \in T^{-1}(\mathcal{P}) \text{ with } \|Tx\| = \alpha. \tag{3.12}$$

$$x \not\leq Tx + Fx \text{ for } x \in T^{-1}(\mathcal{P}) \text{ with } \|Tx\| = \beta. \tag{3.13}$$

Then $T + F$ has a fixed point x in $T^{-1}(\mathcal{P})$ such that $r \leq \|Tx\| \leq R$.

Proof 3.1.26 *The conditions (3.30) and (3.11) of Theorem 3.1.23 are satisfied. Indeed, assume the contrary of Condition (3.30). Then there exist $\lambda_0 > 1$ and $x_0 \in T^{-1}(\mathscr{P})$ with $\|Tx_0\| = \alpha$ such that*

$$x_0 = \lambda_0 T x_0 + F x_0.$$

Thus, $Tx_0 = \dfrac{1}{\lambda_0}(x_0 - Fx_0) < x_0 - Fx_0$, that is

$$x_0 > Tx_0 + Fx_0,$$

which contradicts (3.12). Assume the contrary of Condition (3.11). Then there exist $p \in \mathscr{P} \setminus \{0\}$, $\mu_0 > 0$ and $x_1 \in T^{-1}(\mathscr{P})$ with $\|Tx_1\| = \beta$ such that

$$x_1 = Tx_1 + Fx_1 - \mu_0 p.$$

Since $\mu_0 p \in \mathscr{P} \setminus \{0\}$, we obtain

$$x_1 < Tx_1 + Fx_1,$$

which contradicts (3.13).

Corollary 3.1.27 *Let K be a subset of a Banach space X and $\mathscr{P} \subset X$ a cone. Assume that $T : K \to X$ is an expansive mapping with constant $h > 1$ and $F : K \to X$ is a mapping such that $I - F : K \to \mathscr{P}$ is completely continuous one with $\mathscr{P} \subset T(K)$. Let $\alpha, \beta > 0$, $\alpha \neq \beta$, $r := \min\{\alpha, \beta\}$ and $R := \max\{\alpha, \beta\}$.*

Suppose that the following conditions are satisfied:

$$\|x - Fx\| \leq \|Tx\| \text{ for } x \in T^{-1}(\mathscr{P}) \text{ with } \|Tx\| = \alpha. \qquad (3.14)$$

$$\|x - Fx\| \geq \|Tx\| \text{ for } x \in T^{-1}(\mathscr{P}) \text{ with } \|Tx\| = \beta. \qquad (3.15)$$

Then $T + F$ has a fixed point x in $T^{-1}(\mathscr{P})$ such that $r \leq \|Tx\| \leq R$.

Proof 3.1.28 *The conditions (3.30) and (3.11) of Theorem 3.1.23 are satisfied. Indeed, assume the contrary of Condition (3.30). Then there exist $\lambda_0 > 1$ and $x_0 \in T^{-1}(\mathscr{P})$ with $\|Tx_0\| = \alpha$ such that*

$$x_0 = \lambda_0 T x_0 + F x_0.$$

Then $x_0 - Fx_0 = \lambda_0 T x_0$, that is

$$\|x_0 - Fx_0\| = \lambda_0\|Tx_0\| > \|Tx_0\|,$$

which contradicts (3.14). Assume the contrary of Condition (3.11). Then there exist $p \in \mathscr{P} \setminus \{0\}$, $\mu_0 > 0$, and $x_1 \in T^{-1}(\mathscr{P})$ with $\|Tx_1\| = \beta$ such that

$$x_1 = Tx_1 + Fx_1 - \mu_0 p.$$

$x_1 - Fx_1 = Tx_1 - \mu_0 p$ *that is*

$$\|x_1 - Fx_1\| < \|Tx_1\|,$$

which contradicts (3.15).

We end this section with an application of Theorem 3.1.23 to a non-linear boundary value problem associated to a second order differential equation.

Example 3.1.29 *Consider the following nonlinear boundary value problem*

$$-\frac{d^2}{dt^2} f(t, x(t)) = g(t)h(x(t)), \ 0 < t < 1 \tag{3.16}$$

$$x(0) = x(1) = 0,$$

where $f : [0,1] \times \mathbb{R}_+ \to \mathbb{R}_+$ is continuous function defined by

$$f(t, u) = u^3 + a(t)u, \ a \in \mathscr{C}^2([0,1], \mathbb{R}_+), \ with \ \min_{t \in [0,1]} a(t) > 1,$$

$g \in \mathscr{C}([0,1], \mathbb{R}_+)$ *and $h : \mathbb{R}_+ \to \mathbb{R}_+$ is continuous increasing function. Problem (3.16) is equivalent to the integral equation*

$$f(t, x(t)) = \int_0^1 G(t, s)g(s)h(x(s))ds, \ t \in [0, 1], \tag{3.17}$$

where G is the corresponding Green's function defined in $[0,1] \times [0,1]$ by:

$$G(t, s) = \begin{cases} t(1-s), & if \ 0 \le t \le s \le 1, \\ \\ s(1-t), & if \ 0 \le s \le t \le 1. \end{cases} \tag{3.18}$$

The Green function satisfies the following properties:

$$0 \leq G(t,s) \leq G(s,s), \ \forall \, (t,s) \in [0,1] \times [0,1],$$

$$G(t,s) \geq \frac{1}{4}G(s,s), \ \forall (t,s) \in [\frac{1}{4}, \frac{3}{4}] \times [0,1].$$

$$\int_0^1 G(t,s)\,ds \leq \frac{1}{8}, \ \forall t \in [0,1].$$

$$\int_{\frac{1}{4}}^{\frac{3}{4}} G(t,s)\,ds \geq \frac{1}{16}, \ \forall t \in [\frac{1}{4}, \frac{3}{4}].$$

We will set

$$A := \max_{t \in [0,1]} \int_0^1 G(t,s)g(s)\,ds,$$

$$B := \frac{1}{4} \int_{\frac{1}{4}}^{\frac{3}{4}} G(t_0,s)g(s)\,ds, \ \text{ for some } t_0 \in [0,1].$$

We let the conditions

(C_0) $1 < a_0 := \min_{t \in [0,1]} a(t) \leq a^0 := \max_{t \in [0,1]} a(t).$

Assume that the following assumptions hold for some positive reals α, β *with* $\alpha \neq \beta$:

(C_1) $Ah\left(\dfrac{1}{a_0}\alpha\right) \leq \alpha,$

(C_2) $Bh\left(\dfrac{1}{4}\beta_0\right) \geq \beta,$ *where* $\beta_0 = \beta_0(\beta) > 0$ *such that* $\beta_0^3 + a^0\beta_0 = \beta.$

Remark 3.1.30 *From the properties of the Green function, we get*

$$\max_{t \in [0,1]} \int_0^1 G(t,s)g(s)\,ds \leq \frac{1}{8} \max_{t \in [0,1]} g(t)$$

and

$$\min_{t \in [\frac{1}{4}, \frac{3}{4}]} \int_{\frac{1}{4}}^{\frac{3}{4}} G(t,s)g(s)\,ds \geq \frac{1}{16} \min_{t \in [\frac{1}{4}, \frac{3}{4}]} g(t).$$

Then, for the conditions (C_1) and (C_2) to be satisfied it is enough that constants α and β satisfy

$$\frac{1}{8} \max_{t \in [0,1]} g(t) h \left(\frac{1}{a_0} \alpha \right) \leq \alpha \text{ and } \frac{1}{16} \min_{t \in [\frac{1}{4}, \frac{3}{4}]} g(t) h \left(\frac{1}{4} \beta_0 \right) \geq \beta.$$

The main result of this section is

Theorem 3.1.31 *Let assumptions (C_0)–(C_2) be satisfied. Then the non-linear boundary value problem has a solution x which belongs to $\mathscr{C}([0,1], \mathbb{R}_+)$.*

Proof 3.1.32 *Consider the Banach space $X = \mathscr{C}([0,1])$ normed by $\|x\| = \max_{t \in [0,1]} |x(t)|$, the set*

$$K = \{ x \in X \mid x(t) \geq 0, \forall t \in [0,1] \}$$

and the positive cone $\mathscr{P}$

$$\mathscr{P} = \left\{ x \in X : x \geq 0 \text{ on } [0,1] \text{ and } x(t) \geq \frac{1}{4} \|x\| \text{ for } \frac{1}{4} \leq t \leq \frac{3}{4} \right\}.$$

Define the operators $T : K \rightarrow K$ and $F : K \rightarrow X$ by

$$Tx(t) = x(t)^3 + a(t)x(t)$$

$$Fx(t) = x(t) - \int_0^1 G(t,s) g(s) h(x(s)) \, ds,$$

respectively, for $t \in [0,1]$. Then the integral equation (3.17) is equivalent to the operational equation $x = Tx + Fx$. We check that all assumptions of Theorem 3.1.23 are satisfied.

(a) The operator $T : K \rightarrow K$ is surjective and it is expansive with constant $a_0 > 1$.

(b) Using the Ascoli-Arzelà compactness criteria, we can show that $I - F$ maps bounded sets of K into relatively compact sets. In view of the sup-norm and the continuity of functions G, g and h, it is easily checked that $I - F$ is continuous. Therefore, the operator $I - F : K \rightarrow \mathscr{P}$ is completely continuous.

(c) Assume the existence of $x_0 \in T^{-1}(\mathscr{P})$ with $\|Tx_0\| = \alpha$ and $\lambda_0 > 1$ such that

$$x_0 = \lambda_0 T x_0 + F x_0,$$

Then, $\lambda_0 T x_0 = x_0 - F x_0 = \displaystyle\int_0^1 G(.,s)g(s)h(x_0(s))\,ds$ on $[0,1]$.

So

$$\alpha < \lambda_0 \|Tx_0\| = \max_{t \in [0,1]} \int_0^1 G(t,s)g(s)h(x_0(s))\,ds. \tag{3.19}$$

On the other hand, we have

$$\|x_0\| = \|T^{-1}Tx_0\| \le \frac{1}{a_0}\|Tx_0\| = \frac{1}{a_0}\alpha,$$

where $\dfrac{1}{a_0} < 1$ is the Lipschitz constant of T^{-1}, which implies that

$$0 \le x_0(t) \le \frac{1}{a_0}\alpha \ \text{ for } t \in [0,1].$$

Since the function h is increasing, we get

$$0 \le h(x_0(t)) \le h\left(\frac{1}{a_0}\alpha\right) \ \text{ for } t \in [0,1].$$

Thus, for all $t \in [0,1]$, we obtain

$$
\begin{aligned}
\int_0^1 G(t,s)g(s)h(x_0(s))\,ds &\le h\left(\frac{1}{a_0}\alpha\right)\int_0^1 G(t,s)g(s)\,ds \\
&\le \left\|\int_0^1 G(.,s)g(s)\,ds\right\| h\left(\frac{1}{a_0}\alpha\right) \\
&\le Ah\left(\frac{1}{a_0}\alpha\right) \le \alpha.
\end{aligned}
$$

By passage to the maximum, we obtain

$$\max_{t \in [0,1]} \int_0^1 G(t,s)g(s)h(x_0(s))\,ds \le \alpha,$$

which leads to a contradiction with (3.19).

(d) Assume the existence of $x_1 \in T^{-1}(\mathscr{P})$ with $\|Tx_1\| = \beta$ and $\mu_0 > 0$ such that

$$x_1 = Tx_1 + Fx_1 - \mu_0 y_0,$$

where $y_0 \in \mathscr{P}$ with $y_0(t) > 0$ on $[0,1]$. Then

$$\int_0^1 G(.,s)g(s)h(x_1(s))\,ds = x_1 - Fx_1 = Tx_1 - \mu_0 y_0 < Tx_1 \text{ on } [0,1].$$

Since for all $t \in [0,1]$, $(Tx_1)(t) \leq \|Tx_1\| = \beta$, we get

$$\int_0^1 G(t,s)g(s)h(x_1(s))\,ds < (Tx_1)(t) \leq \beta, \ \forall t \in [0,1]. \qquad (3.20)$$

On the other hand, from the property of the Green function G, for all $t \in [\frac{1}{4}, \frac{3}{4}]$, we have

$$\int_0^1 G(t,s)g(s)h(x_1(s))\,ds \ \geq \ \frac{1}{4}\int_{\frac{1}{4}}^{\frac{3}{4}} G(s,s)g(s)h(x_1(s))\,ds$$

$$\geq \ \frac{1}{4}\int_{\frac{1}{4}}^{\frac{3}{4}} G(t_0,s)g(s)h(x_1(s))\,ds.$$

Since $\|Tx_1\| = \beta$, there exists $t_1 \in [0,1]$ such that $(Tx_1)(t_1) = \beta$. That is

$$(x_1(t_1))^3 + a(t_1)x_1(t_1) = \beta \leq (x_1(t_1))^3 + a^0 x_1(t_1),$$

where $a^0 = \max\limits_{t \in [0,1]} a(t)$. Let $\beta_0 = \beta_0(\beta) > 0$ such that $\beta_0^3 + a^0\beta_0 = \beta$. So $x_1(t_1) \geq \beta_0$,

which implies that $\|x_1\| \geq \beta_0$. Hence $x_1(s) \geq \frac{1}{4}\beta_0, \ \forall s \in [\frac{1}{4}, \frac{3}{4}]$, which gives

$$h(x_1(s)) \geq h\left(\frac{1}{4}\beta_0\right).$$

Thus

$$\int_0^1 G(t,s)g(s)h(x_1(s))\,ds \;\geq\; \frac{1}{4}h\left(\frac{1}{4}\beta_0\right)\int_{\frac{1}{4}}^{\frac{3}{4}} G(t_0,s)g(s)\,ds$$

$$= Bh\left(\frac{1}{4}\beta_0\right)$$

$$\geq \beta,$$

which leads to a contradiction with (3.20). Therefore, Theorem 3.1.23 applies and ensures that Problem (3.16) has at least one positive solution $x \in \mathscr{C}([0,1])$ such that

$$r \leq \|Tx\| \leq R,$$

where $r = \min(\alpha,\beta)$ and $R = \max(\alpha,\beta)$.

3.2 Leggett-Williams Fixed Point Theorems Type

In this section, we present an extension of the original version of the Leggett-Williams fixed point theorem for a k-set contraction perturbed by an expansive operator.

Let us consider be a real Banach space $(E, \|.\|)$, $\mathscr{P} \subset E$ a cone, and $\Omega \subset \mathscr{P}$. Let $\alpha : E \to [0,+\infty)$ a continuous concave functional, i.e.,

$$\alpha(\lambda x + (1-\lambda)y) \geq \lambda\alpha(x) + (1-\lambda)\alpha(y), \quad \text{for all } x,y \in E \text{ and } \lambda \in [0,1].$$

For two numbers $0 < a < c$, define

$$\mathscr{P}_c = \{x \in \mathscr{P} : \|x\| \leq c\},$$

$$S(\alpha,a,c) = \{x \in \mathscr{P} : \alpha(x) \geq a \text{ and } \|x\| \leq c\}$$

such that

$$\Omega \cap S(\alpha,a,c) \neq \emptyset.$$

Theorem 3.2.1 *Let $c \geq b > a > 0$ three real numbers and α be a continuous concave positive functional with $\alpha(x) \leq \|x\|$ for all $x \in \mathscr{P}$. Assume that $T : \Omega \to E$ is an h-expansive mapping, $F : \mathscr{P}_c \to E$ is a k-set contraction with $0 \leq k < h - 1$, and there exist $z_0 \in \mathscr{B}(-T0, (h-1)c) \cap \mathscr{P}_c$ such that*

$$tF(\mathscr{P}_c) + (1-t)z_0 \subset (I-T)(\Omega), \text{ for all } t \in [0,1]. \qquad (3.21)$$

If the following conditions are satisfied:

    1. $\{x \in S(\alpha,a,b) : \alpha(x) > a\} \neq \emptyset$ and $\alpha(Tx + Fx) > a$ if $x \in S(\alpha,a,b)$;

    2. $\alpha(Tx + Fx) > a$ and $\alpha(Tx + z_0) \geq a$ for all $x \in S(\alpha,a,c) \cap \Omega$ with $\|Tx + Fx\| > b$;

    3. $\|Tx + z_0\| \leq b$ for all $x \in S(\alpha,a,c) \cap \Omega$, with $\alpha(x) = a$,

then $T + F$ has at least one fixed point $x \in S(\alpha,a,c) \cap \Omega$, with $\alpha(x) > a$.

Proof 3.2.2 *Consider the set $U = \{x \in S(\alpha,a,c) : \alpha(x) > a\}$; U is then the interior of $S(\alpha,a,c)$ in $\mathscr{P}_c$.*
Suppose that $x \in \partial U \cap \Omega$ is a fixed point of $T + F$. Then $\alpha(x) = a$ with or $x \in S(\alpha,a,b)$, or $\|x\| > b$.
If $x \in S(\alpha,a,b)$, we get $\alpha(x) = \alpha(Tx + Fx) > a$, which is a contradiction.
If $\|x\| > b$, we get $\|Tx + Fx\| > b$ and $\alpha(x) = \alpha(Tx + Fx) > a$, leading again to a contradiction with (2).
Consequently, the fixed point index $i_*(T + F, U \cap \Omega, \mathscr{P})$ is well defined and satisfies the properties (a)–(d) of Theorem 2.2.1.
    Consider the homotopic deformation $H : [0,1] \times \mathscr{P}_c \to E$ defined by

$$H(t,x) = tFx + (1-t)z_0.$$

The operator H is continuous and uniformly continuous in t for each x. Moreover, $H(t,.)$ is a k-set contraction for each t and the mapping $T + H(t,.)$ has no fixed point on $\partial U \cap \Omega$ for each t. Otherwise, there would exist some $x_0 \in \partial U \cap \Omega$ and $t_0 \in [0,1]$ such that

$$x_0 = Tx_0 + H(t_0, x_0).$$

We have $\alpha(x_0) = a$ and we may distinguish between two cases:
(i) If $\|Tx_0 + Fx_0\| > b$, the concavity of α and the condition (2) lead

$$
\begin{aligned}
\alpha(x_0) &= \alpha(Tx_0 + t_0 Fx_0 + (1-t_0)z_0) \\[1em]
&= \alpha(t_0(Tx_0 + Fx_0) + (1-t_0)(Tx_0 + z_0)) \\[1em]
&\geq t_0\,\alpha(Tx_0 + Fx_0) + (1-t_0)\alpha(Tx_0 + z_0) \\[1em]
&> a,
\end{aligned}
$$

which is a contradiction.
(ii) If $\|Tx_0 + Fx_0\| \leq b$, the condition (3) leads

$$
\begin{aligned}
\|x_0\| &= \|Tx_0 + t_0 Fx_0 + (1-t_0)z_0\| \\[1em]
&= \|(t_0(Tx_0 + Fx_0) + (1-t_0)(Tx_0 + z_0))\| \\[1em]
&\leq t_0\|Tx_0 + Fx_0\| + (1-t_0)\|Tx_0 + z_0\| \\[1em]
&\leq b.
\end{aligned}
$$

Thus, $x_0 \in S(\alpha, a, b)$ and by the condition (1), we get $\alpha(Tx_0 + Fx_0) > a$, which imply that $\alpha(x_0) > a$ and again we come to a contradiction with $\alpha(x_0) = a$.

By properties (a) and (d) of the index i_* in Theorem 2.2.1, we deduce that

$$i_*(T + F, U \cap \Omega, \mathscr{P}) = i_*(T + z_0, U \cap \Omega, \mathscr{P}) = 1.$$

As a consequence, $T + F$ has a fixed point in $U \cap \Omega$.

In the sequel, we establish an extension of a Bai-Ge multiple fixed point theorem for a k-set contraction perturbed by an expansive operator.

Theorem 3.2.3 *Let $r_2 \geq d > c > r_1 > 0$, $L_2 \geq L_1 > 0$ be constants, $R > M \max(r_2, L_2)$ and $0 \in \Omega \subset \overline{\mathscr{P}}(\alpha, r_2; \beta, L_2)$. Assume that α, β are nonnegative convex functionals satisfying $(\mathscr{A}_1)$ and $(\mathscr{A}_2)$. Let ψ be a nonnegative concave functional on $\mathscr{P}$ such that $\psi(x) \leq \alpha(x)$ for all $x \in \overline{\mathscr{P}}(\alpha, r_2; \beta, L_2)$. Assume that $T : \Omega \to E$ is an expansive mapping with constant $h > 1$ and $F : \overline{\mathscr{P}}(\alpha, r_2; \beta, L_2) \to E$ is a k-set contraction with $k < h - 1$ such that*

$$F\left(\overline{\mathscr{P}}(\alpha, r_2; \beta, L_2)\right) \subset (I - T)(\overline{\mathscr{P}}(\alpha, r_1; \beta, L_1) \cap \Omega). \qquad (3.22)$$

Suppose that:

(C_1) *if $x \in \mathscr{P}$ with $\alpha(x) = r_1$, then $\alpha(Tx + Fx) \neq r_1$;*

(C_2) *if $x \in \mathscr{P}$ with $\beta(x) = L_1$, then $\beta(Tx + Fx) \neq L_1$;*

(C_3) *there exist $z_0 \in \{x \in \overline{\mathscr{P}}(\alpha, d; \beta, L_2; \psi, c) : \psi(x) > c\}$ such that*

$$z_0 \in \mathscr{B}(-T0, (h-1)R), \ \psi((I-T)^{-1}z_0) > c,$$

    and

$$tF\left(\overline{\mathscr{P}}(\alpha, r_2; \beta, L_2)\right) + (1-t)z_0 \subset (I - T)(\Omega), \text{ for all } t \in [0, 1].$$

(C_4) $\psi(Tx + Fx) > c$, $\psi(Tx + z_0) \geq c$ *and* $\alpha(Tx + z_0) \leq d$ *for all* $x \in \overline{\mathscr{P}}(\alpha, d; \beta, L_2; \psi, c) \cap \Omega$;

(C_5) $\psi(Tx + Fx) > c$ *and* $\psi(Tx + z_0) \geq c$ *for all* $x \in \overline{\mathscr{P}}(\alpha, r_2; \beta, L_2; \psi, c) \cap \Omega$ *with* $\alpha(Tx + Fx) > d$.

Then $T + F$ has at least three fixed points x_1, x_2, and x_3 in $\overline{\mathscr{P}}(\alpha, r_2; \beta, L_2) \cap \Omega$ with

$$x_1 \in \mathscr{P}(\alpha, r_1; \beta, L_1), \quad x_2 \in \{x \in \overline{\mathscr{P}}(\alpha, r_2; \beta, L_2; \psi, c) : \psi(x) > c\}$$

and

$$x_3 \in \overline{\mathscr{P}}(\alpha, r_2; \beta, L_2) \setminus \overline{\mathscr{P}}(\alpha, r_2; \beta, L_2; \psi, c) \cup \overline{\mathscr{P}}(\alpha, r_1; \beta, L_1).$$

Proof 3.2.4 *Let*

$$U_1 = \mathscr{P}(\alpha, r_1; \beta, L_1),$$

$$U_2 = \{x \in \overline{\mathscr{P}}(\alpha, r_2; \beta, L_2; \psi, c) : \psi(x) > c\}.$$

By assumptions on α, β, and ψ, U_1 and U_2 are disjoint bounded nonempty open subsets of $\overline{\mathscr{P}}(\alpha, r_2; \beta, L_2)$.

Claim 1. We show that $i_*(T + F, U_1 \cap \Omega, \overline{\mathscr{P}}(\alpha, r_2; \beta, L_2)) = 1$.
We have $Tx + Fx \neq x$ for all $x \in \partial U_1 \cap \Omega$. Otherwise, there exists $x_0 \in \partial U_1 \cap \Omega$ such that $x_0 = Tx_0 + Fx_0$.
If $\alpha(x_0) = r_1$, by condition (C_1), we get

$$r_1 = \alpha(x_0) = \alpha(Tx_0 + Fx_0) \neq r_1,$$

which is a contradiction.
If $\beta(x_0) = L_1$, by condition $(\mathscr{C})$ we get

$$L_1 = \beta(x_0) = \beta(Tx_0 + Fx_0) \neq L_1,$$

which is a contradiction. Therefore, for $X = \overline{\mathscr{P}}(\alpha, r_2; \beta, L_2)$, $X_1 = \overline{U}_1$ and $U = U_1$, Lemma 2.2.6 applies and gives the conclusion.
Claim 2. We show that

$$i_*(T + F, \overline{\mathscr{P}}(\alpha, r_2; \beta, L_2) \cap \Omega, \overline{\mathscr{P}}(\alpha, r_2; \beta, L_2)) = 1.$$

It is easy to see that this claim follows from the condition (3.22) and Lemma 2.2.6.
Claim 3. We show that

$$i_*(T + F, U_2 \cap \Omega, \overline{\mathscr{P}}(\alpha, r_2; \beta, L_2)) = 1.$$

Suppose that $x_0 \in \partial U_2 \cap \Omega$ is a fixed point of $T + F$ and consider two cases
Case (i): $\psi(x_0) = c$ with $\alpha(x_0) > d$,
Case (ii): $\psi(x_0) = c$ with $x_0 \in \overline{\mathscr{P}}(\alpha, d; \beta, L_2; \psi, c)$.
In Case (i), there is $\alpha(Tx_0 + Fx_0) = \alpha(x_0) > d$, which combined with $(\mathscr{C})$ yields
$\psi(x_0) = \psi(Tx_0 + Fx_0) > c$, which is a contradiction.

In Case (ii), $\psi(x_0) = \psi(Tx_0 + Fx_0) > c$, and a contradiction to (C_4) is reached.

Consequently, the fixed point index $i_*\left(T + F, U_2 \cap \Omega, \overline{\mathscr{P}}(\alpha, r_2; \beta, L_2)\right)$ is well defined and satisfying the properties (a)–(d) of Theorem 2.2.1 as well as the properties given in Lemma 2.2.4.

Consider the homotopic deformation $H : [0,1] \times \overline{U}_2 \to E$ defined by

$$H(t,x) = tFx + (1-t)z_0.$$

The operator H is continuous and uniformly continuous in t for each x and from (C_3), we can easily see that $H([0,1] \times \overline{U}_2) \subset (I - T)(\Omega)$. Moreover, $H(t,.)$ is a k-set contraction for each t and the mapping $T + H(t,.)$ has no fixed point on $\partial U_2 \cap \Omega$ for each t. Otherwise, there would exist some $x_0 \in \partial U_2 \cap \Omega$ and $t_0 \in [0,1]$ such that

$$x_0 = Tx_0 + H(t_0, x_0).$$

Since $x_0 \in \partial U_2$, we have $\psi(x_0) = c$, so we may distinguish between two cases:

(i) $\alpha(Tx_0 + Fx_0) > d$ and (ii) $\alpha(Tx_0 + Fx_0) \leq d$.

If $\alpha(Tx_0 + Fx_0) > d$, the concavity of ψ and the condition (C_5) lead to

$$
\begin{aligned}
c = \psi(x_0) &= \psi(Tx_0 + H(t_0, x_0)) \\[2mm]
&= \psi(Tx_0 + t_0 Fx_0 + (1 - t_0)z_0) \\[2mm]
&\geq t_0 \psi(Tx_0 + Fx_0) + (1 - t_0)\psi(Tx_0 + z_0) \\[2mm]
&> c,
\end{aligned}
$$

which is a contradiction.

If $\alpha(Tx_0 + Fx_0) \leq d$, the convexity of α and the condition (C_4) lead to

$$
\begin{aligned}
\alpha(x_0) &= \alpha(Tx_0 + H(t_0, x_0)) \\[2mm]
&= \alpha(t_0(Tx_0 + Fx_0) + (1 - t_0)(Tx_0 + z_0)) \\[2mm]
&\leq t_0 \alpha(Tx_0 + Fx_0) + (1 - t_0)\alpha(Tx_0 + z_0) \\[2mm]
&\leq d,
\end{aligned}
$$

Thus, $x_0 \in \overline{\mathscr{P}}(\alpha, d; \beta, L_2; \psi, c) \cap \Omega$ and by the condition (C_4), we get $\psi(Tx_0 + Fx_0) > c$, which implies that $\psi(x_0) > c$ and again we come to a contradiction with $\psi(x_0) = c$.

According to the homotopy invariance of the index i_*, we have

$$i_* \left(T + F, U_2 \cap \Omega, \overline{\mathscr{P}}(\alpha, r_2; \beta, L_2)\right) = i_* \left(T + z_0, U_2 \cap \Omega, \overline{\mathscr{P}}(\alpha, r_2; \beta, L_2)\right).$$

Since $\overline{\mathscr{P}}(\alpha, r_2; \beta, L_2) \subset \overline{\mathscr{P}}_R$ is closed and convex, so it is a retract of $\overline{\mathscr{P}}_R$ with $\Omega \subset \overline{\mathscr{P}}(\alpha, r_2; \beta, L_2)$, by the preservation property of the index i_* in Lemma 2.2.4, we deduce that

$$i_* \left(T + z_0, U_2 \cap \Omega, \overline{\mathscr{P}}(\alpha, r_2; \beta, L_2)\right) = i_* \left(T + z_0, U_2 \cap \Omega, \overline{\mathscr{P}}_R\right). \quad (3.23)$$

Since $U_2 \subset \mathscr{P}_R$ and $T + z_0$ has no fixed point in $\overline{\mathscr{P}}_R \setminus U_2$ from the condition (C_3) we have $(I - T)^{-1} z_0 \in U_2$), by the excision property of the index i_* in Lemma 2.2.4, we deduce that

$$i_* \left(T + z_0, U_2 \cap \Omega, \overline{\mathscr{P}}_R\right) = i_* \left(T + z_0, \mathscr{P}_R \cap \Omega, \overline{\mathscr{P}}_R\right). \quad (3.24)$$

Then, our claim follow from (3.23), (3.24), the condition (C_3) and the normality property of the index i_*.

Claim 4. We show that

$$i_* \left(T + F, \left(\mathscr{P}(\alpha, r_2; \beta, L_2) \setminus (\overline{U_1 \cup U_2})\right) \cap \Omega, \overline{\mathscr{P}}(\alpha, r_2; \beta, L_2)\right) \neq 0.$$

From the additivity property of the index i_*, we have

$$i_* \left(T + F, \left(\overline{\mathscr{P}}(\alpha, r_2; \beta, L_2) \setminus (\overline{U_1 \cup U_2})\right) \cap \Omega, \overline{\mathscr{P}}(\alpha, r_2; \beta, L_2)\right)$$

$$= \quad i_* \left(T + F, \overline{\mathscr{P}}(\alpha, r_2; \beta, L_2) \cap \Omega, \overline{\mathscr{P}}(\alpha, r_2; \beta, L_2)\right)$$

$$- i_* \left(T + F, U_1 \cap \Omega, \overline{\mathscr{P}}(\alpha, r_2; \beta, L_2)\right)$$

$$- i_* \left(T + F, U_2 \cap \Omega, \overline{\mathscr{P}}(\alpha, r_2; \beta, L_2)\right)$$

$$= \quad 1 - 1 - 1 = -1.$$

Consequently, $T + F$ has at least three fixed points x_1, x_2 and x_3 in $\mathscr{P}(\alpha, r_2; \beta, L_2) \cap \Omega$ such that

$$x_1 \in U_1 \cap \Omega, \quad x_2 \in U_2 \cap \Omega$$

and

$$x_3 \in \left(\overline{\mathscr{P}}(\alpha,r_2;\beta,L_2) \setminus (\overline{U_1 \cup U_2}) \right) \cap \Omega.$$

This completes the proof.

3.3   Fixed Point Theorems on Open Sets of Cones for Special Mappings

In this section, we present fixed point theorems for some special mappings, including 1-set contractions. The first two results follow from Corollary 2.2.24.

Theorem 3.3.1 *Assume that $T : \Omega \subset \mathscr{P} \to E$ is an expansive mapping with constant $h > 1$ and $F : \overline{\mathscr{P}_r} \to E$ is compact. Assume that $F\left(\overline{\mathscr{P}_r}\right) \subset (I-T)(\Omega)$ and*

$$\|Fx + T0\| < (h-1)r, \ \text{for all } x \in \overline{\mathscr{P}_r}.$$

Then the sum operator $T + F$ has at least one solution in $\mathscr{P}_r \cap \Omega$.

Theorem 3.3.2 *Assume that $T : \Omega \subset \mathscr{P} \to E$ is an expansive mapping with constant $h > 2$ and $F : \overline{\mathscr{P}_r} \to E$ is a 1-set contraction such that $F(\overline{\mathscr{P}_r}) \subset (I-T)(\Omega)$. If $0 \in \Omega$, $r > \dfrac{\|T0\|}{h-2}$ and*

$$\|Fx\| \leq \|x\|, \ \text{for all } x \in \overline{\mathscr{P}_r}.$$

Then $T + F$ has at least one fixed point in $\mathscr{P}_r \cap \Omega$.

Proof 3.3.3 *We have the estimates:*

$$\begin{aligned}
\|Fx + T0\| &\leq \|Fx\| + \|T0\| \\
&\leq \|x\| + \|T0\| \\
&\leq r + \|T0\| \\
&\leq (h-1)r,
\end{aligned}$$

for $r > \dfrac{\|T0\|}{h-2}$. By Corollary 2.2.24, $i_*(T+F, \mathscr{P}_r \cap \Omega, \mathscr{P}) = 1$. As a consequence, $T + F$ has a fixed point in $\mathscr{P}_r \cap \Omega$.

More generally, we have

Theorem 3.3.4 *Assume that the mapping* $T : \Omega \subset \mathscr{P} \to E$ *is such that* $(I - T)$ *is Lipschitz invertible with constant* $0 < \gamma < 1$, $F : \overline{\mathscr{P}_r} \to E$ *is a* k-*set contraction with* $0 \le k < \gamma^{-1}$, *and* $F(\overline{\mathscr{P}_r}) \subset (I - T)(\Omega)$. *If* $0 \in \Omega$, $r > \dfrac{\gamma}{1 - \gamma}\|T0\|$ *and*

$$\|Fx\| \le \|x\|, \quad \textit{for all } x \in \overline{\mathscr{P}_r},$$

then $T + F$ *has at least one fixed point in* $\mathscr{P}_r \cap \Omega$.

Proof 3.3.5 *We have the estimates:*

$$
\begin{aligned}
\|Fx + T0\| &\le \|Fx\| + \|T0\| \\[2mm]
&\le r + \|T0\| \\[2mm]
&\le \frac{1}{\gamma} r.
\end{aligned}
$$

Our claim follows from Corollary 2.2.66 and the solvability property of the index i_*.

Recall that an operator L is said to be semi-closed if the identity perturbation $I - L$ is a closed operator.

Theorem 3.3.6 *Let* Ω *be a closed subset of* $\mathscr{P}$. *Assume that* $T : \Omega \to E$ *is a* 2-*expansive mapping and* $F : \overline{\mathscr{P}_r} \to E$ *is a* 1-*set contraction with* $tF(\overline{\mathscr{P}_r}) \subset (I - T)(\Omega)$, *for all* $t \in [0, 1]$. *Assume further that* $0 \in \Omega$, $\|T0\| < r$, *and* $T + F$ *is semi-closed and satisfies*

$$\|Fx\| + \|Tx\| < \|x\|, \quad \textit{for all } x \in \partial \mathscr{P}_r \cap \Omega.$$

Then $T + F$ *has a fixed point in* $\mathscr{P}_r \cap \Omega$.

Proof 3.3.7 *Assume that* $0 \notin \overline{(I - T - F)(\partial \mathscr{P}_r \cap \Omega)}$, *otherwise we are finished. Since* F *and* T *satisfy the assumptions of Proposition 2.2.35, then*

$$0 \in \overline{(I - T - F)(\mathscr{P}_r \cap \Omega)}.$$

So there exists a sequence $(x_n)_n$ in $\mathscr{P}_r \cap \Omega$ such that

$$x_n - Tx_n - Fx_n \to 0, \ as \ n \to +\infty.$$

Since $I - T - F$ is closed, then $0 \in (I - T - F)(\overline{\mathscr{P}_r \cap \Omega})$. Hence there exists $x \in \overline{\mathscr{P}_r \cap \Omega}$ such that $x = Tx + Fx$. Because $0 \notin (I - T - F)(\partial \mathscr{P}_r \cap \Omega)$, we obtain $x \in \mathscr{P}_r \cap \Omega$.

In what follows, we consider the case when the mapping T is a nonlinear expansive map. We first generalize [46, Theorem 3.1].

Corollary 3.3.8 *Let $\Omega \subset E$ be a nonempty, bounded, closed convex subset and suppose that $T, F : \Omega \to E$ are such that T is a nonlinear expansive mapping with ϕ nondecreasing and right continuous. Let F be a strict k-set contraction with $0 \le k < \omega$. Assume further that*

$$z \in F(\Omega) \ implies \ z + T(\Omega) \supset \Omega. \tag{3.25}$$

Then $F + T$ has a fixed point in Ω.

Proof 3.3.9 *We check (3.25) implies that*

$$F(\Omega) \subset (I - T)(\Omega). \tag{3.26}$$

Let $z \in F(\Omega)$ be arbitrary. By condition (3.25), $(z + T)(\Omega) \supset \Omega$. Consequently [46, Theorem 2.3] with $M = \Omega$ provides a (unique) fixed point $x \in \Omega$ of the mapping $z + T : \Omega \to E$, i.e. $z + Tx = x$. Hence $z \in (I - T)(\Omega)$, as claimed. Therefore (3.26) holds. Then $(I - T)^{-1}F(\Omega) \subset \Omega$. Since $(I - T)^{-1}F$ is a strict $\dfrac{k}{\omega}$-set contraction, then $(I - T)^{-1}F$ has a fixed point by Darbo's fixed point theorem.

Now, we are concerned with the sum of a nonlinear expansive operator and an α-convex ($\alpha > 1$), k-set contraction. Let $\alpha \in \mathbb{R}$ and a mapping $F : \mathscr{P} \to \mathscr{P}$. F is said to be α-convex if

$$F(tx) \ge t^\alpha Fx, \ \forall x \in \mathscr{P}, \ t \ge 1.$$

Theorem 3.3.10 *Let E be a Banach space, $\mathscr{P} \subset E$ a normal cone with normal constant N, and $\Omega = \mathscr{P} \cap B_r$. Let $\rho > 0$ be a real constant. Assume further that $T : \Omega \to E$ is a nonlinear expansive mapping, $F : \overline{\mathscr{P}_\rho} \to E$ is α-convex, k-set contraction with $0 \le k < \omega$, $tF(\overline{\mathscr{P}_\rho}) \subset (I - T)(\Omega)$, for all $t \in [0,1]$, and $F(\partial \mathscr{P}_\rho \cap \Omega) \subset \mathscr{P}$. Let $u_0 \in \mathscr{P}^*$ be such that $T(x - \lambda u_0) \in \mathscr{P}$, for all $(x, \lambda) \in \partial \mathscr{P}_\rho \times [0, +\infty)$. If*

$$\sup\{\|Tx + Fx\| : x \in \mathscr{P}, \|x\| = 1\} < \frac{1}{N} \qquad (3.27)$$

and

$$m = \inf\{\|Fx\| : x \in \mathscr{P}, \|x\| = 1\} > 0,$$

then the sum $T + F$ has at least one nonzero fixed point in $\mathscr{P}$.

Proof 3.3.11 *Let the positive constant L be such that $\rho \ge L >$*

$\max\left\{1, \left(\dfrac{N^3}{m}\right)^{\frac{1}{\alpha-1}}\right\}$. *We claim that*

$$Fx \not\ge x - Tx, \forall x \in \partial \mathscr{P}_1 \cap \Omega. \qquad (3.28)$$

$$Fx \not\le x - \lambda u_0, \forall (x, \lambda) \in \partial \mathscr{P}_L \times [0, 1]. \qquad (3.29)$$

Assume that there exists $x_1 \in \mathscr{P}$ with $\|x_1\| = 1$ such that $Fx_1 \ge x_1 - Tx_1$. By (3.27) we get

$$1 = \|x_1\| \le N\|Fx_1 + Tx_1\| < 1,$$

which is a contradiction, hence (3.28) holds. Assume that there exists $x_2 \in \mathscr{P}$ with $\|x_2\| = L$ and $\lambda_2 \ge 0$ such that $Fx_2 \le x_2 - \lambda_2 u_0$. Then

$$
\begin{aligned}
L &= \|x_2\| \\[4pt]
&\ge \frac{1}{N}\|x_2 - \lambda_2 u_0\| \\[4pt]
&\ge \frac{1}{N^2}\|Fx_2\| \\[4pt]
&= \frac{1}{N^2}\left\|F\left(L\frac{x_2}{L}\right)\right\|
\end{aligned}
$$

$$\geq \frac{L^\alpha}{N^3} \| F(\frac{x_2}{L}) \|$$

$$\geq \frac{L^\alpha}{N^3} m,$$

which contradicts $L \leq \left(\dfrac{N^3}{m} \right)^{\frac{1}{\alpha-1}}$ *and so (3.29) holds. Finally, from Proposition 2.2.47 and Proposition 2.2.26, the additivity property of the index yields*

$$i\left(F+T, P_{1,L} \cap \Omega, \mathscr{P}\right) = -1.$$

By the existence property of the index, the sum $F+T$ has at least one nonzero fixed point x^* in the open set $\mathscr{P}_{1,L} \cap \Omega$, i.e. $1 < \|x^*\| < L$.

In the following results, let K be a subset of a Banach space E with a cone $\mathscr{P}$ and the operators $T, F : K \to E$ be such that:

(D1) $T^{-1} : T(K) \to K$ exists and it is a bounded, δ-Lipschitz mapping,

(D2) $I - F : K \to \mathscr{P}$ is a k-set contraction with $0 \leq k < \delta^{-1}$.

Proposition 3.3.12 *Let U be a bounded open subset of $\mathscr{P}$ such that $0 \in U$ and $\overline{U} \subset T(K)$. Assume that $T, F : K \to E$ are two operators satisfying (D1) and (D2). Suppose that the following condition is satisfied:*

$$x \neq \lambda T x + F x \ \text{ for all } \ x \in T^{-1}(\overline{U}), \ Tx \in \partial U \ \text{ and } \ \lambda > 1. \quad (3.30)$$

Then $T + F$ has a fixed point x in K satisfying $Tx \in \overline{U}$.

Proof 3.3.13 *Since the operator $T^{-1} : T(K) \to K$ is a δ-Lipschitz. Then the operator A defined by*

$$
\begin{aligned}
A : \overline{U} &\to \ \mathscr{P} \\
y &\mapsto \ Ay = T^{-1}y - FT^{-1}y
\end{aligned}
$$

is well defined and it is strict set contraction. In fact, A is continuous and bounded; and for any bounded set B in U, we have

$$\alpha(A(B)) = \alpha((I-F)T^{-1}(B)) \leq k\,\alpha(T^{-1}(B)) \leq \delta k\,\alpha(B).$$

We show that Condition (3.30) implies the Leray-Schauder boundary one:

$$Ay \neq \lambda y \text{ for all } y \in \partial U \text{ and } \lambda > 1.$$

On the contrary, assume the existence of $\lambda_0 > 1$ and $y_0 \in \partial U$ such that

$$Ay_0 = \lambda_0 y_0.$$

From the condition $\overline{U} \subset T(K)$, there exists $x_0 \in K$ such that $x_0 = T^{-1}y_0$. Hence

$$x_0 - Fx_0 = \lambda_0 T x_0.$$

Which lead to a contradiction with Condition (3.30).
Assume that $Ay \neq y$ on ∂U, otherwise we are finished. By [30, Theorem 1.3.7], the fixed point index

$$i(A, U, \mathscr{P}) = 1.$$

The existence property of the fixed point index yields that the operator A has at last a fixed point $y \in U$. That is

$$T^{-1}y - FT^{-1}y = y.$$

Let $x := T^{-1}y$. Then $T + F$ has a fixed point x in K such that $Tx \in \overline{U}$.

4

Applications to ODEs

In this chapter, we consider some applications of some fixed point results in Chapters 2 and 3 in the area of ODEs. We prove applications for existence of periodic solutions, existence of classical solutions for some classes of systems ODEs, for some classes nth-order ODEs and existence of solutions for some classes BVPs for ODEs. Some of the results in this chapter can be found in [11, 12, 24–27].

4.1 Periodic Solutions for First Order ODEs

In this section, we will give an application of Proposition 2.2.8 for existence of periodic solutions for a class of first order ODEs. Let $\omega > 0$ be arbitrarily chosen. For a continuous ω-periodic function a, we denote

$$[a] = \frac{1}{\omega} \int_0^\omega a(\tau) d\tau.$$

We will start with the following useful lemma.

Lemma 4.1.1 *Let ε and ϕ be continuous ω-periodic functions and $[\varepsilon] \neq 0$. Then there exists a unique ω-periodic solution to the equation*

$$x' = -\varepsilon(t)x + \phi(t), \quad t \in \mathbb{R}, \tag{4.1}$$

for which the following representation

$$x(t) = \frac{e^{-[\varepsilon]\omega}}{1 - e^{-[\varepsilon]\omega}} \int_0^\omega e^{\int_t^{t+s} \varepsilon(\tau) d\tau} \phi(t+s) ds, \quad t \in \mathbb{R}, \tag{4.2}$$

holds. Also, there exists a unique ω-periodic solution to the equation

$$x' = \varepsilon(t)x - \phi(t), \quad t \in \mathbb{R}, \tag{4.3}$$

DOI: 10.1201/9781003381969-4

for which we have the following representation

$$x(t) = \frac{1}{1 - e^{-[\varepsilon]\omega}} \int_0^\omega e^{\int_{t+s}^t \varepsilon(\tau)d\tau} \phi(t+s)ds, \quad t \in \mathbb{R}. \qquad (4.4)$$

Proof 4.1.2 *Observe that*

$$\int_{t+\omega}^{t+\omega+s} \varepsilon(\tau)d\tau = \int_t^{t+s} \varepsilon(\tau)d\tau,$$

$$\int_0^{t+\omega} \varepsilon(\tau)d\tau = \int_0^\omega \varepsilon(\tau)d\tau + \int_\omega^{t+\omega} \varepsilon(\tau)d\tau$$

$$= \int_0^\omega \varepsilon(\tau)d\tau + \int_0^t \varepsilon(\tau)d\tau, \quad t,s \in \mathbb{R}.$$

*Assume that the equation (4.1) has two ω-periodic solutions x_1 and x_2.
Set*

$$v = x_1 - x_2.$$

Then v is an ω-periodic solution of the equation

$$v' = -\varepsilon(t)v, \quad t \in \mathbb{R}.$$

Hence,

$$v(t) = v(0)e^{-\int_0^t \varepsilon(\tau)d\tau}$$

and

$$v(0)e^{-\int_0^t \varepsilon(\tau)d\tau} = v(t)$$

$$= v(t+\omega)$$

$$= v(0)e^{-\int_0^{t+\omega} \varepsilon(\tau)d\tau}$$

$$= v(0)e^{-\int_0^t \varepsilon(\tau)d\tau}e^{-\int_0^\omega \varepsilon(\tau)d\tau}.$$

*Therefore $[\varepsilon] = 0$, which is a contradiction. Now, we will prove that
x, defined by (4.2), is an ω-periodic solution of the equation (4.1). We
have*

$$x(t+\omega) = \frac{e^{-[\varepsilon]\omega}}{1 - e^{-[\varepsilon]\omega}} \int_0^\omega e^{\int_{t+\omega}^{t+\omega+s} \varepsilon(\tau)d\tau} \phi(t+\omega+s)ds$$

$$= \frac{e^{-[\varepsilon]\omega}}{1-e^{-[\varepsilon]\omega}} \int_0^\omega e^{\int_t^{t+\omega} \varepsilon(\tau)d\tau} \phi(t+s)ds$$

$$= x(t), \quad t \in \mathbb{R},$$

and

$$x(t) = \frac{e^{-[\varepsilon]\omega}}{1-e^{-[\varepsilon]\omega}} e^{-\int_0^t \varepsilon(\tau)d\tau} \int_0^\omega e^{\int_0^{t+s} \varepsilon(\tau)d\tau} \phi(t+s)ds$$

$$= \frac{e^{-[\varepsilon]\omega}}{1-e^{-[\varepsilon]\omega}} e^{-\int_0^t \varepsilon(\tau)d\tau} \int_t^{t+\omega} e^{\int_0^y \varepsilon(\tau)d\tau} \phi(y)dy, \quad t \in \mathbb{R}.$$

From here,

$$x'(t) = \frac{e^{-[\varepsilon]\omega}}{1-e^{-[\varepsilon]\omega}} (-\varepsilon(t)) e^{-\int_0^t \varepsilon(\tau)d\tau} \int_t^{t+\omega} e^{\int_0^y \varepsilon(\tau)d\tau} \phi(y)dy$$

$$+ \frac{e^{-[\varepsilon]\omega}}{1-e^{-[\varepsilon]\omega}} e^{-\int_0^t \varepsilon(\tau)d\tau} \left(e^{\int_0^{t+\omega} \varepsilon(\tau)d\tau} \phi(t+\omega) - e^{\int_0^t \varepsilon(\tau)d\tau} \phi(t) \right)$$

$$= -\varepsilon(t)x(t) + \frac{e^{-[\varepsilon]\omega}}{1-e^{-[\varepsilon]\omega}} \left(e^{\int_0^\omega \varepsilon(\tau)d\tau} - 1 \right) \phi(t)$$

$$= -\varepsilon(t)x(t) + \phi(t), \quad t \in \mathbb{R}.$$

As in above, one can prove that the equation (4.3) has a unique ω-periodic solution given by (4.4). This completes the proof.

Let r, R, M, N, M_0, M_1, M_2, ω, m, and k be positive constants and a be a non-negative continuous ω-periodic function such that

$$R > r > \frac{e^{-[a]\omega}}{1-e^{-[a]\omega}} e^{M\omega} (M\omega r + M_0\omega + M_1\omega r^m + M_2\omega r^k), \quad (4.5)$$

$$a(t) \leq N, \quad t \in \mathbb{R}, \quad [a] \leq M, \quad a \not\equiv 0. \quad (4.6)$$

Consider the equation

$$x' = f(t,x), \quad t \in \mathbb{R}, \tag{4.7}$$

where

(A1) $f : \mathbb{R} \times [0, \infty) \to [0, \infty)$ is a continuous function that it is ω-periodic in t and

$$0 \le f(t,x) \le a_0(t) + a_1(t)x^m + a_2(t)x^k, \quad t \in \mathbb{R}, \quad x \in [0, \infty),$$

and a_i, $i = 0, 1, 2$, are positive continuous ω-periodic functions such that

$$a_i(t) \le N, \quad t \in \mathbb{R}, \quad [a_i] \le M_i, \quad i = 0, 1, 2.$$

Theorem 4.1.3 *Assume that* (A1), (4.5), *and* (4.6) *hold. Then the equation* (4.7) *has at least one positive continuous ω-periodic solution.*

Proof 4.1.4 *Consider the Banach space $E = \mathscr{C}(\omega)$ of all continuous ω-periodic functions, endowed with the maximum norm*

$$\|x\| = \max_{t \in [0,\omega]} |x(t)|,$$

the positive cone

$$\mathscr{P} = \{x \in E : x(t) \ge 0, \quad t \in \mathbb{R}\},$$

and the conical shell $\mathscr{P}_r$. Let

$$R_1 = R + \frac{e^{-[a]\omega}}{1 - e^{-[a]\omega}} e^{M\omega} \left(M\omega r + M_0\omega + M_1\omega r^m + M_2\omega r^k \right).$$

Define $\Omega = \mathscr{P}_{R_1}$. Take $\varepsilon > 0$ arbitrarily. For $x \in E$, define the operators

$$Tx(t) = (1+\varepsilon)x(t),$$

$$Fx(t) = -\varepsilon \frac{e^{-[a]\omega}}{1 - e^{-[a]\omega}} \int_0^\omega e^{\int_t^{t+s} a(\tau)d\tau} \left(a(t+s)x(t+s) \right)$$

$$+f(t+s,x(t+s))\Big)ds, \quad t \in \mathbb{R}.$$

Let $x \in E$ be a fixed point of $T+F$. Then

$$x(t) = Tx(t)+Fx(t)$$

$$= (1+\varepsilon)x(t)-\varepsilon\frac{e^{-[a]\omega}}{1-e^{-[a]\omega}}\int_0^\omega e^{\int_t^{t+s}a(\tau)d\tau}\Big(a(t+s)x(t+s)$$

$$+f(t+s,x(t+s))\Big)ds,$$

whereupon

$$x(t)=\frac{e^{-[a]\omega}}{1-e^{-[a]\omega}}\int_0^\omega e^{\int_t^{t+s}a(\tau)d\tau}\Big(a(t+s)x(t+s)+f(t+s,x(t+s))\Big)ds,$$

$t \in \mathbb{R}$. Hence and Lemma 4.1.1, we conclude that x satisfies (4.7). We will check all conditions of Proposition 2.2.8.

1. We have that $T : \Omega \to E$ and for any $x,y \in \Omega$, we get

$$\|Tx-Ty\| = (1+\varepsilon)\|x-y\|,$$

i.e., $T : \Omega \to E$ is $(1+\varepsilon)$-expansive operator.

2. Now we will show that $F : \overline{\mathscr{P}_r} \to E$ is a 0-set contraction.

$$|Fx(t)| = \left| -\varepsilon\frac{e^{-[a]\omega}}{1-e^{-[a]\omega}}\int_0^\omega e^{\int_t^{t+s}a(\tau)d\tau}\Big(a(t+s)x(t+s)\right.$$

$$\left.+f(t+s,x(t+s))\Big)ds\right|$$

$$\le \varepsilon\frac{e^{-[a]\omega}}{1-e^{-[a]\omega}}\int_0^\omega e^{\int_t^{t+s}a(\tau)d\tau}(a(t+s)x(t+s)$$

$$+f(t+s,x(t+s)))ds$$

$$\le \varepsilon \frac{e^{-[a]\omega}}{1 - e^{-[a]\omega}} e^{M\omega} \int_0^\omega \left(a(t+s)x(t+s) + a_0(t+s) \right.$$

$$\left. + a_1(t+s)(x(t+s))^m + a_2(t+s)(x(t+s))^k \right) ds$$

$$\le \varepsilon \frac{e^{-[a]\omega}}{1 - e^{-[a]\omega}} e^{M\omega} (M\omega r + M_0\omega + M_1\omega r^m + M_2\omega r^k),$$

$t \in [0, \omega]$. *Hence,*

$$\|Fx\| \le \varepsilon \frac{e^{-[a]\omega}}{1 - e^{-[a]\omega}} e^{M\omega} (M\omega r + M_0\omega + M_1\omega r^m + M_2\omega r^k).$$

This implies that $F(\overline{\mathscr{P}_r})$ *is uniformly bounded. Also, we have*

$$Fx(t) = -\varepsilon \frac{e^{-[a]\omega}}{1 - e^{-[a]\omega}} e^{-\int_0^t a(\tau)d\tau} \int_0^\omega e^{\int_0^{t+s} a(\tau)d\tau} \left(a(t+s)x(t+s) \right.$$

$$\left. + f(t+s, x(t+s)) \right) ds$$

$$= -\varepsilon \frac{e^{-[a]\omega}}{1 - e^{-[a]\omega}} e^{-\int_0^t a(\tau)d\tau} \int_t^{t+\omega} e^{\int_0^y a(\tau)d\tau} (a(y)x(y)$$

$$+ f(y, x(y))) dy, \quad t \in [0, \omega],$$

and

$$\frac{d}{dt} Fx(t) = -\varepsilon \frac{e^{-[a]\omega}}{1 - e^{-[a]\omega}} a(t) e^{-\int_0^t a(\tau)d\tau} \int_t^{t+\omega} e^{\int_0^y a(\tau)d\tau} (a(y)x(y)$$

$$+ f(y, x(y))) dy$$

$$+ \varepsilon \frac{e^{-[a]\omega}}{1 - e^{-[a]\omega}} e^{-\int_0^t a(\tau)d\tau}$$

$$\times \left(e^{\int_0^{t+\omega} a(\tau)d\tau}(a(t)x(t)+f(t,x(t))) \right.$$

$$\left. - e^{\int_0^{t} a(\tau)d\tau}(a(t)x(t)+f(t,x(t))) \right)$$

$$= -\varepsilon \frac{e^{-[a]\omega}}{1-e^{-[a]\omega}}a(t)\int_0^{\omega} e^{\int_t^{t+s} a(\tau)d\tau}(a(t+s)x(t+s)$$

$$+ f(t+s,x(t+s)))ds$$

$$+ \varepsilon \frac{e^{-[a]\omega}}{1-e^{-[a]\omega}}\left(e^{[a]\omega}-1\right)(a(t)x(t)+f(t,x(t)))$$

$$= -\varepsilon \frac{e^{-[a]\omega}}{1-e^{-[a]\omega}}a(t)\int_0^{\omega} e^{\int_t^{t+s} a(\tau)d\tau}(a(t+s)x(t+s)$$

$$+ f(t+s,x(t+s)))ds$$

$$+ \varepsilon(a(t)x(t)+f(t,x(t))), \quad t\in[0,\omega],$$

$$\left| \frac{d}{dt}Fx(t) \right|$$

$$= \left| -\varepsilon \frac{e^{-[a]\omega}}{1-e^{-[a]\omega}}a(t)\int_0^{\omega} e^{\int_t^{t+s} a(\tau)d\tau}(a(t+s)x(t+s) \right.$$

$$+ f(t+s,x(t+s)))ds$$

$$\left. + \varepsilon(a(t)x(t)+f(t,x(t))) \right|$$

$$\leq \varepsilon \frac{e^{-[a]\omega}}{1-e^{-[a]\omega}}a(t)\int_0^{\omega} e^{\int_t^{t+s} a(\tau)d\tau}\left(a(t+s)x(t+s) \right.$$

$$+a_0(t+s)+a_1(t+s)(x(t+s))^m+a_2(t+s)(x(t+s))^k\Big)ds$$

$$+\varepsilon\Big(a(t)x(t)+a_0(t)+a_1(t)(x(t))^m+a_2(t)(x(t))^k\Big)$$

$$\leq\varepsilon\frac{e^{-[a]\omega}}{1-e^{-[a]\omega}}Ne^{M\omega}(M\omega r+M_0\omega+M_1\omega r^m+M_2\omega r^k)$$

$$+\varepsilon(Nr+N+Nr^m+Nr^k),\quad t\in[0,\omega].$$

Consequently

$$\|(Fx)'\|\leq\varepsilon\frac{e^{-[a]\omega}}{1-e^{-[a]\omega}}Ne^{M\omega}(M\omega r+M_0\omega+M_1\omega r^m+M_2\omega r^k)$$

$$+\varepsilon(Nr+N+Nr^m+Nr^k).$$

Now, using the Ascoli-Arzelà theorem, we conclude that $F:$ $\overline{\mathscr{P}_r}\to E$ is a completely continuous mapping. Thus $F:\overline{\mathscr{P}_r}\to E$ is 0-set contraction.

3. Let $x\in\overline{\mathscr{P}_r}$ and $\mu\in[0,1]$ be arbitrarily chosen. We have

$$-\mu Fx(t) \;=\; \mu\varepsilon\frac{e^{-[a]\omega}}{1-e^{-[a]\omega}}\int_0^\omega e^{\int_t^{t+s}a(\tau)d\tau}(a(t+s)x(t+s)$$

$$+f(t+s,x(t+s)))ds$$

$$\geq\;0,\quad t\in[0,\omega].$$

Next,

$$-\mu Fx(t) \;=\; \mu\varepsilon\frac{e^{-[a]\omega}}{1-e^{-[a]\omega}}\int_0^\omega e^{\int_t^{t+s}a(\tau)d\tau}(a(t+s)x(t+s)$$

$$+f(t+s,x(t+s)))ds$$

$$\leq\; \mu\varepsilon\frac{e^{-[a]\omega}}{1-e^{-[a]\omega}}e^{M\omega}(M\omega r+M_1\omega r^m+M_2\omega r^k)$$

$$\leq \quad \varepsilon\left(\frac{e^{-[a]\omega}}{1-e^{-[a]\omega}}e^{M\omega}(M\omega r+M_1\omega r^m+M_2\omega r^k)\right)$$

$$= \quad \varepsilon R_1, \quad t\in[0,\omega].$$

Define

$$y(t) = \frac{-\mu Fx(t)}{\varepsilon}, \quad t\in[0,\omega].$$

We have $y\in\Omega$ and

$$(I-T)y(t) \quad = \quad -\varepsilon y(t)$$

$$= \quad \mu Fx(t), \quad t\in[0,\omega].$$

Consequently

$$\mu F(\overline{\mathscr{P}_r}) \subset (I-T)(\Omega)$$

for any $\mu\in[0,1]$.

4. *Assume that there are an $x\in\partial\mathscr{P}_r\cap\Omega$ and a $\mu\geq 1$ such that*

$$Fx = \mu(x-Tx).$$

Let $t_1\in[0,\omega]$ be such that $x(t_1)=r$. We have

$$\mu(x-Tx)(t_1) \quad = \quad -\mu\varepsilon x(t_1)$$

$$= \quad -\mu\varepsilon r$$

$$\leq \quad -\varepsilon r$$

and

$$Fx(t_1) \geq -\varepsilon\left(\frac{e^{-[a]\omega}}{1-e^{-[a]\omega}}e^{M\omega}(M\omega r+M_0\omega+M_1\omega r^m+M_2\omega r^k)\right).$$

Hence,

$$-\varepsilon r \geq -\varepsilon\left(\frac{e^{-[a]\omega}}{1-e^{-[a]\omega}}e^{M\omega}(M\omega r+M_0\omega+M_1\omega r^m+M_2\omega r^k)\right)$$

or

$$r \leq \left(\frac{e^{-[a]\omega}}{1 - e^{-[a]\omega}} e^{M\omega} (M\omega r + M_0 \omega + M_1 \omega r^m + M_2 \omega r^k) \right),$$

which is a contradiction.

5. Note that $0 \in \Omega$ and

$$\|T0\| = 0$$

$$< \varepsilon r.$$

Hence, and Proposition 2.2.13, it follows that the equation (4.7) has a positive continuous ω-periodic solution. This completes the proof.

4.2 Systems of ODEs

In this section, we will give an application of Theorem 3.1.11 for existence of classical solutions to class of systems ODEs. Let $r, R, m, k, M_j, N_j, j \in \{0, 1, 2\}$, be positive constants such that

$$0 < r < \frac{R}{2}, \quad N_0 + N_1 R^m + N_2 R^k < \frac{1}{6} R.$$

Suppose that $y \in \mathscr{C}([0, \infty))$ is chosen so that

$$\frac{r}{3} \leq y(t) \leq \frac{r}{2}, \quad t \in [0, \infty).$$

Set

$$w^0(t) = (w_1^0(t), w_2^0(t)) = (y(t), y(t)), \quad t \in [0, \infty).$$

Consider the IVP

$$
\begin{aligned}
x_1' &= f_1(t, x_1, x_2) \\
x_2' &= f_2(t, x_1, x_2), \quad t \in (0, \infty), \\
x_1(0) &= x_{10} \\
x_2(0) &= x_{20},
\end{aligned}
$$

(4.8)

where

(A2) $\dfrac{R}{2} \le x_{i0} \le \dfrac{2}{3}R, \quad i \in \{1,2\},$

(A3) $f_i \in \mathscr{C}([0,\infty) \times \mathbb{R} \times \mathbb{R}),$

$$0 \le f_i(t,x_1,x_2) \le a_{i0}(t) + a_{i1}(t)x_1^m + a_{i2}(t)x_2^k,$$

$a_{i0} \in \mathscr{C}([0,\infty)),$

$$0 \le a_{ij}(t) \le M_j, \ \int_0^\infty a_{ij}(s)ds \le N_j, \ j \in \{0,1,2\}, \ i \in \{1,2\}, \ t \in [0,\infty).$$

Let

$$E_1 \;=\; E_2 = \mathscr{C}([0,\infty))$$

be endowed with the supremum norm,

$$\mathscr{P}_1 \;=\; \mathscr{P}_2 = \{y \in \mathscr{C}([0,\infty)) : y(t) \ge 0\},$$

$$\mathscr{P}_r \;=\; (\mathscr{P}_1)_r \times (\mathscr{P}_2)_r,$$

$$\mathscr{P}_R \;=\; (\mathscr{P}_1)_R \times (\mathscr{P}_2)_R,$$

$$\Omega_1 \;=\; (\mathscr{P}_1)_{\frac{3}{2}R},$$

$$\Omega_2 \;=\; (\mathscr{P}_2)_{\frac{3}{2}R},$$

$$\Omega \;=\; \Omega_1 \times \Omega_2.$$

Theorem 4.2.1 *Suppose (A2) and (A3). Then the system (4.8) has at least one positive solution* $x = (x_1,x_2) \in \mathscr{C}\left([0,\infty), \mathbb{R}_+^2\right)$ *such that*

$$r \le \|x_i\| \le R, i = 1,2.$$

Proof 4.2.2 *Take $\varepsilon > 0$ arbitrarily. For $x = (x_1, x_2) \in E_1 \times E_2$ define the operators*

$$T_i x_i(t) = (1 + \varepsilon) x_i(t),$$

$$F_i x(t) = -\varepsilon \left(x_{i0} + \int_0^t f_i(s, x_1(s), x_2(s)) ds \right), \quad t \in [0, \infty), \quad i \in \{1, 2\}.$$

1. We have $T_i : \Omega_i \to E_i$ and for $x_i, y_i \in \Omega_i$

$$\|T_i x_i - T_i y_i\| \ge (1 + \varepsilon) \|x_i - y_i\|,$$

i.e., $T_i : \Omega_i \to E_i$ is $(1 + \varepsilon)$-expansive operator, $i \in \{1, 2\}$. Also, $T_i 0 = 0$, $i \in \{1, 2\}$.

2. We have $F_i : (\overline{\mathscr{P}_1})_R \times (\overline{\mathscr{P}_2})_R \to E_i$, $i = 1, 2$. For $x \in (\overline{\mathscr{P}_1})_R \times (\overline{\mathscr{P}_2})_R$, we get

$$|F_i x(t)|$$

$$\le \varepsilon \left(x_{i0} + \int_0^t \left(a_{i0}(s) + a_{i1}(s)(x_1(s))^m + a_{i2}(s)(x_2(s))^k \right) ds \right)$$

$$\le \varepsilon \left(R + \int_0^t a_{i0}(s) ds + R^m \int_0^t a_{i1}(s) ds + R^k \int_0^t a_{i2}(s) ds \right)$$

$$\le \varepsilon (R + N_0 + N_1 R^m + N_2 R^k), \quad t \in [0, \infty), \quad i \in \{1, 2\},$$

$$\left| \frac{d}{dt} F_i x(t) \right| = \varepsilon f_i(t, x_1(t), x_2(t))$$

$$\le \varepsilon \left(a_{i0}(t) + a_{i1}(t)(x_1(t))^m + a_{i2}(t)(x_2(t))^k \right)$$

$$\le \varepsilon \left(M_0 + M_1 R^m + M_2 R^k \right), \quad t \in [0, \infty), \quad i \in \{1, 2\}.$$

Therefore $F_i : (\overline{\mathscr{P}_1})_R \times (\overline{\mathscr{P}_2})_R \to E_i$ is 0-set contraction, $i \in \{1, 2\}$.

3. *Let* $x = (x_1, x_2) \in (\overline{\mathscr{P}_1})_R \times (\overline{\mathscr{P}_2})_R$ *and* $\lambda \in [0,1]$ *be arbitrarily chosen. Take*

$$v_i(t) = \lambda \left(x_{i0} + \int_0^t f_i(s, x_1(s), x_2(s)) ds \right), \quad t \in [0, \infty), \; i \in \{1, 2\}.$$

We have

$$
\begin{aligned}
v_i(t) \;&\leq\; x_{i0} + \int_0^t f_i(s, x_1(s), x_2(s)) ds \\[4pt]
&\leq\; \frac{2}{3} R + N_0 + N_1 R^m + N_2 R^k \\[4pt]
&<\; \frac{2}{3} R + \frac{1}{6} R \\[4pt]
&=\; \frac{5}{6} R, \quad t \in [0, \infty).
\end{aligned}
$$

Then $v_i \in \Omega_i$ *and* $\lambda F_i x = (I_i - T_i) v_i, \quad i \in \{1, 2\}$.

4. *Assume that there exists* $z = (z_1, z_2) \in \partial(\overline{\mathscr{P}_1})_r \times \partial(\overline{\mathscr{P}_2})_r$ *and* $\lambda \geq 0$ *such that* $z_i \in \Omega_i + \lambda w_i^0$ *and*

$$F_i z(t) = (I_i - T_i)(z_i(t) - \lambda w_i^0(t)), \quad t \in [0, \infty), \quad i \in \{1, 2\}.$$

Let $t_1 \in [0, \infty)$ *be such that*

$$z_i(t_1) \geq \frac{2}{3} r, \quad i \in \{1, 2\}.$$

Then

$$x_{i0} + \int_0^{t_1} f_i(s, z_1(s), z_2(s)) ds = z_i(t_1) - \lambda w_i^0(t_1). \tag{4.9}$$

Note that

$$
\begin{aligned}
\frac{2}{3} r - \lambda \frac{r}{2} \;&\leq\; z_i(t_1) - \lambda w_i^0(t_1) \\[6pt]
&\leq\; r - \lambda \frac{r}{3}
\end{aligned}
$$

$$= \left(1 - \frac{\lambda}{3}\right) r.$$

If $\lambda > 3$, then the equation (4.9) is not valid. Let $\lambda \in [0,3]$.
Then

$$\frac{R}{2} \leq x_{i0}$$

$$\leq x_{i0} + \int_0^{t_1} f_i(s, z_1(s), z_2(s))ds$$

$$= z_i(t_1) - \lambda w_i^0(t_1)$$

$$\leq \left(1 - \frac{\lambda}{3}\right) r$$

$$\leq r,$$

which is a contradiction.

5. *Assume that there exists $z = (z_1, z_2) \in \partial(\mathscr{P}_1)_R \times \partial(\mathscr{P}_2)_R$*
and $\lambda \geq 1$ such that

$$F_i z = \lambda(z_i - T_i z_i), \quad i \in \{1,2\}.$$

Then

$$\lambda z_i(t) = x_{i0} + \int_0^t f_i(s, z_1(s), z_2(s))ds, \quad t \in [0,\infty), \quad i \in \{1,2\}.$$

Let $t_1 \in [0,\infty)$ be such that $z_i(t_1) \geq \frac{5}{6}R$, $i \in \{1,2\}$. Then

$$\lambda \frac{5}{6} R \leq \lambda z_i(t_1)$$

$$= x_{i0} + \int_0^{t_1} f_i(s, z_1(s), z_2(s))ds$$

$$\leq \frac{2}{3}R + N_0 + N_1 R^m + N_2 R^k, \quad i \in \{1,2\},$$

or

$$\lambda \leq \frac{\frac{2}{3}R + N_0 + N_1 R^m + N_2 R^k}{\frac{5}{6}R} < 1,$$

which is a contradiction.

Hence, and Theorem 3.1.11, it follows that the system (4.8) has at least
one positive solution in $\mathscr{C}\left([0,\infty), \mathbb{R}_+^2\right)$.

4.3 Existence of Positive Solutions for a Class of BVPs in Banach Spaces

In this section, we will give an application of Proposition 2.2.21 for existence of positive solutions for a class of BVPs in some Banach spaces. Throughout this section, $(E, \|.\|)$ denotes a Banach space and $\mathscr{P}$ is a cone in E. Being given a positive real parameter k and $f \colon \mathbb{R}^+ \times \mathscr{P} \times E \to \mathscr{P}$ a continuous function, we are interested in the study of the existence of bounded positive solutions to the second-order boundary value problem:

$$\begin{cases} -x''(t) + k^2 x(t) = m(t) f(t, x(t), x'(t)), & t \in (0, +\infty). \\ x(0) = 0, \quad \lim_{t \to +\infty} x(t) = 0, \end{cases} \tag{4.10}$$

where the coefficient $m \in \mathscr{C}((0, +\infty), \mathbb{R}^+) \cap L^1((0, +\infty), \mathbb{R}^+)$ may be singular at $t = 0$ and it does not vanish identically on any subinterval of $(0, +\infty)$.

Also, we consider the problem

$$\begin{cases} -y'' + cy' + \lambda y = m(t) g(t, y(t), y'(t)), & t \in (0, +\infty) \\ y(0) = \lim_{t \to +\infty} y(t) = 0, \end{cases} \tag{4.11}$$

where c, λ are positive constants and $g \colon \mathbb{R}^+ \times \mathscr{P} \times E \to \mathscr{P}$ is a continuous function. Letting $k = \sqrt{\lambda + \dfrac{c^2}{4}}$ and $x(t) = y(t) e^{-\frac{c}{2}t}$, the problem (4.11) leads to the problem (4.10) for the new unknown x and modified nonlinear term

$$f(t, x(t), x'(t)) = e^{\frac{-c}{2}t} g\left(t, e^{\frac{c}{2}t} x(t), e^{\frac{c}{2}t} x'(t) + \frac{c}{2} e^{\frac{c}{2}t} x(t)\right).$$

The following lemmas are concerned with the linear problem associated to (4.10). They provide useful estimates of the kernel G and their proofs are omitted.

Lemma 4.3.1 *Let v be a function such that $v \in \mathscr{C}((0, +\infty), E)$ and $\displaystyle\int_0^{+\infty} \|v(t)\| dt$ exists. Then the problem*

$$\begin{cases} -x''(t) + k^2 x(t) = v(t), & t \in (0, +\infty), \\ x(0) = 0, \quad \lim_{t \to +\infty} x(t) = 0 \end{cases}$$

has a unique solution x given by

$$x(t) = \int_0^{+\infty} G(t,s)v(s)\,ds,$$

where G is the Green function of the problem, namely

$$G(t,s) = \frac{1}{2k}\begin{cases} e^{-ks}(e^{kt} - e^{-kt}), & if \quad 0 \le t \le s < \infty, \\ e^{-kt}(e^{ks} - e^{-ks}), & if \quad 0 \le s \le t < \infty. \end{cases} \qquad (4.12)$$

Throughout this section, $0 < \gamma < \delta$ will denote some fixed numbers. The interval $[\gamma, \delta]$ will play a key role in estimating the solutions of the problem (4.10). Let

$$\begin{aligned}
\Lambda_0 &= \min(e^{-k\delta}, e^{k\gamma} - e^{-k\gamma}), \\
\Lambda_1 &= \min\left(\frac{1-k}{1+k}e^{-k\delta}, e^{k\gamma} + \frac{k-1}{k+1}e^{-k\gamma}\right), \qquad (4.13) \\
\Lambda_2 &= \frac{k}{k+1}e^{-k\delta}.
\end{aligned}$$

Obviously, these constants are less than 1. Some fundamental properties of the kernel G are given hereafter. The proofs are omitted.

Lemma 4.3.2 *The Green's function G satisfies the following estimates:*

(a) $G(t,s) \ge 0, \quad \forall t,s \in \mathbb{R}^+.$

(b) $G(t,s) \le G(s,s) \le \dfrac{1}{2k}, \quad \forall t,s \in \mathbb{R}^+.$

(c) $G(t,s)e^{-\mu t} \le G(s,s)e^{-ks}, \quad \forall t,s \in \mathbb{R}^+, \forall \mu \ge k.$

(d) $G(t,s) \ge \Lambda_0 G(s,s)e^{-ks}, \quad \forall t \in [\gamma,\delta], \forall s \in \mathbb{R}^+.$

Remark 4.3.3 *The problem (4.11) is equivalent to the integral equation:*

$$y(x) = \int_0^{+\infty} e^{\frac{c}{2}(x-s)}m(s)G(x,s)f(s,y(s),y'(s))\,ds. \qquad (4.14)$$

The boundary conditions $y(0) = y(+\infty) = 0$ *follow from* $G(0,s) = 0, \forall s \ge 0,$ *and* $\lim_{x \to +\infty} e^{\frac{c}{2}x}G(x,s) = 0, \forall s \ge 0,$ *since* $k > \dfrac{c}{2},$ *where G is given by (4.12).*

We begin by a new representation formula for the measure of noncompact- ness in the space X.

Let $p\colon \mathbb{R}^+ \to (0, +\infty)$ be a continuous function. Denote by X the space consisting of all weighted functions y, continuously differentiable on $\mathbb{R}^+$ which satisfy

$$\sup_{x\in\mathbb{R}^+} \left([\|y(x)\| + \|y'(x)\|]p(x)\right) < \infty.$$

Equipped with a Bielecki's type norm $\|y\|_p = \sup_{x\in\mathbb{R}^+} ([\|y(x)\| + \|y'(x)\|]p(x))$, it is a Banach space.

In the following, we develop a new non-compactness result in order to use it to show that an operator is ℓ-set contraction in the space X.

Lemma 4.3.4 *Let $B \subset X$ be such that the functions belonging in the sets*

$$pB = \{z \mid z(t) = y(t)\,p(t),\ y \in B\},$$
$$pB' = \{z \mid z(t) = y'(t)\,p(t),\ y \in B\},$$

are almost equicontinuous on $\mathbb{R}^+$ and B is a bounded set in the sense of the norm

$$\|y\|_q = \sup_{x\in\mathbb{R}^+} ([\|y(t)\| + \|y'(t)\|]q(t)),$$

where the function q is positive, continuous on $\mathbb{R}^+$ and satisfies

$$\lim_{t\to+\infty} \frac{p(t)}{q(t)} = 0.$$

Then

$$\alpha_X(B) = \max\left(\sup_{t\in\mathbb{R}^+} \alpha_E(B(t)p(t)),\ \sup_{t\in\mathbb{R}^+} \alpha_E((B)'(t)p(t))\right), \quad (4.15)$$

where $B(t) = \{u(t) \mid u \in B\}$ for $t \in \mathbb{R}^+$.

Proof 4.3.5 *Let $B \subset X$ be bounded in the sense of the norm*

$$\|y\|_q = \sup_{t\in\mathbb{R}^+} ([\|y(t)\| + \|y'(t)\|]q(t)).$$

Thus there exists $r > 0$ such that $\|y\|_q \leq r$ for all $y \in B$. Since the function q is positive on $\mathbb{R}^+$ and satisfies $\lim_{t \to +\infty} \dfrac{p(t)}{q(t)} = 0$, for any $\varepsilon > 0$, there exists $T > 0$ such that

$$\|y(t_1)\,p(t_1) - y(t_2)\,p(t_2)\|$$

$$\leq \frac{p(t_1)}{q(t_1)}\,\|y(t_1)\|\,q(t_1) + \frac{p(t_2)}{q(t_2)}\,\|y(t_2)\|\,q(t_2)$$

$$\leq \frac{p(t_1)}{q(t_1)}\,\left(\|y(t_1)\| + \|y'(t_1)\|\right)\,q(t_1)$$

$$\qquad\qquad (4.16)$$

$$+\frac{p(t_2)}{q(t_2)}\,\left(\|y(t_2)\| + \|y'(t_2)\|\right)\,q(t_2) < \varepsilon,$$

and

$$\|y'(t_1)\,p(t_1) - y'(t_2)\,p(t_2)\|$$

$$\leq \frac{p(t_1)}{q(t_1)}\,\|y'(t_1)\|\,q(t_1) + \frac{p(t_2)}{q(t_2)}\,\|y'(t_2)\|\,q(t_2)$$

$$\leq \frac{p(t_1)}{q(t_1)}\,\left(\|y(t_1)\| + \|y'(t_1)\|\right)\,q(t_1)$$

$$\qquad\qquad (4.17)$$

$$+\frac{p(t_2)}{q(t_2)}\,\left(\|y(t_2)\| + \|y'(t_2)\|\right)\,q(t_2) < \varepsilon,$$

uniformly with respect to $y \in B$ as $t_1, t_2 \geq T$.
   We first claim that

$$\alpha_X(B) \leq \max\left(\sup_{t \in \mathbb{R}^+} \alpha_E\left(B(t)p(t)\right),\ \sup_{t \in \mathbb{R}^+} \alpha_E\left(B'(t)p(t)\right)\right).$$

Denote by $B|_{[0,T]}$ and $B'|_{[0,T]}$ the restriction of B and B' on $[0,T]$. Since the sets $B(t)p(t)$ and $B'(t)p(t)$ are equi-continuous on $[0,T]$, we have *that*

$$\alpha_{\mathscr{C}^1}(Bp|_{[0,T]}) = \max\left(\sup_{t\in[0,T]}\alpha_E(B(t)p(t)),\ \sup_{t\in[0,T]}\alpha_E\left(B'(t)p(t)\right)\right)$$

$$\leq \max\left(\sup_{t\in\mathbb{R}^+}\alpha_E(B(t)p(t)),\ \sup_{t\in\mathbb{R}^+}\alpha_E\left(B'(t)p(t)\right)\right),$$

where $Bp|_{[0,T]} = \{y(t)p(t) : y \in B, t \in [0,T]\}$.
By the definition of the MNC $\alpha_{\mathscr{C}^1}$, *there exists* $\{B_i\}_{i=1}^n$ *such that* $B = \cup_{i=1}^n B_i$ *and for* $i = 1,...,n$,

$$\text{diam}_{\mathscr{C}^1}(B_i p|_{[0,T]}) \leq \max\left(\sup_{t\in\mathbb{R}^+}\alpha_E(B(t)p(t)),\ \sup_{t\in\mathbb{R}^+}\alpha_E\left(B'(t)p(t)\right)\right) + \varepsilon,$$

$$(4.18)$$

where $\text{diam}_{\mathscr{C}^1}(.)$ *denotes the diameter of the bounded subsets of* $\mathscr{C}^1([0,T],E)$.
Furthermore, for $i = 1,...,n$, *fixed, for all* y_1, $y_2 \in B_i$ *and* $t \geq T$, *we deduce from* (4.16)–(4.18), *for* $i = 1,...,n$ *the following estimates:*

$$\|(y_1(t) - y_2(t))\|\,p(t)$$

$$\leq \left(\|(y_1(t) + y_1'(t))\| + \|(y_1'(t) - y_2'(t))\| + \|(y_2(t) + y_2'(t))\|\right)p(t)$$

$$\leq \left(\|(y_1(t) + y_1'(t))\|q(t) + \|(y_2(t) + y_2'(t))\|q(t)\right)\frac{p(t)}{q(t)}$$

$$+ \|(y_1'(t) - y_2'(t))\|\,p(t)$$

$$\leq \varepsilon + \|y_1'(t)p(t) - y_1'(T)p(T)\| + \|y_1'(T)p(T) - y_2'(T)p(T)\|$$

$$+ \|y_2'(T)p(T) - y_2'(t)p(t)\|$$

$$\leq 2\varepsilon + \max\left(\sup_{t\in\mathbb{R}^+}\alpha_E(B(t)p(t)),\ \sup_{t\in\mathbb{R}^+}\alpha_E\left(B'(t)p(t)\right)\right) + \varepsilon + \varepsilon,$$

$$(4.19)$$

and

$$\|(y_1'(t) - y_2'(t))\| \, p(t)$$

$$\leq \ \left(\|(y_1(t) + y_1'(t))\| + \|(y_1(t) - y_2(t))\| + \|(y_2(t) + y_2'(t))\| \right) p(t)$$

$$\leq \ \left(\|(y_1(t) + y_1'(t))\| q(t) + \|(y_2(t) + y_2'(t))\| q(t) \right) \frac{p(t)}{q(t)}$$

$$+ \|(y_1(t) - y_2(t))\| p(t)$$

$$\leq \ \varepsilon + \|y_1(t) p(t) - y_1(T) p(T)\| + \|y_1(T) p(T) - y_2(T) p(T)\|$$

$$+ \|y_2(T) p(T) - y_2(t) p(t)\|$$

$$\leq \ 2\varepsilon + \max\left(\sup_{t \in \mathbb{R}^+} \alpha_E(B(t) p(t)), \ \sup_{t \in \mathbb{R}^+} \alpha_E(B'(t) p(t)) \right) + \varepsilon + \varepsilon. \tag{4.20}$$

Therefore (4.18), (4.19), and (4.20) guarantee that

$$\mathrm{diam}_X(B_i) \leq \max\left(\sup_{t \in \mathbb{R}^+} \alpha_E(B(t) p(t)), \ \sup_{t \in \mathbb{R}^+} \alpha_E(B'(t) p(t)) \right) + 4\varepsilon$$

Noting that $B = \cup_{i=1}^n B_i$, we infer

$$\alpha_X(B) \leq \max\left(\sup_{t \in \mathbb{R}^+} \alpha_E(B(t) p(t)), \ \sup_{t \in \mathbb{R}^+} \alpha_E(B'(t) p(t)) \right) + 4\varepsilon,$$

and ε being arbitrary, we deduce that

$$\alpha_X(B) \leq \max\left(\sup_{t \in \mathbb{R}^+} \alpha_E(B(t) p(t)), \ \sup_{t \in \mathbb{R}^+} \alpha_E(B'(t) p(t)) \right).$$

Conversely, we prove that

$$\max\left(\sup_{t \in \mathbb{R}^+} \alpha_E(B(t) p(t)), \ \sup_{t \in \mathbb{R}^+} \alpha_E(B'(t) p(t)) \right) \leq \alpha_X(B).$$

Given $\varepsilon > 0$, there exists $\{B_i\}_{i=1}^{n}$ such that $B = \cup_{i=1}^{n} B_i$ and $\text{diam}_X(B_i) \leq \alpha_X(B) + \varepsilon$. Thus, for fixed i, for every $t \in \mathbb{R}^+$ and all $y_1, y_2 \in B_i$, we have

$$\|(y_1(t) - y_2(t))\| p(t)\| \leq \|y_1 - y_2\|_\omega < \alpha_X(B) + \varepsilon,$$

and

$$\|(y_1'(t) - y_2'(t))\| p(t)\| \leq \|y_1 - y_2\|_\omega < \alpha_X(B) + \varepsilon.$$

Since $B(t) = \cup_{i=1}^{n} B_i(t)$, we have $\alpha_E(B(t)p(t)) \leq \alpha_X(B) + \varepsilon$. Now ε being arbitrary, we deduce that $\sup\limits_{t \in \mathbb{R}^+} \alpha_E(B(t)p(t)) \leq \alpha_X(B)$. In accordance with $B'(t) = \cup_{i=1}^{n} B_i'(t)$, we get $\sup\limits_{t \in \mathbb{R}^+} \alpha_E(B'(t)p(t)) \leq \alpha_X(B)$, where $H'(t) = \{y'(t) | \ y \in H\}$, $t \in \mathbb{R}^+$, whence the reversed inequality and then the desired result.

Let $\omega > 0$ be a given real parameter. Consider the Banach space with weight function $e^{-\omega t}$

$$X = \left\{ x \in \mathscr{C}^1(\mathbb{R}^+, E) : \sup\limits_{t \in \mathbb{R}^+} \left((\|x(t)\| + \|x'(t)\|) e^{-\omega t} \right) < \infty \right\},$$

endowed with the norm

$$\|x\|_\omega = \sup\limits_{t \in \mathbb{R}^+} \left((\|x(t)\| + \|x'(t)\|) e^{-\omega t} \right).$$

Define the cone

$$\mathscr{K} = \{ x \in X : \quad x \geq 0 \quad \text{on} \quad \mathbb{R}^+ \}.$$

Take $\varepsilon \in (0,1)$ and $p,q > 0$ arbitrarily. Let A, B_1, B_2, B_3, R be positive constants such that

$$G(t,s) + |G_t(t,s)| \leq A, \quad t,s \in [0,\infty),$$

and

$$A(B_1 + B_2 R^p + B_3 R^q) < R.$$

Define the conical shell

$$\mathscr{K}_R = \{ x \in \mathscr{K} : \|x\|_\omega < R \}.$$

With $\widetilde{\mathscr{K}_R}$ we will denote the set of all equi-continuous families of $\mathscr{K}_R$ with respect to the norm $\| \cdot \|_\omega$. Assume that

(A4) $f \in \mathscr{C}(\mathbb{R}^+ \times \mathscr{P} \times E, \mathscr{P})$ be such that

$$\|f(t,x,y)\| \leq a_0(t) + a_1(t)\|x\|^p + a_2(t)\|y\|^q$$

for any $(t,x,y) \in \mathbb{R}^+ \times \mathscr{P} \times E$, where $a_i \in \mathscr{C}(\mathbb{R}^+, \mathbb{R}^+)$, $i \in \{0,1,2\}$,

$$\int_0^\infty m(s)a_0(s)ds \leq B_1, \quad \int_0^\infty m(s)a_1(s)e^{\omega sp}ds \leq B_2,$$

$$\int_0^\infty m(s)a_2(s)e^{\omega sq}ds \leq B_3.$$

Theorem 4.3.6 *Assume* (A4). *Then the problem* (4.10) *has at least one positive solution* x *in* $\mathscr{K}$ *such that*

$$\sup_{t \in \mathbb{R}^+} \left(\left(\|x(t)\| + \|x'(t)\| \right) e^{-\omega t} \right) < R.$$

Proof 4.3.7 *For* $x \in X$ *define the operators*

$$Tx(t) = (1+\varepsilon)x(t),$$

$$Fx(t) = -\varepsilon \int_0^\infty G(t,s)m(s)f(s,x(s),x'(s))ds, \quad t \in (0,\infty).$$

1. *Note that* $T : \widetilde{\mathscr{K}_R} \to X$ *is an* $(1+\varepsilon)$-*expansive operator.*

2. *Now we will prove that the operator* $F : \widetilde{\mathscr{K}_R} \to X$ *is continuous. From the assumption* (A4), *we can show that*

$$\sup_{t \in \mathbb{R}^+} e^{-\omega t} \left(\|Fx(t)\| + \|(Fx)'(t)\| \right) < \infty,$$

which imply that

$$F(\widetilde{\mathscr{K}_R}) \subset X.$$

Let $\{x_n\}_{n \in \mathbb{N}}$, $\{x\} \subset \widetilde{\mathscr{K}_R}$ *with* $\|x_n - x\|_w \to 0$, *as* $n \to \infty$. *Hence,* $\{x_n\}_{n \in \mathbb{N}}$ *is bounded in* $\widetilde{\mathscr{K}_R}$. *Then there exists a positive constant* r *such that* $\max\{\|x_n\|_\omega, n \in \mathbb{N}, \|x\|_\omega\} \leq r$. *We have*

$$\int_0^\infty e^{-\omega t} G(t,s)m(s)\|f(s,x_n(s),x_n'(s)) - f(s,x(s),x'(s))\|ds$$

$$\leq \int_0^\infty e^{-\omega t} G(t,s) m(s) \Big(\|f(s,x_n(s),x_n'(s))\|$$

$$+ \|f(s,x(s),x'(s))\| \Big) ds$$

$$\leq \int_0^\infty e^{-\omega t} G(t,s) m(s) \Big(2a_0(s) + a_1(s) \left(\|x_n(s)\|^p + \|x(s)\|^p \right)$$

$$+ a_2(s) \left(\|x_n'(s)\|^q + \|x'(s)\|^q \right) \Big) ds$$

$$\leq 2B_1 A + AB_2 \left(\|x_n\|_\omega^p + \|x\|_\omega^p \right) + AB_3 \left(\|x_n\|_\omega^q + \|x\|_\omega^q \right)$$

$$\leq 2A \left(B_1 + B_2 r^p + B_3 r^q \right), \quad t \in (0,\infty),$$

and

$$\int_0^\infty e^{-\omega t} |G_t(t,s)| m(s) \|f(s,x_n(s),x_n'(s)) - f(s,x(s),x'(s))\| ds$$

$$\leq \int_0^\infty e^{-\omega t} |G_t(t,s)| m(s) \Big(\|f(s,x_n(s),x_n'(s))\|$$

$$+ \|f(s,x(s),x'(s))\| \Big) ds$$

$$\leq \int_0^\infty e^{-\omega t} |G_t(t,s)| m(s) \Big(2a_0(s) + a_1(s) \left(\|x_n(s)\|^p + \|x(s)\|^p \right)$$

$$+ a_2(s) \left(\|x_n'(s)\|^q + \|x'(s)\|^q \right) \Big) ds$$

$$\leq 2B_1 A + AB_2 \left(\|x_n\|_\omega^p + \|x\|_\omega^p \right) + AB_3 \left(\|x_n\|_\omega^q + \|x\|_\omega^q \right)$$

$$\leq 2A \left(B_1 + B_2 r^p + B_3 r^q \right), \quad t \in (0,\infty).$$

Thus, the Lebesgue dominated convergence theorem both with the continuity of f imply

$$\sup_{t \in \mathbb{R}^+} \left(e^{-\omega t} \|(Fx_n)(t) - (Fx)(t)\| \right) \to 0, \quad as \quad n \to \infty,$$

and

$$\sup_{t \in \mathbb{R}^+} \left(e^{-\omega t} \|(Fx_n)'(t) - (Fx)'(t)\| \right) \to 0, \quad as \quad n \to \infty.$$

As a result,

$$\|Fx_n - Fx\|_\omega \to 0, \quad as \quad n \to \infty,$$

i.e., the operator F is continuous.

3. We have $F : \overline{\widetilde{\mathcal{K}_R}} \to X$ *and for* $x \in \overline{\widetilde{\mathcal{K}_R}}$, *we get*

$$\left(\|Fx(t)\| + \|(Fx)'(t)\| \right) e^{-\omega t}$$

$$\leq \varepsilon e^{-\omega t} \int_0^\infty \left(G(t,s) + |G_t(t,s)| \right) m(s) \|f(s,x(s),x'(s))\| ds$$

$$\leq \varepsilon e^{-\omega t} A \int_0^\infty m(s) \left(a_0(s) + a_1(s)\|x(s)\|^p + a_2(s)\|x'(s)\|^q \right) ds$$

$$\leq \varepsilon e^{-\omega t} A \int_0^\infty m(s) \left(a_0(s) + a_1(s)e^{\omega p s} R^p + a_2(s)e^{\omega q s} R^q \right) ds$$

$$\leq \varepsilon e^{-\omega t} A \left(B_1 + B_2 R^p + B_3 R^q \right)$$

$$\leq \varepsilon A \left(B_1 + B_2 R^p + B_3 R^q \right)$$

$$< R, \quad t \in (0, \infty).$$

Hence,

$$\|Fx\|_\omega \leq R.$$

Therefore, $F : \overline{\widetilde{\mathcal{K}_R}} \to X$ *is a 0-set contraction.*

4. Let $y \in \widetilde{\mathscr{K}_R}$ be arbitrarily chosen. Set

$$z(t) = \int_0^\infty G(t,s)m(s)f(s,y(s),y'(s))ds, \quad t \in (0,\infty).$$

We have that $z \in \mathscr{K}$ and using the above computations, we have

$$\left(\|z(t)\| + \|z'(t)\| \right) e^{-\omega t}$$

$$\leq e^{-\omega t} \int_0^\infty \left(G(t,s) + |G_t(t,s)| \right) m(s) \|f(s,y(s),y'(s))\| ds$$

$$\leq A \left(B_1 + B_2 R^p + B_3 R^q \right)$$

$$\leq A \left(B_1 + B_2 R^p + B_3 R^q \right)$$

$$< R,$$

so, $\|z\|_\omega < R$. Therefore $z \in \widetilde{\mathscr{K}_R}$. Also,

$$(I - T)z(t) = -\varepsilon z(t)$$

$$= -\varepsilon \int_0^\infty G(t,s)m(s)f(s,y(s),y'(s))ds$$

$$= Fy(t), \quad t \in (0,\infty).$$

Thus

$$F\left(\widetilde{\mathscr{K}_R} \right) \subset (I - T)(\widetilde{\mathscr{K}_R}).$$

5. Note that $0 \in \widetilde{\mathscr{K}_R}$ and for any $x \in \partial \widetilde{\mathscr{K}_R}$ we have

$$\|Fx + T0\|_\omega = \|Fx\|_\omega$$

$$\leq \varepsilon A \left(B_1 + B_2 R^p + B_3 R^q \right)$$

$$\leq \quad \varepsilon R$$

$$= \quad \varepsilon \|x\|_{\omega}.$$

Assume that there exists $x \in \partial \widetilde{\mathscr{H}_R}$ such that

$$Fx + Tx = x.$$

Then

$$\left(\|x(t)\| + \|x'(t)\| \right) e^{-\omega t}$$

$$\geq \quad (1+\varepsilon) e^{-\omega t} \left(\|x(t)\| + \|x'(t)\| \right)$$

$$-\varepsilon e^{-\omega t} \int_0^{\infty} \left(G(t,s) + |G_t(t,s)| \right) m(s) \|f(s,x(s),x'(s))\| ds$$

$$\geq \quad (1+\varepsilon) e^{-\omega t} \left(\|x(t)\| + \|x'(t)\| \right)$$

$$-\varepsilon A \left(B_1 + B_2 R^p + B_3 R^q \right), \quad t \in (0,\infty),$$

or

$$\varepsilon A \left(B_1 + B_2 R^p + B_3 R^q \right) \geq \varepsilon e^{-\omega t} \left(\|x(t)\| + \|x'(t)\| \right), \quad t \in (0,\infty),$$

whereupon
$$A \left(B_1 + B_2 R^p + B_3 R^q \right) \geq R.$$

This is a contradiction.

By 1, 2, 3, 4, 5 and Proposition 2.2.21, we conclude that the operator $T + F$ has a fixed point $x \in \widetilde{\mathscr{H}_R}$, which is a solution of the problem (4.10). This completes the proof.

Remark 4.3.8 *In the general case when we consider $\mathscr{H}_R$ and do not consider $\widetilde{\mathscr{H}_R}$, additional conditions on the nonlinearity f are needed to show that the operator F is a l-set contraction.*

To overcome the problem pointed out in the remark we consider the conditions (A5) and (A6).

(A5) For every $r > 0$ and all subinterval $[a,b] \subset \mathbb{R}^+$, the nonlinearity f is uniformly continuous on $[a,b] \times B_E(0,r) \times B_E(0,r)$, where $B_E(0,r) = \{x \in E : \|x\| \le r\}$.

(A6) There exist a positive functions $l_1, l_2 \in L^1(\mathbb{R}^+)$ such that

$$\alpha(f(t, B_1, B_2)) \le l_1(t)\alpha(B_1) + l_2(t)\alpha(B_2), \quad t \in \mathbb{R}^+,$$

for every bounded subsets $B_1, B_2 \subset E$, where

$$A \int_0^{+\infty} m(t)(l_1(t) + l_2(t))dt < 1.$$

And we present the following theorem.

Theorem 4.3.9 *Assume* $(A4) - (A6)$. *Then the problem* (4.10) *has at least one positive solution* x *in* $\mathcal{K}$ *such that*

$$\sup_{t \in \mathbb{R}^+} \left(\left(\|x(t)\| + \|x'(t)\| \right) e^{-\omega t} \right) < R.$$

Proof 4.3.10 *The proof of this theorem is similar to that of Theorem 4.3.6, we will only show how the operator F is a ℓ-set contraction with $\ell < \varepsilon$ under conditions* $(A5)$ *and* $(A6)$. *Firstly, using Lemma 4.3.4 for $p(t) = e^{-\omega t}$ and $q(t) = e^{-\mu t}$ with $\mu < \omega$, we get the following result.*

Lemma 4.3.11 *Assume that* $(A4)$ *holds. If V be a bounded subset of* $\overline{\mathcal{K}_R}$, *then*

$$\alpha_X(FV) = \max \left(\sup_{t \in \mathbb{R}^+} \alpha_E \left(e^{-\omega t} FV(t) \right), \; \sup_{t \in \mathbb{R}^+} \alpha_E \left(e^{-\omega t} (FV)'(t) \right) \right).$$

Proof 4.3.12 *Let $V \subset \overline{\mathcal{K}_R}$ be arbitrary.*
    (a) $F(V) \subset X$ is a uniformly bounded set with respect to the norm $\|.\|_\mu$. Indeed, as in Theorem 4.3.6, we obtain

$$\|x\|_\mu \le \varepsilon A(B_1 + B_2\|x\|_\mu^p + B_3\|x\|_\mu^q), \forall x \in V.$$

    (b) The families $\{e^{-\omega t}(FV(t))\}_{t \in \mathbb{R}^+}$ and $\{e^{-\omega t}(FV)'(t))\}_{t \in \mathbb{R}^+}$ are almost equicontinuous on $\mathbb{R}^+$.

Now, suppose that $V \subset \overline{\mathcal{K}_R}$; we prove that there exists a constant $0 \le \ell < \varepsilon$ such that $\alpha_X(FV) \le \ell\alpha_X(V)$. Lemma 4.3.11 tells us that it is enough to verify that

$$\max\left(\sup_{t\in\mathbb{R}^+} \alpha_E\left(e^{-\omega t}FV(t)\right), \sup_{t\in\mathbb{R}^+} \alpha_E\left(e^{-\omega t}(FV)'(t)\right)\right) \le \ell\alpha_X(V).$$

(4.21)

Let $x \in V$. we introduce, for each $n \ge 1$, the approximating operator F_n by

$$F_n x(t) = -\varepsilon \int_0^n G(t,s)m(s)f(s,x(s),x'(s))ds.$$

Step 1. From (A4)and (A6), for every $t \in (0,\infty)$, we have that

$$e^{-\omega t}\|Fx(t) - F_n x(t)\|$$

$$\le \int_n^{+\infty} e^{-\omega t}G(t,s)m(s)\|f(s,x(s),x'(s))\|ds$$

$$\le \varepsilon A\left(\int_n^{+\infty} m(s)a_0(s)ds + \|x\|_\omega^p \int_n^{+\infty} e^{p\omega s}m(s)a_1(s)ds\right.$$
$$\left. + \|x\|_\omega^q \int_n^{+\infty} e^{q\omega s}m(s)a_2(s)ds\right).$$

Similarly, we also have

$$e^{-\omega t}\|(F_n x)'(t) - (Fx)'(t)\|$$

$$\le \varepsilon A\left(\int_n^{+\infty} m(s)a_0(s)ds + \|x\|_\omega^p \int_n^{+\infty} e^{p\omega s}m(s)a_1(s)ds\right.$$
$$\left. + \|x\|_\omega^q \int_n^{+\infty} e^{q\omega s}m(s)a_2(s)ds\right).$$

As a consequence, we get

$$\|Fx - F_n x\|_\omega$$
$$= \sup_{t\in\mathbb{R}^+} \{e^{-\omega t}\left(\|Fx(t) - F_n x(t)\| + \|(F_n x)'(t) - (Fx)'(t)\|\right)\}$$
$$\le 2\varepsilon A\left(\int_n^{+\infty} m(s)a_0(s)ds + \|x\|_\omega^p \int_n^{+\infty} e^{p\omega s}m(s)a_1(s)ds\right.$$
$$\left. + \|x\|_\omega^q \int_n^{+\infty} e^{q\omega s}m(s)a_2(s)ds\right).$$

The convergence of the integrals guarantee that

$$\lim_{n \to +\infty} \int_n^{+\infty} m(s)a_0(s)ds = 0,$$

$$\lim_{n \to +\infty} \int_n^{+\infty} e^{p\omega s} m(s)a_1(s)ds = 0,$$

$$\lim_{n \to +\infty} \int_n^{+\infty} e^{q\omega s} m(s)a_2(s)ds = 0.$$

Then, for all $x \in V$ and $t \in (0, \infty)$, we have

$$d(e^{-\omega t}(F_n x)(t), e^{-\omega t}(FV)(t))$$

$$= \inf_{y \in B} \{ e^{-\omega t} \left(\|F_n x(t) - Fy(t)\| + \|(F_n x)'(t) - (Fx)'(t)\| \right) \}$$

$$\leq e^{-\omega t} \left(\|F_n x(t) - Fx(t)\| + \|(F_n x)'(t) - (Fx)'(t)\| \right)$$

$$\to 0, \text{ as } n \to \infty,$$

hence for every $t \in (0, \infty)$

$$\sup_{x \in V} d(e^{-\omega t}(F_n x)(t), e^{-\omega t}(FV)(t)) \to 0, \text{ as } n \to \infty.$$

Similarly, for every $t \in (0, \infty)$

$$\sup_{x \in V} d(e^{-\omega t}(F_n V)(t), e^{-\omega t}(Fx)(t)) \to 0, \text{ as } n \to \infty.$$

Then the Hausdorff distance

$$H_d(e^{-\omega t}FV(t), e^{-\omega t}F_n V(t))$$

tends to 0, as n tends to $+\infty$ for all t in $(0, \infty)$. The Lipschitz property of the MNC α guarantees

$$\lim_{n \to +\infty} \alpha \left(e^{-\omega t}F_n V(t) \right) = \alpha \left(e^{-\omega t}FV(t) \right), \ \forall t \in (0, +\infty), \qquad (4.22)$$

and

$$\lim_{n \to +\infty} \alpha \left(e^{-\omega t}(F_n V)'(t) \right) = \alpha \left(e^{-\omega t}(FV)'(t) \right), \ \forall t \in (0, +\infty), \quad (4.23)$$

Step 2. In what follows, we estimate $\alpha\left(e^{-\omega t}F_nV(t)\right)$. *Using Assumption (A6), Lemma 4.3.4 and the properties of the Green function lead to estimations:*

$$\alpha\left(e^{-\omega t}FV(t)\right) = \lim_{n\to+\infty}\alpha\left(e^{-\omega t}(F_nV)(t)\right)$$

$$= \varepsilon\lim_{n\to+\infty}\alpha\left(\{e^{-\omega t}\int_0^n G(t,s)m(s)f(s,x(s),x'(s))ds\},\right.$$

$$\left. \times x\in V\right)$$

$$\leq \varepsilon A\lim_{n\to+\infty}\int_0^n m(s)\alpha\left(f(s,x(s),x'(s)),\, x\in V\right)ds$$

$$\leq \varepsilon A\lim_{n\to+\infty}\int_0^n m(s)\left(l_1(s)\alpha\left(e^{-\omega s}V(s)\right)\right.$$

$$\left. + l_2(s)\alpha\left(e^{-\omega s}V'(s)\right)\right)ds$$

$$\leq \alpha_X(V)\varepsilon A\lim_{n\to+\infty}\int_0^n m(s)(l_1(s)+l_2(s))ds$$

$$\leq \alpha_X(V)\varepsilon A\int_0^{+\infty} m(s)(l_1(s)+l_2(s))ds.$$

Since t is arbitrary

$$\sup_{t\in\mathbb{R}^+}\alpha\left(e^{-\omega t}(FV)(t)\right)\leq \ell\alpha_X(V),$$

Similarly, we find that

$$\sup_{t\in\mathbb{R}^+}\alpha\left(e^{-\omega t}(FV)'(t)\right)\leq \ell\alpha_X(V),$$

where, $\ell = \varepsilon A\int_0^{+\infty} m(s)(l_1(s)+l_2(s))ds.$
From Lemma 4.3.11, we immediately deduce that

$$\alpha_X(FV)\leq \ell\,\alpha_X(V),$$

meaning that $F:\overline{\mathscr{K}_R}\to X$ *is a* ℓ*-set contraction with* $\ell < \varepsilon$.

Example 4.3.13 *Consider the following nonlinear boundary value problem for system of n scalar differential equations in the Banach space $E = \mathbb{R}^n$ with the Euclidean norm $\|x\| = \left(\sum\limits_{i=1}^{n} x_i^2\right)^{\frac{1}{2}}$ with $x = (x_1,\ldots,x_n) \mid x_i \in \mathbb{R}, i = 1,\ldots,n$ and let $0 < p, q < 1$:*

$$
\begin{cases}
-x_i''(t) + k^2 x_i(t) = \dfrac{e^{-p\omega t}}{t^2 \sqrt{t}} \dfrac{1 - \cos t}{t+1}(1 + (x_i(t))^p + (x_i'(t))^q), \, t > 0, \\[2mm]
x_i(0) = 0, \quad \lim_{t \to +\infty} x_i(t) = 0, \quad i = 1,2,\ldots,n.
\end{cases}
$$

$$(4.24)$$

Let $\mathscr{P} = \{x = (x_1,x_2,\ldots,x_n) \in \mathbb{R}^n \mid x_i \geq 0, \, i = 1,2,\ldots,n\}$. Then $\mathscr{P}$ is a cone in $\mathbb{R}^n$ and clearly System (4.37) can be rewritten in the form (4.10) in E. In this case,
$x = (x_1,\ldots,x_n)$, $y = (y_1,\ldots,y_n)$, $f = (f^{(1)},\ldots,f^{(n)})$ *where for any $i \in \{1,\ldots,n\}$, $f^{(i)}$ is defined by*

$$
f^{(i)}(t,x_1,\ldots,x_n,y_1,\ldots,y_n) = \frac{1 - \cos t}{t+1}(1 + x_i^p + y_i^q), \quad \text{for } t \geq 0.
$$

Then, we have $f^{(i)}$ is uniformly continuous on $[a,b] \times B_E(0,r)$, for all $[a,b] \subset I$ and $r > 0$. The singular coefficient is given by $m(t) = \dfrac{e^{-p\omega t}}{t^2 \sqrt{t}}$ for $t > 0$. Then

$$
\begin{aligned}
\|f(t,x,y)\|^2 &= \sum_{i=1}^{n} (f^{(i)}(t,x,y))^2 \\
&\leq 4\frac{(1-\cos t)^2}{(t+1)^2}\left(n + \sum_{i=1}^{n} x_i^{2p} + \sum_{i=1}^{n} y_i^{2q}\right) \\
&\leq 4\frac{(1-\cos t)^2}{(t+1)^2}\left(n + (\sum_{i=1}^{n} x_i^2)^p + (\sum_{i=1}^{n} y_i^2)^q\right) \\
&\leq 4\frac{(1-\cos t)^2}{(t+1)^2}\left(n + \|x\|^{2p} + \|y\|^{2q}\right).
\end{aligned}
$$

Hence,

$$
\|f(t,x,y)\| \leq a_1(t)\|x\|^p + a_2(t)\|y\|^q + a_0(t),
$$

where $a_1(t) = a_2(t) = 2\dfrac{1-\cos t}{t+1}$ and $a_2(t) = 2\sqrt{n}\dfrac{1-\cos t}{t+1}$.

Moreover, since in the vicinity the origin, $\dfrac{1-\cos s}{s^2\sqrt{s}} \sim \dfrac{1}{2\sqrt{s}}$ and for

any $\alpha > 0$, $\displaystyle\int_0^{+\infty} \dfrac{e^{-\alpha s}}{(s+1)\sqrt{s}}\,dx < \infty$, we deduce the convergence of the integrals

$$\int_0^{+\infty} e^{p\omega s} m(s)a_1(s)ds = 2\int_0^{+\infty} \frac{1-\cos s}{s^2\sqrt{s}(s+1)}ds,$$

$$\int_0^{+\infty} m(s)a_0(s)ds = 2\sqrt{n}\int_0^{+\infty} \frac{(1-\cos s)e^{-p\omega s}}{s^2\sqrt{s}(s+1)}ds.$$

Also, the integral

$$\int_0^{+\infty} e^{q\omega s} m(s)a_2(s)ds = 2\int_0^{+\infty} \frac{(1-\cos s)e^{(q-p)\omega s}}{s^2\sqrt{s}(s+1)}ds$$

is convergent provided $p > q$.

Here the real numbers p,q satisfy $0 < p,q < 1$, then there exists $R > 0$ such that $A(B_1 + B_2 R^p + B_3 R^q) < R$.

Finally, for every bounded subsets $D_1, D_2 \subset E$ and for all $t \in \mathbb{R}^+$, $x \in D_1$, $y \in D_2$, we have

$$\|f(t,x,y)\| \leq 2\frac{1-\cos t}{t+1}\left(n + \|x\|^{2p} + \|y\|^{2q}\right) \leq 4\left(n + \|x\|^{2p} + \|y\|^{2q}\right).$$

Moreover, for all $0 < t_1 < t_2 < +\infty$, $x \in D_1$, and $y \in D_2$, we have

$$\lim_{t_1\to t_2} |f^{(i)}(t_1,x,y) - f^{(i)}(t_2,x,y)|$$

$$\leq \lim_{t_1\to t_2} \left|\frac{1-\cos t_1}{t_1+1}(1+x_i^p+y_i^q) - \frac{1-\cos t_2}{t_2+1}(1+x_i^p+y_i^q)\right|$$

$$\leq \lim_{t_1\to t_2}(1+\|x\|_\infty^p+\|x\|_\infty^q)\left|\frac{1-\cos t_1}{t_1+1} - \frac{1-\cos t_2}{t_2+1})\right| = 0,$$

$\forall i = 1,\ldots,n$. Then $\lim_{t_1\to t_2}\|f(t_1,x,y) - f(t_2,x,y)\| = 0$ and

$$\lim_{t\to+\infty}|f^{(i)}(t,x,y) - \lim_{s\to+\infty}f^{(i)}(s,x,y)|$$

$$\leq \lim_{t\to+\infty}\left|\frac{1-\cos t}{t+1}(1+x_i^p+y_i^q) - 0\right| = 0, \forall i = 1,\ldots,n.$$

Hence, $\lim\limits_{t \to +\infty} \|f^{(i)}(t,x,y) - \lim\limits_{s \to +\infty} f^{(i)}(s,x,y)\| = 0.$

As a consequence, Corduneanu's compactness criterion ensures that $f(t,D_1,D_2)$ *is relatively compact in* $\mathbb{R}^n$. *So,* $\alpha(f(t,D_1,D_2)) = 0$, *for all* $t \in \mathbb{R}^+$ *and all bounded subset* $D_1, D_2 \subset E$.

Theorem 4.3.9 ensures the sub-linear singular problem (4.37) has a bounded positive solution for every constants k and all $0 < p, q < 1$.

4.4   A Nonlinear IVP

In this section, we will give an application of Proposition 2.2.55 for existence of non-negative bounded solutions for a class of ODEs. We investigate the IVP

$$x' = f(t,x), \quad t > 0,$$

$$x(0) = x_0,$$

(4.25)

where $x_0 \in \mathbb{R}$ is a given constant, $f : [0,\infty) \times \mathbb{R} \to \mathbb{R}$ is a given function. Let $l \in \mathbb{N}$ and x_0, s, r, A_j, $j \in \{0,1,\dots,l\}$, are positive constants such that

(A7)

$$x_0 + \sum_{j=0}^{l} \left(\frac{r}{2}\right)^j A_j < \frac{r}{2},$$

(A8) $f \in \mathscr{C}([0,\infty) \times \mathbb{R})$ and

$$0 \le f(y,x) \le \sum_{j=0}^{l} a_j(y)|x|^j, \quad y \in [0,\infty), \quad x \in \mathbb{R},$$

where $a_j \in \mathscr{C}([0,\infty))$, $a_j \ge 0$ on $[0,\infty)$ and

$$\int_0^\infty a_j(y)\,dy \le A_j, \quad j \in \{0,1,\dots,l\}.$$

Theorem 4.4.1 *Assume (A7) and (A8). Then the IVP (4.25) has a solution $x \in \mathscr{C}^1([0,\infty))$ such that $0 \le x(t) < \dfrac{r}{2}$, $t \in [0,\infty)$.*

Proof 4.4.2 Case 1. *Let $t \in [0,1]$. Consider the IVP*

$$x' = f(t,x), \quad t \in (0,1],$$
$$x(0) = x_0. \tag{4.26}$$

Take $\varepsilon > 0$ arbitrarily. Let $E_1 = \mathscr{C}([0,1])$ be endowed with the maximum norm and

$$\mathscr{P}_1 = \{x \in E_1 : x(t) \ge 0, \quad t \in [0,1]\},$$

$$\Omega_1 = \mathscr{P}_{1r} = \{x \in \mathscr{P}_1 : \|x\| < r\},$$

$$U_1 = \mathscr{P}_{1\frac{r}{2}} = \left\{x \in \mathscr{P}_1 : \|x\| < \frac{r}{2}\right\}.$$

For $x \in E_1$, define the operators

$$T_1 x(t) = (1+\varepsilon)x(t),$$

$$F_1 x(t) = -\varepsilon\left(x_0 + \int_0^t f(y,x(y))dy\right), \quad t \in [0,1].$$

Note that for any fixed point $x \in E_1$ of the operator $T_1 + F_1$, we have that $x \in \mathscr{C}^1([0,1])$ and it is a solution of the IVP (4.26).

1. *For $x,y \in E_1$, we have*

$$\|(I - T_1)^{-1}x - (I - T_1)^{-1}y\| = \frac{1}{\varepsilon}\|x - y\|,$$

   i.e., $(I - T_1) : E_1 \to E_1$ is Lipschitz invertible with constant $\dfrac{1}{\varepsilon}$.
2. *For $x \in \overline{U}_1$ and $t \in [0,1]$, we have*

$$|F_1 x(t)| = \varepsilon\left(x_0 + \int_0^t f(y,x(y))dy\right)$$

$$\leq \ \varepsilon \left(x_0 + \int_0^t \sum_{j=0}^{l} a_j(y)(x(y))^j dy \right)$$

$$\leq \ \varepsilon \left(x_0 + \sum_{j=0}^{l} \left(\frac{r}{2} \right)^j \int_0^t a_j(y) dy \right)$$

$$\leq \ \varepsilon \left(x_0 + \sum_{j=0}^{l} \left(\frac{r}{2} \right)^j A_j \right)$$

and

$$|(F_1 x)'(t)| \ = \ \varepsilon f(t, x(t))$$

$$\leq \ \varepsilon \sum_{j=0}^{l} a_j(y)(x(y))^j$$

$$\leq \ \varepsilon \sum_{j=0}^{l} \left(\frac{r}{2} \right)^j a_j(y)$$

$$\leq \ \varepsilon \sum_{j=0}^{l} \left(\frac{r}{2} \right)^j B_j$$

Thus,

$$\|F_1 x\| \leq \varepsilon \left(x_0 + \sum_{j=0}^{l} \left(\frac{r}{2} \right)^j A_j \right)$$

and

$$\|(F_1 x)'\| \leq \varepsilon \sum_{j=0}^{l} \left(\frac{r}{2} \right)^j B_j.$$

Hence, using the Ascoli-Arzelà theorem, we conclude that
$F_1 : \overline{U}_1 \to E$ is a completely continuous mapping.
Therefore, $F_1 : \overline{U}_1 \to E$ is a 0-set contraction.

3. *Let $\lambda \in [0,1]$ and $x \in \overline{U}_1$ be arbitrarily chosen. Then*

$$z(t) = \lambda \left(x_0 + \int_0^t f(y, x(y)) ds \right) \in E_1$$

and

$$z(t) \leq \lambda \left(x_0 + \int_0^\infty f(y, x(y)) dy \right)$$

$$\leq \lambda \left(x_0 + \sum_{j=0}^l \int_0^\infty a_j(y)(x(y))^j dy \right)$$

$$\leq \lambda \left(x_0 + \sum_{j=0}^l \left(\frac{r}{2} \right)^j A^j \right)$$

$$< \lambda \frac{r}{2}$$

$$\leq \frac{r}{2}, \quad t \in [0, 1],$$

i.e., $z \in \Omega_1$. Next,

$$\lambda F_1 x(t) = -\lambda \varepsilon \left(x_0 + \int_0^t f(y, x(y)) dy \right)$$

$$= -\varepsilon z(t)$$

$$= (I - T_1) z(t), \quad t \in [0, 1].$$

Thus, $\lambda F_1(\overline{U_1}) \subset (I - T_1)(\Omega_1)$.

4. *Note that*
$$(I - T_1)^{-1} 0 = 0 \in U_1.$$

5. *Assume that there are $x \in \partial U_1 \cap \Omega_1$ and $\lambda \in [0, 1]$ such that*

$$x - T_1 x = \lambda F_1 x.$$

If $\lambda = 0$, then

$$0 = x - T_1 x = -\varepsilon x \quad on \quad [0, 1],$$

whereupon $x(t) = 0$, $t \in [0, 1]$. This is a contradiction because $x \in \partial U_1$. Therefore $\lambda \in (0, 1]$. Let $t_1 \in [0, 1]$ be such that $x(t_1) = \frac{r}{2}$. Then

$$(I - T_1) x(t_1) = -\varepsilon x(t_1)$$

$$= -\varepsilon \frac{r}{2}$$

$$= -\lambda \varepsilon \left(x_0 + \int_0^{t_1} f(y, x(y)) dy \right),$$

whereupon

$$\frac{r}{2} = \lambda \left(x_0 + \int_0^{t_1} f(y, x(y)) dy \right)$$

$$\leq \lambda \left(x_0 + \int_0^{\infty} f(y, x(y)) dy \right)$$

$$\leq \lambda \left(x_0 + \sum_{j=0}^{l} \int_0^{\infty} a_j(y) (x(y))^j dy \right)$$

$$\leq \lambda \left(x_0 + \sum_{j=0}^{l} A_j \left(\frac{r}{2} \right)^j \right)$$

$$< \lambda \frac{r}{2}$$

$$\leq \frac{r}{2},$$

i.e., $\dfrac{r}{2} < \dfrac{r}{2}$, *which is a contradiction.*

By 1, 2, 3, 4, 5 and Proposition 2.2.55, it follows that the operator $T_1 + F_1$ *has a fixed point in* U_1. *Denote it by* x_1. *We have*

$$0 \leq x_1(t) < \frac{r}{2}, \quad t \in [0, 1],$$

and $x_1 \in \mathscr{C}^1([0, 1])$ *is a solution of the IVP (4.26).*

Case 2. *Let* $t \in [1, 2]$. *Consider the IVP*

$$x' = f(t, x), \quad t \in (1, 2],$$

$$x(1) = x_1(1).$$

(4.27)

Take $\varepsilon > 0$ arbitrarily. Let $E_2 = \mathscr{C}([1,2])$ be endowed with the maximum norm and

$$\mathscr{P}_2 \quad = \quad \{x \in E_2 : x(t) \geq 0, \quad t \in [1,2]\},$$

$$\Omega_2 \quad = \quad \mathscr{P}_{2r} = \{x \in \mathscr{P}_2 : \|x\| < r\},$$

$$U_2 \quad = \quad \mathscr{P}_{2\frac{r}{2}} = \left\{x \in \mathscr{P}_2 : \|x\| < \frac{r}{2}\right\}.$$

For $x \in E_2$ define the operators

$$T_2 x(t) \quad = \quad (1+\varepsilon)x(t),$$

$$F_2 x(t) \quad = \quad -\varepsilon \left(x_1(1) + \int_1^t f(s,x(s))ds \right), \quad t \in [1,2].$$

Note that for $x \in U_2$, we have

$$
\begin{aligned}
x_1(1) + \int_1^t f(s,x(s))ds \quad &= \quad x_0 + \int_0^t f(y,x(y))dy \\
&\leq \quad x_0 + \int_0^\infty f(y,x(y))dy \\
&\leq \quad x_0 + \sum_{j=0}^l a_j(y)(x(y))^j dy \\
&\leq \quad x_0 + \sum_{j=0}^l A_j r^j \\
&< \quad \frac{r}{2}, \quad t \in [1,2].
\end{aligned}
$$

As in Case 1 we prove that the operator $T_2 + F_2$ has a fixed point $x_2 \in U_2$. We have that

$$0 \leq x_2(t) < \frac{r}{2}, \quad t \in [1,2], \quad x_2 \in \mathscr{C}^1([1,2]).$$

Note that

$$x_1(1) \quad = \quad x_2(1),$$

$$x_1'(1) = f(1, x_1(1))$$

$$= f(1, x_2(1))$$

$$= x_2'(1).$$

Thus,

$$x(t) = \begin{cases} x_1(t) & t \in [0,1] \\ x_2(t) & t \in [1,2] \end{cases}$$

is a solution to the IVP

$$x' = f(t,x), \quad t \in (0,2],$$

$$x(0) = x_0.$$

Case 3. *Consider the IVP*

$$x' = f(t,x), \quad t \in (2,3],$$

$$x(2) = x_2(2).$$

And so on, the function

$$x(t) = \begin{cases} x_1(t) & t \in [0,1] \\ x_2(t) & t \in [1,2] \\ x_3(t) & t \in [2,3] \\ x_4(t) & t \in [3,4] \\ \cdots \end{cases}$$

is a solution to the IVP (4.25). This completes the proof.

4.5   Existence of Positive Solutions for a Class of *n*th-Order ODEs

In this section, we will give an application of Proposition 2.2.61 for existence of positive solutions for a class of *n*th-order ODEs. We investigate the equation

$$y^{(n)} = f(t, y, y', \ldots, y^{(n-1)}), \quad t \in \mathbb{R}, \tag{4.28}$$

for existence of positive ω-periodic solutions, where the period $\omega > 0$ is arbitrarily chosen and fixed, and $n \in \mathbb{N}, n \geq 2$. Our main assumptions in this section are as follows.

(A9) $f \in \mathscr{C}(\mathbb{R}^{n+1})$,

$$f(t + \omega, v_1, v_2, \ldots, v_n) = f(t, v_1, v_2, \ldots, v_n),$$

$$0 \leq f(t, v_1, v_2, \ldots, v_n) \leq a(t) + \sum_{j=1}^{m} \sum_{k=1}^{l_j} a_{kj}(t) |v_j|^{p_{kj}}, \quad t \in \mathbb{R},$$

$v = (v_1, v_2, \ldots, v_n) \in \mathbb{R}^n$, $a, a_{kj} \in \mathscr{C}(\mathbb{R})$ are non-negative ω-periodic functions, $0 \leq a, a_{kj} \leq r$ on $[0, \omega]$, $r > 0$ is a constant that satisfies the condition

$$4 \left(r + \left(r + \sum_{j=1}^{m} \sum_{k=1}^{l_j} r^{p_{kj}+1} \right) \omega \right) < 1, \tag{4.29}$$

$p_{kj} \geq 0$, $k \in \{1, \ldots, l_j\}$, $l_j \in \mathbb{N}$, $j \in \{1, \ldots, m\}$, $m \leq n$.

Theorem 4.5.1 *Let* $\omega > 0$ *be arbitrarily chosen and fixed. Suppose that (A9) holds. Then the equation (4.28) has a non-negative ω-periodic solution.*

Proof 4.5.2 *Take* $q > 1$ *arbitrarily and set*

$$R = r + \left(r + \sum_{j=1}^{m} \sum_{k=1}^{l_j} r^{p_{kj}+1} \right) \frac{\omega}{q}.$$

We choose $\varepsilon \in (1,2)$ close enough to 2 so that

$$R < \frac{\varepsilon r}{(2-\varepsilon)q}.$$ (4.30)

Firstly, note that the equation (4.28) can be rewritten in the form

$$y_1' = -\frac{q}{\omega}y_1 + \frac{q}{\omega}y_1 + y_2$$

$$y_2' = -\frac{q}{\omega}y_2 + \frac{q}{\omega}y_2 + y_3$$

$$\vdots$$

$$y_n' = -\frac{q}{\omega}y_n + \frac{q}{\omega}y_n + f(t,y_1,y_2,\ldots,y_n), \quad t \in \mathbb{R}.$$

Now, using (4.2) and (4.4), the last system has a ω-periodic solution if and only if $(y_1,y_2,\ldots,y_n)$ satisfies the system

$$y_1(t) = \frac{e^{-q}}{1-e^{-q}} \int_t^{t+\omega} e^{\frac{q}{\omega}(s-t)} \left(\frac{q}{\omega}y_1(s) + y_2(s)\right) ds$$

$$y_2(t) = \frac{e^{-q}}{1-e^{-q}} \int_t^{t+\omega} e^{\frac{q}{\omega}(s-t)} \left(\frac{q}{\omega}y_2(s) + y_3(s)\right) ds$$

(4.31)

$$\vdots$$

$$y_n(t) = \frac{e^{-q}}{1-e^{-q}} \int_t^{t+\omega} e^{\frac{q}{\omega}(s-t)}$$
$$\times \left(\frac{q}{\omega}y_n(s) + f(s,y_1(s),y_2(s),\ldots,y_n(s))\right) ds,$$

$t \in [0,\omega]$. Let

$$E_1 = \{u \in \mathscr{C}(\mathbb{R}) : u(t+\omega) = u(t), \quad t \in \mathbb{R}\}$$

be endowed with the norm

$$\|u\|_1 = \max_{t \in [0,\omega]} |u(t)|.$$

Let also, $E = E_1^n$ be endowed with the norm

$$\|v\| = \max_{j \in \{1,\dots,n\}} \|v_j\|_1, \quad v = (v_1, v_2, \dots, v_n) \in E.$$

Denote

$$\mathscr{P} = \{v \in E : v = (v_1, v_2, \dots, v_n), \quad v_j \geq 0, \quad j \in \{1, \dots, n\}\},$$

$$\Omega = \{v \in \mathscr{P} : \|v\| \leq R\},$$

$$U = \{v \in \mathscr{P} : \|v\| \leq r, \quad \min_{t \in [0,\omega]} v_j(t) \geq \frac{1}{q}\|v\|, \quad j \in \{1, \dots, n\}\}.$$

For $y \in E$, $y = (y_1, y_2, \dots, y_n)$, define the operators

$$T_1 y(t) = (1 - \varepsilon) y_1(t),$$

$$T_2 y(t) = (1 - \varepsilon) y_2(t),$$

$$\vdots$$

$$T_n y(t) = (1 - \varepsilon) y_n(t),$$

$$T y(t) = (T_1 y(t), T_2 y(t), \dots, T_n y(t)),$$

$$S_1 y(t) = \varepsilon \frac{e^{-q}}{1 - e^{-q}} \int_t^{t+\omega} e^{\frac{q}{\omega}(s-t)} \left(\frac{q}{\omega} y_1(s) + y_2(s) \right) ds$$

$$S_2 y(t) = \varepsilon \frac{e^{-q}}{1 - e^{-q}} \int_t^{t+\omega} e^{\frac{q}{\omega}(s-t)} \left(\frac{q}{\omega} y_2(s) + y_3(s) \right) ds$$

$$\vdots$$

$$S_n y(t) = \varepsilon \frac{e^{-q}}{1 - e^{-q}}$$

$$\times \int_t^{t+\omega} e^{\frac{q}{\omega}(s-t)} \left(\frac{q}{\omega}y_n(s) + f(s, y_1(s), y_2(s), \ldots, y_n(s))\right) ds,$$

$$Sy(t) = (S_1 y(t), S_2 y(t), \ldots, S_n y(t)),$$

$t \in [0, \omega]$. *Observe that any fixed point $v \in E$ of the operator $T + S$ is a solution of the system (4.31) and hence, it is a solution of the equation (4.28).*

1. *For $v \in \Omega$, we have*

$$|(I - T_j)v(t)| = \varepsilon|v_j(t)|, \quad t \in [0, \omega], \quad j \in \{1, \ldots, n\}.$$

Then

$$\frac{\varepsilon}{2}\|v\| \le \|(I - T)v\| \le 2\varepsilon\|v\|$$

and $I - T : \Omega \to E$ is Lipschitz invertible with a constant $\gamma \in \left[\frac{1}{2\varepsilon}, \frac{2}{\varepsilon}\right]$.

2. *For $u \in \overline{U}$, we find*

$$
\begin{aligned}
|S_j u(t)| &= \varepsilon \frac{e^{-q}}{1 - e^{-q}} \int_t^{t+\omega} e^{\frac{q}{\omega}(s-t)} \left(\frac{q}{\omega}u_j(s) + u_{j+1}(s)\right) ds \\
&= \varepsilon \frac{e^{-q}}{1 - e^{-q}} \left(\frac{q}{\omega}\right) \int_t^{t+\omega} e^{\frac{q}{\omega}(s-t)} u_j(s) ds \\
&\quad + \varepsilon \frac{e^{-q}}{1 - e^{-q}} \int_t^{t+\omega} e^{\frac{q}{\omega}(s-t)} u_{j+1}(s) ds \\
&\le \varepsilon r \left(\frac{q}{\omega}\right) \frac{e^{-q}}{1 - e^{-q}} \int_t^{t+\omega} e^{\frac{q}{\omega}(s-t)} ds \\
&\quad + \varepsilon r \frac{e^{-q}}{1 - e^{-q}} \int_t^{t+\omega} e^{\frac{q}{\omega}(s-t)} ds \\
&= \varepsilon r \left(1 + \frac{\omega}{q}\right), \quad t \in [0, \omega],
\end{aligned}
$$

$j \in \{1, \ldots, n-1\}$, *and*

$$|S_n u(t)| = \varepsilon \frac{e^{-q}}{1 - e^{-q}} \int_t^{t+\omega} e^{\frac{q}{\omega}(s-t)} \left(\frac{q}{\omega}u_n(s)\right.$$

$$+ f(s, u_1(s), u_2(s), \ldots, u_n(s)) \Big) ds$$

$$= \varepsilon \left(\frac{q}{\omega} \right) \frac{e^{-q}}{1 - e^{-q}} \int_t^{t+\omega} e^{\frac{q}{\omega}(s-t)} u_n(s) ds$$

$$+ \varepsilon \frac{e^{-q}}{1 - e^{-q}} \int_t^{t+\omega} e^{\frac{q}{\omega}(s-t)} f(s, u_1(s), u_2(s), \ldots, u_n(s)) ds$$

$$\leq \varepsilon r \left(\frac{q}{\omega} \right) \frac{e^{-q}}{1 - e^{-q}} \int_t^{t+\omega} e^{\frac{q}{\omega}(s-t)} ds$$

$$+ \varepsilon \frac{e^{-q}}{1 - e^{-q}}$$

$$\times \int_t^{t+\omega} e^{\frac{q}{\omega}(s-t)} \left(a(s) + \sum_{j=1}^{m} \sum_{k=1}^{l_j} a_{kj}(s)(u_j(s))^{p_{kj}} \right) ds$$

$$\leq \varepsilon r + \varepsilon \frac{e^{-q}}{1 - e^{-q}} \left(r + \sum_{j=1}^{m} \sum_{k=1}^{l_j} r^{p_{kj}+1} \right) \int_t^{t+\omega} e^{\frac{q}{\omega}(s-t)} ds$$

$$= \varepsilon r + \varepsilon \left(r + \sum_{j=1}^{m} \sum_{k=1}^{l_j} r^{p_{kj}+1} \right) \frac{\omega}{q}, \quad t \in [0, \omega].$$

Consequently

$$\|Su\| \leq \varepsilon r + \varepsilon \left(r + \sum_{j=1}^{m} \sum_{k=1}^{l_j} r^{p_{kj}+1} \right) \frac{\omega}{q} \tag{4.32}$$

$$= \varepsilon R.$$

Next,

$$\left| \frac{d}{dt} S_j u(t) \right| = \left| -\frac{q}{\omega} S_j u(t) + \varepsilon \left(\frac{q}{\omega} u_j(t) + u_{j+1}(t) \right) \right|$$

$$\leq \frac{q}{\omega} S_j u(t) + \varepsilon \left(\frac{q}{\omega} u_j(t) + u_{j+1}(t) \right)$$

$$\leq \varepsilon r \frac{q}{\omega} \left(1 + \frac{\omega}{q} \right) + \varepsilon \left(\frac{q}{\omega} r + r \right), \quad t \in [0, \omega],$$

$j \in \{1, \ldots, n-1\}$, *and*

$$\left| \frac{d}{dt} S_n u(t) \right| = \left| -\frac{q}{\omega} S_n u(t) \right.$$

$$\left. + \varepsilon \left(\frac{q}{\omega} u_n(t) + f(t, u_1(t), u_2(t), \ldots, u_n(t)) \right) \right|$$

$$\leq \frac{q}{\omega} S_n u(t)$$

$$+ \varepsilon \left(\frac{q}{\omega} u_n(t) + f(t, u_1(t), u_2(t), \ldots, u_n(t)) \right)$$

$$\leq \frac{q}{\omega} S_n u(t)$$

$$+ \varepsilon \left(\frac{q}{\omega} u_n(t) + a(t) + \sum_{j=1}^{m} \sum_{k=1}^{l_j} a_{kj}(t) u_j^{p_{kj}}(t) \right)$$

$$\leq \varepsilon \frac{q}{\omega} \left(r + \left(r + \sum_{j=1}^{m} \sum_{k=1}^{l_j} r^{p_{kj}+1} \right) \frac{\omega}{q} \right)$$

$$+ \varepsilon \left(\frac{q}{\omega} r + r + \sum_{j=1}^{m} \sum_{k=1}^{l_j} r^{p_{kj}+1} \right),$$

$t \in [0, \omega]$. *Thus,*

$$\| (Su)' \| \leq \varepsilon \frac{q}{\omega} \left(r + \left(r + \sum_{j=1}^{m} \sum_{k=1}^{l_j} r^{p_{kj}+1} \right) \frac{\omega}{q} \right)$$

$$+\varepsilon\left(\frac{q}{\omega}r+r+\sum_{j=1}^{m}\sum_{k=1}^{l_j}r^{p_{kj}+1}\right).$$

Now, using the Ascoli-Arzelà theorem, we conclude that $S :$ $\overline{U} \to E$ is a completely continuous mapping. Therefore, $S :$ $\overline{U} \to E$ is a 0-set contraction.

3. Let $u \in \overline{U}$ be arbitrarily chosen. Take $z = \dfrac{Su}{\varepsilon}$. We have that $z \in \mathscr{P}$ and using (4.32), we obtain

$$\|z\| = \frac{\|Su\|}{\varepsilon} \le R,$$

i.e., $z \in \Omega$. Note that $(I - T)z = Su$. Therefore $S(\overline{U}) \subset (I - T)(\Omega)$.

4. Assume that there are $u \in \partial U$ and $\lambda \ge 1$ so that

$$Su = (I - T)(\lambda u) \quad and \quad \lambda u \in \Omega.$$

Since $u \in \partial U$ and $\lambda u \in \Omega$, we have

$$\|u\| = r \quad and \quad \lambda r = \lambda \|u\| \le R.$$

Thus, $\lambda \le \dfrac{R}{r}$. Next,

$$\lambda u = Su + T(\lambda u)$$

$$= Su + (1 - \varepsilon)\lambda u.$$

Hence,

$$R \ge \lambda r$$

$$= \lambda \|u\|$$

$$= \|Su + (1 - \varepsilon)\lambda u\|$$

$$\geq \quad \|Su\| - (1-\varepsilon)\lambda\|u\|$$

$$\geq \quad \varepsilon\frac{e^{-q}}{1-e^{-q}}\int_t^{t+\omega} e^{\frac{q}{\omega}(s-t)}\frac{q}{\omega}u_1(s)ds - (1-\varepsilon)\lambda r$$

$$\geq \quad \left(\frac{\varepsilon}{q}\right)\frac{e^{-q}}{1-e^{-q}}\int_t^{t+\omega} e^{\frac{q}{\omega}(s-t)}\frac{q}{\omega}\|u\|ds - (1-\varepsilon)\lambda r$$

$$= \quad \left(\frac{\varepsilon r}{q}\right)\frac{e^{-q}}{1-e^{-q}}\int_t^{t+\omega} e^{\frac{q}{\omega}(s-t)}\frac{q}{\omega}ds - (1-\varepsilon)\lambda r$$

$$= \quad \frac{\varepsilon r}{q} - (1-\varepsilon)\lambda r,$$

whereupon

$$(2-\varepsilon)\lambda r \geq \frac{\varepsilon r}{q}$$

and

$$(2-\varepsilon)R \quad \geq \quad (2-\varepsilon)\lambda r$$

$$\geq \quad \varepsilon\frac{r}{q}$$

or

$$R \geq \frac{\varepsilon r}{(2-\varepsilon)q}.$$

The last inequality contradicts with (4.30). Hence and Proposition 2.2.61, we conclude that the operator $T+S$ has a fixed point in U which is a positive ω-periodic solution of the equation (4.28). This completes the proof of the main result.

Example 4.5.3 *Consider the equation*

$$y''' = \frac{1}{200} + \frac{\sin^2(t)}{100}\,(y'')^{\frac{8}{3}} + \frac{|\cos(t)|}{300}\,(y')^{\frac{14}{5}} + \frac{1}{100}\,(y)^{\frac{22}{7}}, \quad t\in\mathbb{R}.$$
$$(4.33)$$

Here

$$f(t,y_1,y_2,y_3) = \frac{1}{200} + \frac{1}{100}\,(y_1)^{\frac{22}{7}} + \frac{|\cos(t)|}{300}\,(y_2)^{\frac{14}{5}} + \frac{\sin^2(t)}{100}\,(y_3)^{\frac{8}{3}},$$

t ∈ ℝ. Then, the inequality (4.29) in assumption (A9) is satisfied for

$$n = 3, \ m = 3, \ l_1 = l_2 = l_3 = 1, \ p_{11} = \frac{22}{7}, \ p_{12} = \frac{14}{5}, \ p_{13} = \frac{8}{3}, \ \omega =$$

$$2\pi, \ r = \frac{1}{100} \ and \ q = 3. \ In \ fact,$$

$$4\left(r + \left(r + \sum_{j=1}^{m} \sum_{k=1}^{l_j} r^{p_{kj}+1} \right) \omega \right) < 0.298 < 1.$$

Moreover, $R = r + \left(r + \sum_{j=1}^{m} \sum_{k=1}^{l_j} r^{p_{kj}+1} \right) \dfrac{\omega}{q} < 0.00315, \ so \ for \ \varepsilon = \dfrac{3}{2},$

we get,

$$R < \frac{\varepsilon r}{(2 - \varepsilon)q} = 0.01 \cdot$$

Therefore, the equation (4.33) has a non-negative 2π-periodic solution.

Example 4.5.4 *Consider the equation*

$$\frac{|\sin(t)|}{100} + \frac{1}{150}(y^{\frac{22}{3}} + \cos^2(t) y^4) + \frac{1}{900} (y'')^2 + \frac{|\sin(t)|}{500} (y''')^6 - y^{(5)} = 0,$$

$$(4.34)$$

t ∈ ℝ. This equation is of the form $y^{(5)} = f(t, y, y', y'', y^{(3)}, y^{(4)})$, t ∈ ℝ, where

$$f(t, y_1, y_2, y_3, y_4, y_5) = \frac{|\sin(t)|}{100} + \frac{1}{150}(y_1^{\frac{22}{3}} + \cos^2(t) y_1^4)$$

$$+ \frac{1}{900} y_3^2 + \frac{|\sin(t)|}{500} y_4^6, \ t \in \mathbb{R}.$$

The inequality (4.29) in Assumption (A9) is satisfied for $n = 5$, $m =$ 4, $l_1 = 2$, $l_2 = l_3 = l_4 = l_5 = 1$, $p_{11} = \dfrac{22}{3}$, $p_{21} = 4$, $p_{12} = 0$ $p_{13} =$ 2, $p_{14} = 6$, $\omega = 2\pi$, $r = \dfrac{1}{100}$ and $q = 5$. In fact,

$$4\left(r + \left(r + \sum_{j=1}^{m} \sum_{k=1}^{l_j} r^{p_{kj}+1} \right) \omega \right) < 0.2964 < 1.$$

Moreover, $R = r + \left(r + \sum_{j=1}^{m} \sum_{k=1}^{l_j} r^{p_{kj}+1} \right) \dfrac{\omega}{q} < 0.0036,$ *so for* $\varepsilon = \dfrac{19}{10},$

we get,

$$R < \frac{\varepsilon r}{(2-\varepsilon)q} = 0.038.$$

Therefore, the equation (4.34) *has a non-negative* 2π-*periodic solution.*

4.6 Existence of Positive Solutions for a Class of BVPs with p-Laplacian in Banach Spaces

Let E be a Banach space with a norm $\|\cdot\|$ and zero element θ. Let also,

$$\mathscr{P} = \{u \in E: \quad u \geq \theta\}.$$

With $\mathscr{P}^*$ we will denote the dual cone of the cone $\mathscr{P}$. The space $(\mathscr{C}([0,1],E), \|\cdot\|_c)$ is a Banach space with $\|x\|_c = \max_{t \in [0,1]} \|x(t)\|$.

In this section, we investigate the following boundary value problem (BVP for short):

$$\begin{aligned} \left(\phi_p\left(u'(t)\right) \right)' + f(u(t)) &= \theta, \quad 0 < t < 1, \\ u'(0) = u(1) &= \theta, \end{aligned}$$

(4.35)

where

(A10) $\phi_p(s) = |s|^{p-2}s$, $p > 1$ $\phi_p^{-1} = \phi_q$, $\dfrac{1}{p} + \dfrac{1}{q} = 1$,

(A11) $f \in \mathscr{C}(\mathscr{P}, \mathscr{P})$ and

$$\sup\{\|f(u(t))\| : u \in Q \cap B_r\} \leq M < \infty,$$

where $M > 0$ is a given constant such that

$$\frac{M^{q-1}}{q} < 1.$$

(4.36)

Our main result is as follows.

Theorem 4.6.1 *Suppose* (A10) *and* (A11). *Then the BVP* (4.35) *has at least one positive bounded solution.*

Proof 4.6.2 *Take $\varepsilon > 0$. Set $R = \dfrac{M^{q-1}}{q}$ and*

$$\Omega \;=\; \{u \in Q : \|u\|_c \le R\},$$

$$U \;=\; \{u \in Q : \|u\|_c \le r\}.$$

For $u \in Q$, define the operators

$$Tu(t) \;=\; (1-\varepsilon)u(t),$$

$$Su(t) \;=\; \varepsilon \int_t^1 \phi_q \left(\int_0^s f(u(\tau))d\tau \right) ds, \quad t \in [a,b].$$

Note that any fixed point $u \in Q$ of the operator $T + S$ is a solution of the BVP (4.35).

1. *For, $u \in \Omega$, we have that*

$$\|(I-T)u(t)\| = \varepsilon\|u(t)\|, \quad t \in [a,b].$$

Therefore $I - T : \Omega \to \mathscr{C}([a,b],E)$ is Lipschitz invertible with a constant $\gamma = \dfrac{1}{\varepsilon}$.

2. *For $u \in \overline{U}$, we have*

$$
\begin{aligned}
\|Su(t)\| &= \varepsilon \left\| \int_t^1 \phi_q \left(\int_0^s f(u(\tau))d\tau \right) ds \right\| \\
&\le \varepsilon \int_t^1 \phi_q \left(\int_0^s \|f(u(\tau))\|d\tau \right) ds \\
&\le M^{q-1}\varepsilon \int_t^1 s^{q-1}ds
\end{aligned}
$$

$$\leq \ \varepsilon \frac{M^{q-1}}{q}, \quad t \in [a,b],$$

and

$$\|Su\|_c \leq \varepsilon \frac{M^{q-1}}{q}.$$

Next,

$$\left\| \frac{d}{dt} Su(t) \right\| = \varepsilon \left\| -\phi_q \left(\int_0^t f(u(s))ds \right) \right\|$$

$$\leq \varepsilon \phi_q \left(\int_0^1 \|f(u(s))\|ds \right)$$

$$\leq \varepsilon M^{q-1}, \quad t \in [0,1].$$

Then,

$$\|(Su)'\|_c \leq \varepsilon M^{q-1}.$$

Hence and the Ascoli-Arzelà theorem, we conclude that S : $\overline{U} \to \mathscr{C}([a,b],E)$ is a completely continuous mapping. Therefore $S : \overline{U} \to \mathscr{C}([a,b],E)$ is a 0-set contraction.

3. *Let $u \in \overline{U}$ be arbitrarily chosen. Take $v = \dfrac{Su}{\varepsilon}$. We have $v \in Q$ and*

$$\|v\|_c = \frac{\|Su\|_c}{\varepsilon} \leq \frac{M^{q-1}}{q},$$

i.e., $v \in \Omega$. Note that $(I-T)v = Su$. Therefore $S(\overline{U}) \subset (I-T)(\Omega)$.

4. *Assume that there are $u \in \partial U$ and $\lambda \geq 1$ so that*

$$Su = (I-T)(\lambda u) \quad and \quad \lambda u \in \Omega.$$

We have

$$Su = \varepsilon \lambda u, \quad \|u\|_c = r,$$

and

$$\varepsilon \frac{M^{q-1}}{q} \geq \|Su\|_c = \varepsilon \lambda \|u\|_c \geq \varepsilon r,$$

whereupon

$$r \leq \frac{M^{q-1}}{q} < 1.$$

This is a contradiction, because $r > 1$.

Hence and Proposition 2.2.61, it follows that the operator $T + S$ has a fixed point $u \in Q$, which is a bounded solution of the BVP (4.35). This completes the proof of the main result.

Example 4.6.3 *For $m, k \geq 0$, consider the BVP:*

$$\left(|u'|^3 u'\right)'(t) + (u^m(t) + \ln(u^k(t) + 1)) = 0, \quad 0 < t < 1,$$
$$(4.37)$$
$$u'(0) = 0, \quad u(1) = 0.$$

Here $E = \mathbb{R}$, $\mathscr{P} = \mathbb{R}^+$, $\phi_p(s) = |s|^3 s$ $(p = 5, q = \dfrac{5}{4})$, and $f(y) = y^m + \ln(y^k + 1)$. Clearly, the function f is positive continuous and bounded when y is bounded.
Moreover, for some $r > 1$, (A11) is satisfied for all constants m and k satisfying $r^m + \ln(r^k + 1) < (\dfrac{5}{4})^4$.
Therefore, the problem (4.37) has a bounded positive solution.

Example 4.6.4 *Consider the following BVP of infinite system of scalar differential equations in the infinite-dimensional Banach space $E = l^\infty = \{u = (u_1, \ldots, u_n, \ldots) \mid \sup_n |u_n| < +\infty\}$ with the sup-norm $\|u\| = \sup_n |u_n|$:*

$$\left(|u_n'|u_n'\right)'(t) + \frac{1}{100}(|\sin u_{n+1}(t)| + 5 u_n^2(t)) = 0, \quad 0 < t < 1,$$

$$x_n'(0) = 0, \quad x_n(1) = 0, \quad n = 1, 2, \ldots$$
$$(4.38)$$

Let $\mathscr{P} = \{x = (x_n) \in l^\infty \mid x_n \geq 0, \ n = 1, 2, \ldots\}$. It is easy to see that $\mathscr{P}$ is a cone in E. System (4.38) can be regarded as a BVP of the form

(4.35) in l^∞ with $\phi_p(s) = |s|s$ $(p = 3, q = \dfrac{3}{2})$, $u = (u_1,\ldots,u_n,\ldots)$, $f = (f_1,\ldots,f_n,\ldots)$,

$$f_n(u(t)) = \frac{1}{100}(|\sin u_{n+1}(t)| + 5u_n^2(t)), \quad for \ n = 1,2,\ldots.$$

Then $f \in \mathscr{C}(\mathscr{P},\mathscr{P})$. Furthermore, for any $r > 1$ satisfying

$$\frac{1}{100}(1 + 5r^2) < \frac{9}{4},$$

$$\sup\{\|f(u(t))\| : u \in Q \cap B_r\} \leq M = \frac{1}{100}(1 + 5r^2) < \infty,$$

and

$$\frac{M^{q-1}}{q} < 1.$$

Therefore, the system (4.38) has a bounded positive solution.

4.7   A Three-Point Fourth-Order Eigenvalue BVP

Here we will give an application of Theorem 3.2.3. In this section, we will investigate the following eigenvalue three-point boundary value problem:

$$
\begin{aligned}
u^{(4)} \quad &+ \quad \lambda g(t)f(u) = 0, \quad 0 < t < 1, \\
u(0) \quad &= \quad u'(1) = u''(0) = u''(p) - u''(1) = 0,
\end{aligned}
\tag{4.39}
$$

where

(A12) $\lambda > 0$, $\lambda \neq 1$, $0 < p < 1$.

(A13) $g \in \mathscr{C}([0,1])$ is a non-negative function such that

$$0 \leq g(t) \leq N, \quad t \in [0,1],$$

N is a positive constant.

(A14) $f \in \mathscr{C}([0,\infty))$ is a non-negative function,

$$f(0) = 0, \quad |f(x) - f(y)| \leq b|x - y|, \quad x, y \in [0,\infty),$$

b is a positive constant.

Define

$$G_1(t,s) = \begin{cases} t & \text{if } t \leq s \\[2mm] s & \text{if } s \leq t, \end{cases}$$

$$G_2(t,s) = \begin{cases} t & \text{if } t \leq s \leq p \\[2mm] s & \text{if } s \leq t \text{ and } s \leq p \\[2mm] \dfrac{1-s}{1-p}t & \text{if } s \geq p \text{ and } t \leq s \\[2mm] s + \dfrac{p-s}{1-p}t & \text{if } t \geq s \geq p, \end{cases}$$

$$J(t,s) = \int_0^1 G_1(t,v)G_2(v,s)dv, \quad t,s \in [0,1].$$

We have $J(0,s) = 0$, $s \in [0,1]$, and

$$0 \leq G_1(t,s), G_2(t,s) \leq 1, \quad t,s \in [0,1].$$

Thus,

$$0 \leq J(t,s) \leq 1, \quad t,s \in [0,1].$$

When $\lambda \neq 1$ we have that $J(\cdot,\cdot)$ is the Green function for the BVP (4.39). Then the solutions of the BVP (4.39) can be represented in the form

$$u(t) = -\lambda \int_0^1 J(t,s)g(s)f(u(s))ds, \quad t \in [0,1].$$

Let $E = \mathscr{C}([0,1])$ be endowed with the maximum norm. Below, suppose that $z_0, r_1, r_2, L_1, L_2, b, N, d, c, m, \varepsilon$ and R are positive constants that satisfy the following inequalities

(A15)

$$r_2 \ = \ L_2 \geq d > \frac{d}{\varepsilon} \geq z_0 > c > r_1,$$

$$(4.40)$$

$$m \ > \ 1, \quad r_1 = L_1, \quad \lambda = \frac{\varepsilon}{\varepsilon + 1 - \frac{1}{m}},$$

$$0 < A_1 < 1, \quad \varepsilon > 2, \quad \varepsilon(1 - A_1) > 1, \qquad (4.41)$$

$$z_0 \ < \ (\varepsilon(1-A_1)-1)R, \quad \frac{r_2}{m(\varepsilon+1-\varepsilon A_1)} \leq r_1. \qquad (4.42)$$

Here $A_1 = bN$.

After the proof of the main result in this section, we will give an example for such constants and functions f and g that satisfy $(A12)$–$(A15)$. Our main result in this section is as follows.

Theorem 4.7.1 *Suppose $(A12)$-$(A15)$. Then the BVP (4.39) has at least three non-negative solutions.*

Proof 4.7.2 *For $u \in E$, define the operators*

$$T_1 u(t) \ = \ \int_0^1 J(t,s)g(s)f(u(s))ds,$$

$$T_2 u(t) \ = \ -\varepsilon u(t) - \varepsilon T_1 u(t),$$

$$T u(t) \ = \ T_2 u(t) - z_0,$$

$$F_1 u(t) \ = \ \frac{1}{m}u(t),$$

$$F u(t) \ = \ F_1 u(t) + z_0 \quad t \in [0,1].$$

Define the functional $\alpha : E \to \mathbb{R}$ as follows

$$\alpha(u) = \begin{cases} z_0 & \text{if } u(0) = 0 \\ |u(0)| & \text{if } u(0) \neq 0. \end{cases}$$

Next, for $u \in E$, define the functionals

$$\psi(u) = \alpha(u), \quad \beta(u) = \|u\|.$$

Let

$$\mathscr{P} \; = \; \{u \in E : u(t) \geq 0, \quad t \in [0,1]\}.$$

Note that any fixed point $u \in \mathscr{P}$ of the operator $T + F$ is a solution to the eigenvalue BVP (4.39). We have that α and β are convex functionals on $\mathscr{P}$ and ψ is a concave functional on $\mathscr{P}$ and

$$\psi(u) = \alpha(u), \quad \|u\| \leq \max(\alpha(u), \beta(u)) \quad u \in \mathscr{P}.$$

Now, let $r > 0$ and $L > 0$ be arbitrarily chosen. Let also, $\widetilde{L} = \min\{r, L\}$. Then $\dfrac{\widetilde{L}}{2} \in \mathscr{P}$ and

$$\alpha\left(\frac{\widetilde{L}}{2}\right) = \frac{\widetilde{L}}{2} < r, \quad \beta\left(\frac{\widetilde{L}}{2}\right) = \frac{\widetilde{L}}{2} < L.$$

Therefore $\dfrac{\widetilde{L}}{2} \in \mathscr{P}(\alpha, r; \beta, L)$ and $\mathscr{P}(\alpha, r; \beta, L) \neq \emptyset$. So, $(\mathscr{A}_2)$ holds. Let $\Omega = \overline{\mathscr{P}}\left(\alpha, \dfrac{d}{\varepsilon}; \beta, \dfrac{d}{\varepsilon}\right)$. We have that $0 \in \Omega$ and $\Omega \subset \overline{\mathscr{P}}(\alpha, r_2; \beta, L_2)$. For $u, u_1, u_2 \in \mathscr{P}$, we have

$$T_1 u(t) \; \leq \; \int_0^1 J(t,s)g(s)f(u(s))ds$$

$$\leq \; bN \int_0^1 u(s)ds$$

$$\leq \; bN\|u\|$$

$$= \; A_1\|u\|, \quad t \in [0,1],$$

$$\|T_1 u_1 - T_1 u_2\| \; = \; \max_{t \in [0,1]} \left| \int_0^1 J(t,s)g(s)(f(u_1(s)) - f(u_2(s)))ds \right|$$

$$\leq \int_0^1 g(s)|f(u_1(s)) - f(u_2(s))|ds$$

$$\leq bN \int_0^1 |u_1(s) - u_2(s)|ds$$

$$\leq A_1\|u_1 - u_2\|.$$

1. Let $u_1, u_2 \in \Omega$ be arbitrarily chosen. Then

$$\|Tu_1 - Tu_2\| = \|\varepsilon(u_1 - u_2) + \varepsilon(T_1u_1 - T_1u_2)\|$$

$$\geq \varepsilon\|u_1 - u_2\| - \varepsilon\|T_1u_1 - T_1u_2\|$$

$$\geq \varepsilon\|u_1 - u_2\| - \varepsilon A_1\|u_1 - u_2\|$$

$$= \varepsilon(1 - A_1)\|u_1 - u_2\|.$$

Therefore $T : \Omega \to E$ is an expansive operator with a constant $h = \varepsilon(1 - A_1) > 1$.

2. We have $F : \overline{\mathscr{P}}(\alpha, r_2; \beta, L_2) \to E$ is a completely continuous operator and then it is a 0-set contraction.

3. Let $u \in \overline{\mathscr{P}}(\alpha, r_2; \beta, L_2)$ be arbitrarily chosen. Then

$$\|(I - T_2)u\| = \|(1 + \varepsilon)u + \varepsilon T_1u\|$$

$$\geq (1 + \varepsilon)\|u\| - \varepsilon\|T_1u\|$$

$$\geq (\varepsilon + 1 - \varepsilon A_1)\|u\|$$

and

$$\|(I - T_2)u\| = \|(1 + \varepsilon)u + \varepsilon T_1u\|$$

$$\leq (\varepsilon + 1)\|u\| + \varepsilon\|T_1u\|$$

$$\leq \ (\varepsilon + 1 + \varepsilon A_1) \|u\|.$$

Take

$$u_1 = (I - T_2)^{-1} F_1 u.$$

Then $u_1 \in \mathscr{P}$ and

$$\|u_1\| \ = \ \|(I - T_2)^{-1} F_1 u\|$$

$$\leq \ \frac{1}{m(\varepsilon + 1 - \varepsilon A_1)} \|u\|$$

$$\leq \ \frac{r_2}{m(\varepsilon + 1 - \varepsilon A_1)}$$

$$\leq \ r_1$$

and

$$(I - T)u_1 \ = \ (I - T_2)u_1 + z_0$$

$$= \ F_1 u + z_0$$

$$= \ F u.$$

Therefore $u_1 \in \Omega$ and

$$F(\overline{\mathscr{P}}(\alpha, r_2; \beta, L_2)) \subset (I - T)(\overline{\mathscr{P}}(\alpha, r_1; \beta, L_1) \cap \Omega).$$

4. *Let $u \in \mathscr{P}$. Then*

$$Tu + Fu \ = \ -\left(\varepsilon - \frac{1}{m}\right)u - \varepsilon T_1 u,$$

$$(Tu + Fu)(0) \ = \ -\left(\varepsilon - \frac{1}{m}\right)u(0),$$

$$\beta(Tu + Fu) \ = \ \beta\left(-\left(\varepsilon - \frac{1}{m}\right)u - \varepsilon T_1 u\right)$$

$$= \beta \left(\left(\varepsilon - \frac{1}{m} \right) u + \varepsilon T_1 u \right)$$

$$\geq \left(\varepsilon - \frac{1}{m} \right) \beta(u).$$

Let $\alpha(u) = r_1$. If $u(0) = 0$, then

$$\alpha(Tu + Fu) = z_0 > r_1.$$

If $u(0) \neq 0$, then $r_1 = \alpha(u) = |u(0)|$ and

$$\alpha(Tu + Fu) = \left(\varepsilon - \frac{1}{m} \right) u(0) = \left(\varepsilon - \frac{1}{m} \right) r_1 > r_1.$$

If $\beta(u) = L_1$, then

$$\beta(Tu + Fu) \geq \left(\varepsilon - \frac{1}{m} \right) L_1 > L_1.$$

5. We have $z_0 \in \{ x \in \overline{\mathscr{P}}(\alpha, d; \beta, L_2; \psi, c) : \psi(x) > c \}$,

$z_0 < (\varepsilon(1 - A_1) - 1)R$ and $\psi \left((I - T)^{-1} z_0 \right) = \psi(0) = z_0 > c$.

Let now, $u \in \overline{\mathscr{P}}(\alpha, r_2; \beta, L_2)$ and $\mu \in [0, 1]$ be arbitrarily chosen. Take

$$u_2 = (I - T_2)^{-1}(\mu F_1 u).$$

We have $u_2 \in \mathscr{P}$ and

$$\|u_2\| \leq \frac{\|u\|}{m(\varepsilon + 1 - \varepsilon A_1)}$$

$$\leq \frac{r_2}{m(\varepsilon + 1 - \varepsilon A_1)}$$

$$\leq \frac{d}{\varepsilon}$$

and

$$(I - T)u_2 = (I - T_2)u_2 + z_0$$

$$= \mu F_1 u + \mu z_0 + (1 - \mu)z_0$$

$$= \mu F u + (1 - \mu)z_0.$$

Therefore $u_2 \in \Omega$ and

$$\mu F(\overline{\mathscr{P}}(\alpha, r_2; \beta, L_2)) + (1 - \mu)z_0 \subset (I - T)(\Omega), \quad \mu \in [0, 1].$$

6. *Let $u \in \overline{\mathscr{P}}(\alpha, d; \beta, L_2; \psi, c) \cap \Omega$ be arbitrarily chosen.*
Then $\alpha(u) \leq \dfrac{d}{\varepsilon}$ and

$$\psi(Tu + Fu) = \psi\left(\left(\varepsilon - \frac{1}{m}\right)u\right)$$

$$= \begin{cases} \left(\varepsilon - \dfrac{1}{m}\right)u(0) > c & \text{if } u(0) \neq 0 \\[4mm] z_0 > c & \text{if } u(0) = 0, \end{cases}$$

$$\psi(Tu + z_0) = \psi(T_2 u) = \psi(\varepsilon u)$$

$$= \begin{cases} \varepsilon u(0) > c & \text{if } u(0) \neq 0 \\[4mm] z_0 > c & \text{if } u(0) = 0, \end{cases}$$

$$\alpha(Tu + z_0) = \alpha(T_2 u) = \alpha(\varepsilon u)$$

$$= \begin{cases} \varepsilon u(0) \leq d & \text{if } u(0) \neq 0 \\[4mm] z_0 \leq d & \text{if } u(0) = 0. \end{cases}$$

7. *Let $u \in \{\overline{\mathscr{P}}(\alpha, r_2; \beta, L_2; \psi, c) \cap \Omega : \alpha(Tu + Fu) > d\}$.*
Then $\alpha(u) \leq \dfrac{d}{\varepsilon}$ and

$$\psi(Tu + Fu) = \psi\left(\left(\varepsilon - \frac{1}{m}\right)u\right)$$

$$= \begin{cases} \left(\varepsilon - \dfrac{1}{m}\right) u(0) > c & \text{if} \quad u(0) \neq 0 \\[2ex] z_0 > c & \text{if} \quad u(0) = 0, \end{cases}$$

$$\psi(Tu + z_0) = \psi(T_2 u) = \psi(\varepsilon u)$$

$$= \begin{cases} \varepsilon u(0) > c & \text{if} \quad u(0) \neq 0 \\[2ex] z_0 > c & \text{if} \quad u(0) = 0, \end{cases}$$

$$\alpha(Tu + z_0) = \alpha(T_2 u) = \alpha(\varepsilon u)$$

$$= \begin{cases} \varepsilon u(0) \leq r_2 & \text{if} \quad u(0) \neq 0 \\[2ex] z_0 \leq r_2 & \text{if} \quad u(0) = 0. \end{cases}$$

Thus, the condition (C5) in Theorem 3.2.3 holds.

Hence and Theorem 3.2.3, it follows that the problem (4.39) has at least three solutions u_1, u_2, u_3 such that $u_1 \in \mathscr{P}(\alpha, r_1; \beta, L_1)$,

$$u_2 \in \{x \in \overline{\mathscr{P}}(\alpha, r_2; \beta, L_2; \psi, c) : \psi(x) > c\}$$

and

$$u_3 \in \overline{\mathscr{P}}(\alpha, r_2; \beta, L_2) \backslash \left(\overline{\mathscr{P}}(\alpha, r_2; \beta, L_2; \psi, c) \cup \overline{\mathscr{P}}(\alpha, r_1; \beta, L_1) \right).$$

This completes the proof.

Example 4.7.3 *Let*

$$r_2 = L_2 = d = b = 1, \quad N = \frac{1}{2}, \quad \varepsilon = \frac{1994}{3},$$

$$z_0 = \frac{1}{1994}, \quad R = 1, \quad m = 1994, \quad L_1 = r_1 = \frac{3}{1994 \cdot 10^3},$$

$$\lambda = \frac{1994^2}{1994 \cdot 1997 - 3}, \quad c = \frac{3}{4 \cdot 1994}.$$

1. We have

$$\varepsilon + 1 - \frac{1}{m} = \frac{1994}{3} + 1 - \frac{1}{1994} = \frac{1994 \cdot 1997 - 3}{3 \cdot 1994},$$

$$\frac{\varepsilon}{\varepsilon + 1 - \frac{1}{m}} = \frac{\frac{1994}{3}}{\frac{1994 \cdot 1997 - 3}{3 \cdot 1994}} = \frac{1994^2}{1994 \cdot 1997 - 3} = \lambda$$

and $\dfrac{d}{\varepsilon} = \dfrac{3}{1994}$ *and*

$$r_2 = L_2 = d > \frac{d}{\varepsilon} > z_0 > c > r_1, \quad m > 1, \quad r_1 = L_1,$$

i.e., (4.40) holds.

2. Note that $A_1 = bN = \dfrac{1}{2}$ *and*

$$0 < A_1 < 1, \quad \varepsilon > 2, \quad \varepsilon(1 - A_1) = \frac{997}{3} > 1,$$

i.e., (4.41) holds.

3. We have

$$(\varepsilon(1 - A_1) - 1)R = \frac{997}{3} - 1 = \frac{994}{3} > z_0$$

and

$$\frac{d}{10^3 \varepsilon} = \frac{1}{10^3 \cdot \frac{1994}{3}} = \frac{3}{1994 \cdot 10^3} = r_1,$$

$$\varepsilon + 1 - \varepsilon A_1 = \varepsilon(1 - A_1) + 1 = \frac{997}{3} + 1 = \frac{10^3}{3},$$

$$\frac{r_2}{m(\varepsilon + 1 - \varepsilon A_1)} = \frac{1}{1994 \cdot \frac{10^3}{3}} = \frac{3}{1994 \cdot 10^3} = r_1,$$

i.e., (4.42) holds.

Now, consider the BVP

$$u^{(4)} + \frac{1994^2}{1994 \cdot 1997 - 3} \left(\frac{1}{2(1+t^2)} \right) \left(\frac{u}{1+u} \right) = 0, \quad t \in (0,1),$$

$$u(0) = u'(1) = u''(0) = u'' \left(\frac{1}{2} \right) - u''(1) = 0. \tag{4.43}$$

Here

$$\lambda = \frac{1994^2}{1994 \cdot 1997 - 3}, \quad g(t) = \frac{1}{2(1+t^2)}, \quad f(u) = \frac{u}{1+u},$$

$t \in [0,1]$, $u \in [0,\infty)$. *Then*

$$0 \leq g(t) \leq \frac{1}{2}, \quad t \in [0,1],$$

and

$$f(0) = 0, \quad |f(x_1) - f(x_2)| \leq |x_1 - x_2|, \quad x_1, x_2 \in [0,\infty).$$

Hence, and Theorem 4.7.1, it follows that the problem (4.43) has at least three non-negative solutions.

Now, suppose that z_0, r_1, r_2, L_1, L_2, b, N, d, c, m, ε, and R are positive constants that satisfy the following inequalities

(**A16**)

$$r_2 = L_2 \geq d > \frac{d}{\varepsilon + 1} \geq z_0 > c > r_1,$$

$$\tag{4.44}$$

$$1 > \varepsilon > \frac{1}{m}, \quad r_1 = L_1, \quad \lambda = \frac{\varepsilon}{\varepsilon - \frac{1}{m}}, \quad 0 < A_1 < 1,$$

$$z_0 < \varepsilon(1 - A_1)R, \quad \frac{r_2}{m\varepsilon(1 - A_1)} \leq r_1. \tag{4.45}$$

Here $A_1 = bN$.

Our next main result is as follows.

Theorem 4.7.4 *Suppose* (A12)-(A14) *and* (A16). *Then the BVP* (4.39) *has at least three non-negative solutions.*

Proof 4.7.5 *For* $u \in E$, *define the operators*

$$T_3 u(t) \;=\; (\varepsilon + 1) u(t) + \varepsilon T_1 u(t),$$

$$T_4 u(t) \;=\; T_3 u(t) - z_0,$$

$$F_2 u(t) \;=\; -F_1 u(t) + z_0 \quad t \in [0,1],$$

where F_1 *and* T_1 *are as in the proof of Theorem 4.7.1. Let* $\mathscr{P}$, α, β, *and* ψ *are as in the proof of Theorem 4.7.1. Any fixed point* $u \in \mathscr{P}$ *of the operator* $T_4 + F_2$ *is a solution to the eigenvalue BVP* (4.39). *Let* $\Omega = \overline{\mathscr{P}}(\alpha, \dfrac{d}{1+\varepsilon}; \beta, \dfrac{d}{1+\varepsilon})$. *We have that* $0 \in \Omega$ *and* $\Omega \subset \overline{\mathscr{P}}(\alpha, r_2; \beta, L_2)$.

1. Let $u_1, u_2 \in \Omega$ *be arbitrarily chosen. Then*

$$
\begin{aligned}
\|T_4 u_1 - T_4 u_2\| &= \|(\varepsilon + 1)(u_1 - u_2) + \varepsilon(T_1 u_1 - T_1 u_2)\| \\[1mm]
&\geq (\varepsilon + 1)\|u_1 - u_2\| - \varepsilon\|T_1 u_1 - T_1 u_2\| \\[1mm]
&\geq (\varepsilon + 1)\|u_1 - u_2\| - \varepsilon A_1\|u_1 - u_2\| \\[1mm]
&= (\varepsilon(1 - A_1) + 1)\|u_1 - u_2\|.
\end{aligned}
$$

Therefore, $T_4 : \Omega \to E$ *is an expansive operator with a constant* $h = \varepsilon(1 - A_1) + 1 > 1$.

2. We have $F_2 : \overline{\mathscr{P}}(\alpha, r_2; \beta, L_2) \to E$ *is a completely continuous operator and then it is a 0-set contraction.*

3. Let $u \in \overline{\mathscr{P}}(\alpha, r_2; \beta, L_2)$ *be arbitrarily chosen. Then*

$$
\begin{aligned}
\|(I - T_3) u\| &= \|\varepsilon u + \varepsilon T_1 u\| \\[1mm]
&\geq \varepsilon\|u\| - \varepsilon\|T_1 u\| \\[1mm]
&\geq \varepsilon(1 - A_1)\|u\|
\end{aligned}
$$

and

$$\|(I - T_3)u\| = \|\varepsilon u + \varepsilon T_1 u\|$$

$$\leq \varepsilon\|u\| + \varepsilon\|T_1 u\|$$

$$\leq \varepsilon(1 + A_1)\|u\|.$$

Take

$$u_1 = -(I - T_3)^{-1} F_1 u.$$

Then $u_1 \in \mathscr{P}$ and

$$\|u_1\| = \|(I - T_3)^{-1} F_1 u\|$$

$$\leq \frac{1}{m\varepsilon(1 - A_1)}\|u\|$$

$$\leq \frac{r_2}{m\varepsilon(1 - A_1)}$$

$$\leq r_1$$

and

$$(I - T_4)u_1 = (I - T_3)u_1 + z_0$$

$$= -F_1 u + z_0$$

$$= F_2 u.$$

Therefore, $u_1 \in \Omega$ and

$$F_2(\overline{\mathscr{P}}(\alpha, r_2; \beta, L_2)) \subset (I - T_4)(\overline{\mathscr{P}}(\alpha, r_1; \beta, L_1) \cap \Omega).$$

4. *Let $u \in \mathscr{P}$. Then*

$$T_4 u + F_2 u = \left(\varepsilon + 1 - \frac{1}{m}\right)u + \varepsilon T_1 u,$$

$$(T_4u + F_2u)(0) = \left(\varepsilon + 1 - \frac{1}{m}\right)u(0),$$

$$\beta(T_4u + F_2u) = \beta\left(\left(\varepsilon + 1 - \frac{1}{m}\right)u + \varepsilon T_1 u\right)$$

$$= \beta\left(\left(\varepsilon + 1 - \frac{1}{m}\right)u + \varepsilon T_1 u\right)$$

$$\geq \left(\varepsilon + 1 - \frac{1}{m}\right)\beta(u).$$

Let $\alpha(u) = r_1$. If $u(0) = 0$, then

$$\alpha(T_4u + F_2u) = z_0 > r_1.$$

If $u(0) \neq 0$, then

$$\alpha(T_4u + F_2u) = \left(\varepsilon + 1 - \frac{1}{m}\right)u(0) = \left(\varepsilon + 1 - \frac{1}{m}\right)r_1 > r_1.$$

If $\beta(u) = L_1$, then

$$\beta(T_4u + F_2u) \geq \left(\varepsilon + 1 - \frac{1}{m}\right)L_1 > L_1.$$

5. We have $z_0 \in \{x \in \overline{\mathscr{P}}(\alpha, d; \beta, L_2; \psi, c) : \psi(x) > c\}$,

$$z_0 < \varepsilon(1 - A_1)R \quad and \quad \psi\left((I - T_4)^{-1}z_0\right) = \psi(0) = z_0 > c.$$

Let now, $u \in \overline{\mathscr{P}}(\alpha, r_2; \beta, L_2)$ and $\mu \in [0, 1]$ be arbitrarily chosen. Take

$$u_2 = -(I - T_3)^{-1}(\mu F_1 u).$$

We have $u_2 \in \mathscr{P}$ and

$$\|u_2\| \leq \frac{\|u\|}{m\varepsilon(1 - A_1)}$$

$$\leq \frac{r_2}{m\varepsilon(1 - A_1)}$$

$$\leq \frac{d}{\varepsilon + 1}$$

and

$$(I - T_4)u_2 = (I - T_3)u_2 + z_0$$

$$= -\mu F_1 u + \mu z_0 + (1 - \mu)z_0$$

$$= \mu F_2 u + (1 - \mu)z_0.$$

Therefore $u_2 \in \Omega$ and

$$\mu F_2(\overline{\mathscr{P}}(\alpha, r_2; \beta, L_2)) + (1 - \mu)z_0 \subset (I - T_4)(\Omega), \quad \mu \in [0, 1].$$

6. *Let $u \in \overline{\mathscr{P}}(\alpha, d; \beta, L_2; \psi, c) \cap \Omega$ be arbitrarily chosen. Then $\alpha(u) \leq \dfrac{d}{1 + \varepsilon}$ and*

$$\psi(T_4 u + F_2 u) = \psi\left(\left(\varepsilon + 1 - \frac{1}{m}\right)u\right)$$

$$= \begin{cases} \left(\varepsilon + 1 - \dfrac{1}{m}\right)u(0) > c & \text{if} \quad u(0) \neq 0 \\ z_0 > c & \text{if} \quad u(0) = 0, \end{cases}$$

$$\psi(T_4 u + z_0) = \psi(T_3 u) = \psi((1 + \varepsilon)u)$$

$$= \begin{cases} (\varepsilon + 1)u(0) > c & \text{if} \quad u(0) \neq 0 \\ z_0 > c & \text{if} \quad u(0) = 0, \end{cases}$$

$$\alpha(T_4 u + z_0) = \alpha(T_3 u)$$

$$= \begin{cases} (\varepsilon + 1)u(0) \leq d & \text{if} \quad u(0) \neq 0 \\ z_0 \leq d & \text{if} \quad u(0) = 0. \end{cases}$$

7. *Let $u \in \{\overline{\mathscr{P}}(\alpha, r_2; \beta, L_2; \psi, c) \cap \Omega : \alpha(T_4 u + F_2 u) > d\}$.*

Then $\alpha(u) \leq \dfrac{d}{1+\varepsilon}$ *and*

$$\psi(T_4u + F_2u) = \psi\left(\left(\varepsilon + 1 - \frac{1}{m}\right)u\right)$$

$$= \begin{cases} \left(\varepsilon + 1 - \dfrac{1}{m}\right)u(0) > c & \text{if } u(0) \neq 0 \\[2mm] z_0 > c & \text{if } u(0) = 0, \end{cases}$$

$$\psi(T_4u + z_0) = \psi(T_3u) = \psi((1+\varepsilon)u)$$

$$= \begin{cases} (\varepsilon + 1)u(0) > c & \text{if } u(0) \neq 0 \\[2mm] z_0 > c & \text{if } u(0) = 0, \end{cases}$$

$$\alpha(T_4u + z_0) = \alpha(T_3u)$$

$$= \begin{cases} (\varepsilon + 1)u(0) \leq r_2 & \text{if } u(0) \neq 0 \\[2mm] z_0 \leq r_2 & \text{if } u(0) = 0. \end{cases}$$

Hence, and Theorem 3.2.3, it follows that the problem (4.39) has at least three solutions u_1, u_2, u_3 such that $u_1 \in \mathscr{P}(\alpha, r_1; \beta, L_1)$,

$$u_2 \in \{x \in \overline{\mathscr{P}}(\alpha, r_2; \beta, L_2; \psi, c) : \psi(x) > c\}$$

and

$$u_3 \in \overline{\mathscr{P}}(\alpha, r_2; \beta, L_2) \setminus \left(\overline{\mathscr{P}}(\alpha, r_2; \beta, L_2; \psi, c) \cup \overline{\mathscr{P}}(\alpha, r_1; \beta, L_1)\right).$$

This completes the proof.

Example 4.7.6 *Let*

$$r_2 = L_2 = d = b = 1, \quad N = \varepsilon = A_1 = \frac{1}{2}, \quad m = 40,$$

$$z_0 = \frac{1}{5}, \quad c = \frac{1}{9}, \quad r_1 = \frac{1}{10}, \quad \lambda = \frac{20}{19}, \quad R = 1.$$

Then

$$A_1 = \frac{1}{2}, \quad \varepsilon + 1 = \frac{3}{2}, \quad \frac{d}{\varepsilon + 1} = \frac{2}{3}, \quad \varepsilon - \frac{1}{m} = \frac{1}{2} - \frac{1}{40} = \frac{19}{40}.$$

1. We have

$$r_2 = L_2 = d > 1 > \frac{2}{3} = \frac{d}{\varepsilon + 1} > \frac{1}{5} = z_0 > \frac{1}{9} = c > \frac{1}{10} = r_1,$$

and

$$1 > \frac{1}{2} = \varepsilon > \frac{1}{40} = \frac{1}{m},$$

$$\frac{\varepsilon}{\varepsilon - \frac{1}{m}} = \frac{\frac{1}{2}}{\frac{19}{40}} = \frac{20}{19} = \lambda, \quad 0 < A_1 < 1,$$

i.e., (4.44) holds.

2. We have

$$\varepsilon(1 - A_1)R = \frac{1}{4} > \frac{1}{5} = z_0,$$

$$m\varepsilon(1 - A_1) = 40 \cdot \frac{1}{2} \cdot \frac{1}{2} = 10, \quad \frac{r_2}{m\varepsilon(1 - A_1)} = r_1,$$

i.e., (4.45).

Now, consider the BVP

$$u^{(4)} + \frac{10}{19} \left(\frac{1}{1 + t^2} \right) \left(\frac{u}{1 + u} \right) = 0, \quad t \in (0, 1),$$

$$u(0) = u'(1) = u''(0) = u'' \left(\frac{1}{2} \right) - u''(1) = 0. \tag{4.46}$$

Here

$$\lambda = \frac{20}{19}, \quad g(t) = \frac{1}{2(1 + t^2)}, \quad f(u) = \frac{u}{1 + u},$$

$t \in [0, 1], u \in [0, \infty)$. *Then*

$$0 \le g(t) \le \frac{1}{2}, \quad t \in [0, 1],$$

and

$$f(0) = 0, \quad |f(x_1) - f(x_2)| \le |x_1 - x_2|, \quad x_1, x_2 \in [0, \infty).$$

Hence, and Theorem 4.7.4, it follows that the problem (4.46) has at least three non-negative solutions.

5

Applications to Parabolic Equations

In this chapter, we will investigate applications of some results in Chapters 2 and 3 for classes parabolic equations. We give criteria for existence of classical solutions for classes IBVPs for nonlinear parabolic equations. Next, we give an application for existence of classical solutions for the Burgers-Fisher equation. Some of the results in this chapter can be found in [27, 28].

5.1 Existence of Solutions of a Class IBVPs for Parabolic Equations

In this section, we will give an application of Proposition 2.3.1 for existence of classical solutions for a class of IBVPs for parabolic equations. Let the positive constants m, k, s, A, T, R, and $a \geq 0$ satisfy the following conditions

$$3A + aTRA + AR^m + AR^k < \frac{2}{3}R, \quad 0 < s < \frac{1}{2}, \tag{5.1}$$

$$R + 3A + aTR + AR^m + AR^k < s. \tag{5.2}$$

Consider the IBVP

$$
\begin{aligned}
u_t &= \Delta u + f(t, x, u), \quad (t, x) \in [0, T] \times \Omega, \\
u(0, x) &= \phi(x), \quad x \in \Omega, \\
\frac{\partial u}{\partial n} &= au + g \quad \text{on} \quad [0, T] \times \partial\Omega,
\end{aligned}
\tag{5.3}
$$

where $\Omega \subset \mathbb{R}^n$ is a bounded open set, $\partial\Omega$ is smoothed,

$$\int_{\partial\Omega} d\sigma \leq A,$$

DOI: 10.1201/9781003381969-5

$d\sigma$ is a vector element of $\partial\Omega$,

(B1) $f : [0,T] \times \overline{\Omega} \times \mathbb{R} \to [0,\infty)$, $f \in \mathscr{C}([0,T] \times \overline{\Omega} \times \mathbb{R})$,

$$f(t,x,u) \le a_0(t,x) + a_1(t,x)u^m + a_2(t,x)u^k, \quad (t,x) \in [0,T] \times \overline{\Omega},$$

$u \in [0,\infty)$, $a_j \in \mathscr{C}([0,T] \times \overline{\Omega})$, $j = 0,1,2$,

$$\int_0^T \int_\Omega a_j(\tau,x)dxd\tau \ \le \ A, \quad j = 0,1,2,$$

(B2) $g : [0,T] \times \partial\Omega \to [0,\infty)$, $g \in \mathscr{C}([0,T] \times \partial\Omega)$,

$$\int_0^T \int_{\partial\Omega} g(\tau,x)d\sigma d\tau \le A,$$

(B3) $\phi \in \mathscr{C}(\overline{\Omega})$, $B_0 \le \phi(x) \le A$, $x \in \overline{\Omega}$.

Definition 5.1.1 *We say that $u \in \mathscr{C}([0,T] \times \overline{\Omega})$ is a solution of the IBVP (5.3), if it satisfies the integral equation*

$$u(t,x) \ = \ \phi(x) + a\int_0^t \int_{\partial\Omega} u(\tau,x)d\sigma d\tau + \int_0^t \int_{\partial\Omega} g(\tau,x)d\sigma d\tau$$

$$+ \int_0^t \int_\Omega f(\tau,y,u(\tau,y))dyd\tau, \quad (t,x) \in [0,T] \times \overline{\Omega}.$$

Let $E = \mathscr{C}([0,T] \times \overline{\Omega})$ be endowed with the norm

$$\|u\| = \max_{(t,x) \in [0,T] \times \overline{\Omega}} |u(t,x)|.$$

Define

$$\mathscr{P} = \{u \in E : u(t,x) \ge 0, \quad (t,x) \in [0,T] \times \overline{\Omega}\}.$$

We take $\theta \in \mathscr{P}$ so that

$$0 \le \theta(t,x) < \frac{R}{3}, \quad (t,x) \in [0,R] \times \overline{\Omega}.$$

Let $\mathscr{K} = \mathscr{P} + \theta$ and $\widetilde{\mathscr{K}}$ be the set of all equi-continuous families in $\widetilde{\mathscr{K}}$. Let also, $\widetilde{\mathscr{K}} = \widetilde{\widetilde{\mathscr{K}}}$. Take $\varepsilon > 0$ arbitrarily. For $u \in E$ define the operators

$$Tu(t,x) = (1+\varepsilon)u(t,x) - \varepsilon R,$$

$$Fu(t,x) = \varepsilon R - \varepsilon \left(\phi(x) + a \int_0^t \int_{\partial \Omega} u(\tau,x) d\sigma d\tau + \int_0^t \int_{\partial \Omega} g(\tau,x) d\sigma d\tau \right.$$

$$\left. + \int_0^t \int_{\Omega} f(\tau,y,u(\tau,y)) dy d\tau \right), \quad (t,x) \in [0,T] \times \overline{\Omega}.$$

Note that any fixed point $u \in E$ of the operator $T + F$ is a solution of the IBVP (5.3).

Theorem 5.1.2 *Suppose* (5.1), (5.2), (B1)-(B3). *Then the IBVP* (5.3) *has at least one positive solution* $u \in \mathscr{P}$.

Proof 5.1.3 *Define*

$$U = \{y \in \widetilde{\mathscr{K}} : \|y\| < R\},$$

$$\Omega_1 = \{y \in \widetilde{\mathscr{K}} : \|y\| < \frac{3}{2}R\}.$$

We have

$$U \subset \Omega_1 \subset \mathscr{K}.$$

Note that $\theta \in U$.

1. Let $u, v \in \Omega_1$. *Then*

$$\|Tu - Tv\| = (1+\varepsilon)\|u - v\|.$$

Therefore $T : \Omega_1 \to E$ *is an* $(\varepsilon + 1)$-*expansive operator.*

2. For $u \in \overline{U}$, we have

$$|Fu(t,x)| \leq \varepsilon R + \varepsilon \phi(x) + \varepsilon a \int_0^T \int_{\partial \Omega} u(\tau,x) d\sigma d\tau$$

$$+\varepsilon \int_0^T \int_{\partial \Omega} g(\tau,x) d\sigma d\tau$$

$$+\varepsilon \int_0^T \int_{\Omega} f(\tau,y,u(\tau,y)) dy d\tau$$

$$\leq \varepsilon R + \varepsilon A + \varepsilon a T R A + \varepsilon A$$

$$+\varepsilon \int_0^T \int_{\Omega} a_0(\tau,y) dy d\tau$$

$$+\varepsilon \int_0^T \int_{\Omega} a_1(\tau,y)(u(\tau,y))^m dy d\tau$$

$$+\varepsilon \int_0^T \int_{\Omega} a_2(\tau,y)(u(\tau,y))^k dy d\tau$$

$$\leq \varepsilon R + \varepsilon A + \varepsilon a T R A + \varepsilon A + \varepsilon A + \varepsilon A R^m + \varepsilon A R^k$$

$$= \varepsilon(R + 3A + a T R A + A R^m + A R^k)$$

$$< s\varepsilon, \quad (t,x) \in [0,T] \times \overline{\Omega}.$$

Therefore, $F : \overline{U} \to E$ is a 0-set contraction.

3. Let $u \in \overline{U}$ be arbitrarily chosen. Take

$$z(t,x) = \phi(x) + a \int_0^T \int_{\partial \Omega} u(\tau,x) d\sigma d\tau$$

$$+ \int_0^T \int_{\partial \Omega} g(\tau,x) d\sigma d\tau + \int_0^T \int_{\Omega} f(\tau,y,u(\tau,y)) dy d\tau,$$

$(t,x) \in [0,T] \times \overline{\Omega}$. We have

$$z(t,x) \leq 3A + a T R A + A R^m + A R^k$$

$$< \frac{2}{3}R, \quad (t,x) \in [0,T] \times \overline{\Omega},$$

i.e., $z \in \Omega_1$, *and*

$$
\begin{aligned}
(I-T)z(t,x) &= -\varepsilon z(t,x) + \varepsilon R \\
&= \varepsilon R - \varepsilon \left(\phi(x) + a \int_0^T \int_{\partial\Omega} u(\tau,x) d\sigma d\tau \right. \\
&\quad + \int_0^T \int_{\partial\Omega} g(\tau,x) d\sigma d\tau \\
&\quad \left. + \int_0^T \int_{\Omega} f(\tau,y,u(\tau,y)) dy d\tau \right) \\
&= Fu(t,x), \quad (t,x) \in [0,T] \times \overline{\Omega}.
\end{aligned}
$$

Therefore

$$F(\overline{U}) \subset (I-T)(\Omega_1).$$

4. *Assume that there exist an* $u \in \partial U$ *and a* $\lambda \geq 1$ *such that*

$$\lambda u + (1 - \lambda)\theta \in \Omega_1$$

and

$$Fu(t,x) = (I-T)(\lambda u + (1-\lambda)\theta)(t,x), \quad (t,x) \in [0,T] \times \overline{\Omega},$$

or

$$
\begin{aligned}
\varepsilon R - \varepsilon &\left(\phi(x) + a \int_0^t \int_{\partial\Omega} u(\tau,x) d\sigma d\tau \right. \\
&+ \int_0^t \int_{\partial\Omega} g(\tau,x) d\sigma d\tau \\
&\left. + \int_0^t \int_{\Omega} f(\tau,y,u(\tau,y)) dy d\tau \right)
\end{aligned}
$$

$$= -\varepsilon\lambda u(t,x) - \varepsilon(1-\lambda)\theta(t,x) + \varepsilon R,$$

or

$$\phi(x) + a\int_0^t \int_{\partial\Omega} u(\tau,x)\,d\sigma d\tau + \int_0^t \int_{\partial\Omega} g(\tau,x)\,d\sigma d\tau$$

$$+ \int_0^t \int_{\Omega} f(\tau,y,u(\tau,y))\,dy d\tau$$

$$= \lambda u(t,x) + (1-\lambda)\theta(t,x), \quad (t,x) \in [0,T] \times \overline{\Omega}.$$

Let $(t_1,x_1) \in [0,T] \times \overline{\Omega}$ be such that

$$u(t_1,x_1) = \frac{3}{2}R \geq R.$$

Since $\lambda \geq 1$, we get

$$(1-\lambda)\theta(t_1,x_1) \geq (1-\lambda)R$$

and

$$\lambda u(t_1,x_1) + (1-\lambda)\theta(t_1,x_1) \geq R.$$

Hence,

$$R \leq \lambda u(t_1,x_1) + (1-\lambda)\theta(t_1,x_1)$$

$$= \phi(x_1) + a\int_0^{t_1} \int_{\partial\Omega} u(\tau,x)\,d\sigma d\tau + \int_0^{t_1} \int_{\partial\Omega} g(\tau,x)\,d\sigma d\tau$$

$$+ \int_0^{t_1} \int_{\Omega} f(\tau,y,u(\tau,y))\,dy d\tau$$

$$\leq 3A + aTRA + AR^m + AR^k$$

$$< \frac{2}{3}R,$$

which is a contradiction.

Hence, and Proposition 2.3.1, it follows that the IBVP (5.3) has at least one positive solution $u \in U$. This completes the proof.

5.2   Existence of Classical Solutions for Burgers-Fisher Equation

In this section, we give an application of Proposition 2.2.61 for existence of classical solutions for the Burgers-Fisher equation. Consider the IVP for Burgers-Fisher equation

$$u_t - u_{xx} + \alpha(t)uu_x = \beta(t)u(1-u), \quad t > 0, \quad x \geq 0, \tag{5.4}$$

$$u(0,x) = u_0(x), \quad x \geq 0, \tag{5.5}$$

where

(B4) $u_0 \in \mathscr{C}^2([0,\infty))$, $r_1 \geq u_0 \geq \dfrac{r_1}{2}$ on $[0,\infty)$, where $r_1 \in \left(0, \dfrac{1}{2}\right)$ is a given constant,

(B5) $\alpha, \beta \in \mathscr{C}([0,\infty))$, $\alpha < 0$, $\beta \geq 0$ on $[0,\infty)$, $A \in (0,1)$ is a constant and g is a positive continuous function on $(0,\infty) \times (0,\infty)$ such that

$$1 - (1 + 2r_1)A > 0, \quad \left(4 + \frac{3}{2}r_1\right)A < \frac{1}{2},$$

and

$$120 \left(1 + t + t^2 + t^3 + t^4\right) \left(1 + x + x^2 + x^3 + x^4 + x^5 + x^6\right)$$

$$\times \int_0^t \int_0^x g(t_1, x_1) \left(1 + \int_0^{t_1} (\beta(t_2) - \alpha(t_2))dt_2\right) dx_1 dt_1 \leq A,$$

$t \geq 0$, $x \geq 0$.

Let $E = \mathscr{C}^1([0,\infty), \mathscr{C}^2([0,\infty)))$ be endowed with the norm

$$\|u\| = \Bigg\{ \sup_{(t,x)\in[0,\infty)\times[0,\infty)} |u(t,x)|, \quad \sup_{(t,x)\in[0,\infty)\times[0,\infty)} \left|\frac{\partial}{\partial t}u(t,x)\right|,$$

$$\sup_{(t,x)\in[0,\infty)\times[0,\infty)} \left|\frac{\partial}{\partial x}u(t,x)\right|, \quad \sup_{(t,x)\in[0,\infty)\times[0,\infty)} \left|\frac{\partial^2}{\partial x^2}u(t,x)\right| \Bigg\}.$$

Lemma 5.2.1 *Suppose (B4) and (B5). If a function $u \in E$ is a solution of the integral equation*

$$0 = \int_0^t \int_0^x (t-t_1)^4 (x-x_1)^4 g(t_1,x_1) \int_0^{t_1} \int_0^{x_1} (x_1-x_2)\beta(t_2)$$

$$\times u(t_2,x_2)(1-u(t_2,x_2))dx_2dt_2dx_1dt_1$$

$$-\frac{1}{2}\int_0^t \int_0^x (t-t_1)^4 (x-x_1)^4 g(t_1,x_1) \int_0^{t_1} \int_0^{x_1} \alpha(t_2)(u(t_2,x_2))^2$$

$$\times dx_2dt_2dx_1dt_1$$

$$+\int_0^t \int_0^x (t-t_1)^4 (x-x_1)^4 g(t_1,x_1) \int_0^{t_1} u(t_2,x_1)dt_2dx_1dt_1$$

$$+\int_0^t \int_0^x (t-t_1)^4 (x-x_1)^4 g(t_1,x_1) \int_0^{x_1} (x_1-x_2)$$

$$\times(u_0(x_2)-u(t_1,x_2))dx_2dx_1dt_1,$$

$(t,x) \in [0,\infty) \times [0,\infty)$, *then it is a solution to the IVP (5.4)-(5.5).*

Proof 5.2.2 *We differentiate the considered integral equation five times in t and five times in x, we get*

$$0 = g(t,x)\int_0^t \int_0^x \int_0^{x_1} \beta(t_1)u(t_1,x_2)(1-u(t_1,x_2))dx_2dx_1dt_1$$

$$-\frac{1}{2}g(t,x)\int_0^t \int_0^x \alpha(t_1)(u(t_1,x_1))^2 dx_1dt_1$$

$$+g(t,x)\int_0^t u(t_1,x)dt_1$$

$$+g(t,x)\int_0^x \int_0^{x_1} (u_0(x_2)-u(t_1,x_2))dx_2dx_1, \quad (t,x) \in [0,\infty) \times [0,\infty),$$

whereupon

$$0 = \int_0^t \int_0^x \int_0^{x_1} \beta(t_1)u(t_1,x_2)(1-u(t_1,x_2))dx_2dx_1dt_1$$

$$-\frac{1}{2}\int_0^t \int_0^x \alpha(t_1)(u(t_1,x_1))^2 dx_1 dt_1$$

$$+\int_0^t u(t_1,x)dt_1$$

$$+\int_0^x \int_0^{x_1} (u_0(x_2) - u(t_1,x_2))dx_2 dx_1, \quad (t,x) \in [0,\infty) \times [0,\infty).$$

The last equation we differentiate twice in x and we get

$$
0 = \int_0^t \beta(t_1)u(t_1,x)(1 - u(t_1,x))dt_1
$$

$$
-\int_0^t \alpha(t_1)u(t_1,x)u_x(t_1,x)dt_1 + \int_0^t u_{xx}(t_1,x)dt_1 \qquad (5.6)
$$

$$
+u_0(x) - u(t,x), \quad (t,x) \in [0,\infty) \times [0,\infty),
$$

which we differentiate in t and we obtain

$$
0 = \beta(t)u(t,x)(1 - u(t,x)) - \alpha(t)u(t,x)u_x(t,x)
$$

$$
+u_{xx}(t,x) - u_t(t,x), \quad (t,x) \in [0,\infty) \times [0,\infty),
$$

i.e., u satisfies (5.4). Now we put $t = 0$ in (5.6) and we get

$$u(0,x) = u_0(x), \quad x \in [0,\infty).$$

This completes the proof.

For $u \in E$, define the operators

$$
F_1 u(t,x) = \int_0^t \int_0^x (t - t_1)^4 (x - x_1)^4 g(t_1,x_1) \int_0^{t_1} \int_0^{x_1} (x_1 - x_2)\beta(t_2)
$$

$$
\times (u(t_2,x_2))^2 dx_2 dt_2 dx_1 dt_1
$$

$$
+\int_0^t \int_0^x (t - t_1)^4 (x - x_1)^4 g(t_1,x_1) \int_0^{x_1} (x_1 - x_2)
$$

$$
\times u(t_1,x_2)dx_2 dx_1 dt_1,
$$

$$F_2u(t,x) = \int_0^t \int_0^x (t-t_1)^4(x-x_1)^4 g(t_1,x_1) \int_0^{t_1} \int_0^{x_1} (x_1-x_2)\beta(t_2)$$

$$\times u(t_2,x_2)dx_2dt_2dx_1dt_1$$

$$-\frac{1}{2}\int_0^t \int_0^x (t-t_1)^4(x-x_1)^4 g(t_1,x_1) \int_0^{t_1}$$

$$\times \int_0^{x_1} \alpha(t_2)(u(t_2,x_2))^2 dx_2 dt_2 dx_1 dt_1$$

$$+\int_0^t \int_0^x (t-t_1)^4(x-x_1)^4 g(t_1,x_1) \int_0^{t_1} u(t_2,x_1)dt_2dx_1dt_1$$

$$+\int_0^t \int_0^x (t-t_1)^4(x-x_1)^4 g(t_1,x_1) \int_0^{x_1} (x_1-x_2)$$

$$\times u_0(x_2)dx_2dx_1dt_1,$$

$(t,x) \in [0,\infty) \times [0,\infty)$. Note that if $u \in E$ is a fixed point of the operator $F_2 - F_1$, then it is a solution of the IVP (5.4)-(5.5).

Lemma 5.2.3 *Suppose* (B4), (B5) *and* $r > 0$. *If* $u \in E$ *and* $\|u\| \le r$, *then*

$$\|F_1u\| \le (1+r)A\|u\|, \quad \|F_2u\| \le \left(3+\frac{r}{2}\right)rA$$

and $F_2 : \{u \in E : \|u\| \le r\} \to E$ *is a completely continuous operator. Moreover,*

$$\|F_1u_1 - F_1u_2\| \le (2r+1)A\|u_1 - u_2\|$$

for any $u_1,u_2 \in \{u \in E : \|u\| \le r\}$.

Proof 5.2.4 *Take* $u \in \{E : \|u\| \le r\}$ *arbitrarily. Then*

$$|F_1u(t,x)| \le \int_0^t \int_0^x (t-t_1)^4(x-x_1)^4 g(t_1,x_1) \int_0^{t_1} \int_0^{x_1} (x_1-x_2)\beta(t_2)$$

$$\times (u(t_2,x_2))^2 dx_2 dt_2 dx_1 dt_1$$

$$+ \int_0^t \int_0^x (t-t_1)^4 (x-x_1)^4 g(t_1,x_1) \int_0^{x_1} (x_1-x_2)$$

$$\times |u(t_1,x_2)| dx_2 dx_1 dt_1$$

$$\leq r\|u\| \int_0^t \int_0^x x_1^2 (t-t_1)^4 (x-x_1)^4 g(t_1,x_1) \int_0^{t_1} \beta(t_2) dt_2 dx_1 dt_1$$

$$+ \|u\| \int_0^t \int_0^x x_1^2 (t-t_1)^4 (x-x_1)^4 g(t_1,x_1) dx_1 dt_1$$

$$\leq r\|u\| t^4 x^6 \int_0^t \int_0^x g(t_1,x_1) \int_0^{t_1} \beta(t_2) dt_2 dx_1 dt_1$$

$$+ \|u\| t^4 x^6 \int_0^t \int_0^x g(t_1,x_1) dx_1 dt_1$$

$$\leq (1+r) A \|u\|, \quad t \geq 0, \quad x \geq 0,$$

and

$$\left| \frac{\partial}{\partial t} F_1 u(t,x) \right| \leq 4 \int_0^t \int_0^x (t-t_1)^3 (x-x_1)^4 g(t_1,x_1) \int_0^{t_1} \int_0^{x_1} (x_1-x_2) \beta(t_2)$$

$$\times (u(t_2,x_2))^2 dx_2 dt_2 dx_1 dt_1$$

$$+ 4 \int_0^t \int_0^x (t-t_1)^3 (x-x_1)^4 g(t_1,x_1) \int_0^{x_1} (x_1-x_2)$$

$$\times |u(t_1,x_2)| dx_2 dx_1 dt_1$$

$$\leq 4r\|u\| \int_0^t \int_0^x x_1^2 (t-t_1)^3 (x-x_1)^4 g(t_1,x_1)$$

$$\times \int_0^{t_1} \beta(t_2)dt_2 dx_1 dt_1$$

$$+4\|u\| \int_0^t \int_0^x x_1^2(t-t_1)^3(x-x_1)^4 g(t_1,x_1)dx_1 dt_1$$

$$\leq 4r\|u\|t^3 x^6 \int_0^t \int_0^x g(t_1,x_1) \int_0^{t_1} \beta(t_2)dt_2 dx_1 dt_1$$

$$+4\|u\|t^3 x^6 \int_0^t \int_0^x g(t_1,x_1)dx_1 dt_1$$

$$\leq (1+r)A\|u\|, \quad t \geq 0, \quad x \geq 0,$$

and

$$\left| \frac{\partial}{\partial x}F_1 u(t,x) \right| \leq 4 \int_0^t \int_0^x (t-t_1)^4(x-x_1)^3 g(t_1,x_1) \int_0^{t_1} \int_0^{x_1} (x_1-x_2)\beta(t_2)$$

$$\times (u(t_2,x_2))^2 dx_2 dt_2 dx_1 dt_1$$

$$+4 \int_0^t \int_0^x (t-t_1)^4(x-x_1)^3 g(t_1,x_1) \int_0^{x_1} (x_1-x_2)$$

$$\times |u(t_1,x_2)|dx_2 dx_1 dt_1$$

$$\leq 4r\|u\| \int_0^t \int_0^x x_1^2(t-t_1)^4(x-x_1)^3 g(t_1,x_1)$$

$$\times \int_0^{t_1} \beta(t_2)dt_2 dx_1 dt_1$$

$$+4\|u\| \int_0^t \int_0^x x_1^2(t-t_1)^4(x-x_1)^3 g(t_1,x_1)dx_1 dt_1$$

$$\leq 4r\|u\|t^4x^5 \int_0^t \int_0^x g(t_1,x_1) \int_0^{t_1} \beta(t_2)dt_2dx_1dt_1$$

$$+4\|u\|t^4x^5 \int_0^t \int_0^x g(t_1,x_1)dx_1dt_1$$

$$\leq (1+r)A\|u\|, \quad t \geq 0, \quad x \geq 0,$$

and

$$\left| \frac{\partial^2}{\partial x^2} F_1 u(t,x) \right| \leq 12 \int_0^t \int_0^x (t-t_1)^4(x-x_1)^2 g(t_1,x_1)$$

$$\times \int_0^{t_1} \int_0^{x_1} (x_1-x_2)\beta(t_2)$$

$$\times(u(t_2,x_2))^2 dx_2dt_2dx_1dt_1$$

$$+12 \int_0^t \int_0^x (t-t_1)^4(x-x_1)^2 g(t_1,x_1) \int_0^{x_1} (x_1-x_2)$$

$$\times|u(t_1,x_2)|dx_2dx_1dt_1$$

$$\leq 12r\|u\| \int_0^t \int_0^x x_1^2(t-t_1)^4(x-x_1)^2 g(t_1,x_1) \int_0^{t_1} \beta(t_2)dt_2dx_1dt_1$$

$$+12\|u\| \int_0^t \int_0^x x_1^2(t-t_1)^4(x-x_1)^2 g(t_1,x_1)dx_1dt_1$$

$$\leq 12r\|u\|t^4x^4 \int_0^t \int_0^x g(t_1,x_1) \int_0^{t_1} \beta(t_2)dt_2dx_1dt_1$$

$$+12\|u\|t^4x^4 \int_0^t \int_0^x g(t_1,x_1)dx_1dt_1$$

$$\leq (1+r)A\|u\|, \quad t \geq 0, \quad x \geq 0.$$

Consequently

$$\|F_1 u\| \leq (1+r)A\|u\|.$$

Next,

$$|F_2u(t,x)| \leq \int_0^t \int_0^x (t-t_1)^4(x-x_1)^4 g(t_1,x_1) \int_0^{t_1} \int_0^{x_1} (x_1-x_2)\beta(t_2)$$

$$\times |u(t_2,x_2)| dx_2 dt_2 dx_1 dt_1$$

$$-\frac{1}{2} \int_0^t \int_0^x (t-t_1)^4(x-x_1)^4 g(t_1,x_1)$$

$$\times \int_0^{t_1} \int_0^{x_1} \alpha(t_2)(u(t_2,x_2))^2$$

$$\times dx_2 dt_2 dx_1 dt_1$$

$$+\int_0^t \int_0^x (t-t_1)^4(x-x_1)^4 g(t_1,x_1) \int_0^{t_1} |u(t_2,x_1)| dt_2 dx_1 dt_1$$

$$+\int_0^t \int_0^x (t-t_1)^4(x-x_1)^4 g(t_1,x_1) \int_0^{x_1} (x_1-x_2)$$

$$\times u_0(x_2) dx_2 dx_1 dt_1$$

$$\leq r \int_0^t \int_0^x x_1^2 (t-t_1)^4 (x-x_1)^4 g(t_1,x_1) \int_0^{t_1} \beta(t_2) dt_2 dx_1 dt_1$$

$$-\frac{1}{2} r^2 \int_0^t \int_0^x x_1 (t-t_1)^4 (x-x_1)^4 g(t_1,x_1) \int_0^{t_1} \alpha(t_2) dt_2 dx_1 dt_1$$

$$+r \int_0^t \int_0^x t_1 (t-t_1)^4 (x-x_1)^4 g(t_1,x_1) dx_1 dt_1$$

$$+r \int_0^t \int_0^x x_1^2 (t-t_1)^4 (x-x_1)^4 g(t_1,x_1) dx_1 dt_1$$

$$\leq r t^4 x^6 \int_0^t \int_0^x g(t_1,x_1) \int_0^{t_1} \beta(t_2) dt_2 dx_1 dt_1$$

$$-\frac{1}{2} r^2 t^4 x^5 \int_0^t \int_0^x g(t_1,x_1) \int_0^{t_1} \alpha(t_2) dt_2 dx_1 dt_1$$

$$+ rt^5 x^4 \int_0^t \int_0^x g(t_1, x_1) dx_1 dt_1$$

$$+ rt^4 x^6 \int_0^t \int_0^x g(t_1, x_1) dx_1 dt_1$$

$$\leq \left(3 + \frac{r}{2}\right) rA, \quad t \geq 0, \quad x \geq 0,$$

and

$$\left| \frac{\partial}{\partial t} F_2 u(t, x) \right| \leq 4 \int_0^t \int_0^x (t - t_1)^3 (x - x_1)^4 g(t_1, x_1) \int_0^{t_1} \int_0^{x_1} (x_1 - x_2) \beta(t_2)$$

$$\times |u(t_2, x_2)| dx_2 dt_2 dx_1 dt_1$$

$$- 2 \int_0^t \int_0^x (t - t_1)^3 (x - x_1)^4 g(t_1, x_1)$$

$$\times \int_0^{t_1} \int_0^{x_1} \alpha(t_2) (u(t_2, x_2))^2$$

$$\times dx_2 dt_2 dx_1 dt_1$$

$$+ 4 \int_0^t \int_0^x (t - t_1)^3 (x - x_1)^4 g(t_1, x_1) \int_0^{t_1} |u(t_2, x_1)| dt_2 dx_1 dt_1$$

$$+ 4 \int_0^t \int_0^x (t - t_1)^3 (x - x_1)^4 g(t_1, x_1) \int_0^{x_1} (x_1 - x_2)$$

$$\times u_0(x_2) dx_2 dx_1 dt_1$$

$$\leq 4r \int_0^t \int_0^x x_1^2 (t - t_1)^3 (x - x_1)^4 g(t_1, x_1) \int_0^{t_1} \beta(t_2) dt_2 dx_1 dt_1$$

$$- 2r^2 \int_0^t \int_0^x x_1 (t - t_1)^3 (x - x_1)^4 g(t_1, x_1) \int_0^{t_1} \alpha(t_2) dt_2 dx_1 dt_1$$

$$+4r \int_0^t \int_0^x t_1(t-t_1)^3(x-x_1)^4 g(t_1,x_1)dx_1 dt_1$$

$$+4r \int_0^t \int_0^x x_1^2(t-t_1)^3(x-x_1)^4 g(t_1,x_1)dx_1 dt_1$$

$$\leq 4rt^3 x^6 \int_0^t \int_0^x g(t_1,x_1) \int_0^{t_1} \beta(t_2)dt_2 dx_1 dt_1$$

$$-2r^2 t^3 x^5 \int_0^t \int_0^x g(t_1,x_1) \int_0^{t_1} \alpha(t_2)dt_2 dx_1 dt_1$$

$$+4rt^4 x^4 \int_0^t \int_0^x g(t_1,x_1)dx_1 dt_1$$

$$+4rt^3 x^6 \int_0^t \int_0^x g(t_1,x_1)dx_1 dt_1$$

$$\leq \left(3+\frac{r}{2}\right) rA, \quad t \geq 0, \quad x \geq 0,$$

and

$$\left| \frac{\partial^2}{\partial t^2} F_2 u(t,x) \right|$$

$$\leq 12 \int_0^t \int_0^x (t-t_1)^2(x-x_1)^4 g(t_1,x_1) \int_0^{t_1} \int_0^{x_1} (x_1-x_2)\beta(t_2)$$

$$\times |u(t_2,x_2)|dx_2 dt_2 dx_1 dt_1$$

$$-6 \int_0^t \int_0^x (t-t_1)^2(x-x_1)^4 g(t_1,x_1) \int_0^{t_1}$$

$$\times \int_0^{x_1} \alpha(t_2)(u(t_2,x_2))^2 dx_2 dt_2 dx_1 dt_1$$

$$+12 \int_0^t \int_0^x (t-t_1)^2(x-x_1)^4 g(t_1,x_1) \int_0^{t_1} |u(t_2,x_1)|dt_2 dx_1 dt_1$$

$$+ 12 \int_0^t \int_0^x (t-t_1)^2 (x-x_1)^4 g(t_1, x_1) \int_0^{x_1} (x_1 - x_2)$$

$$\times u_0(x_2) dx_2 dx_1 dt_1$$

$$\leq 12r \int_0^t \int_0^x x_1^2 (t-t_1)^2 (x-x_1)^4 g(t_1, x_1) \int_0^{t_1} \beta(t_2) dt_2 dx_1 dt_1$$

$$-6r^2 \int_0^t \int_0^x x_1 (t-t_1)^2 (x-x_1)^4 g(t_1, x_1) \int_0^{t_1} \alpha(t_2) dt_2 dx_1 dt_1$$

$$+12r \int_0^t \int_0^x t_1 (t-t_1)^2 (x-x_1)^4 g(t_1, x_1) dx_1 dt_1$$

$$+12r \int_0^t \int_0^x x_1^2 (t-t_1)^2 (x-x_1)^4 g(t_1, x_1) dx_1 dt_1$$

$$\leq 12rt^2 x^6 \int_0^t \int_0^x g(t_1, x_1) \int_0^{t_1} \beta(t_2) dt_2 dx_1 dt_1$$

$$-6r^2 t^2 x^5 \int_0^t \int_0^x g(t_1, x_1) \int_0^{t_1} \alpha(t_2) dt_2 dx_1 dt_1$$

$$+12rt^3 x^4 \int_0^t \int_0^x g(t_1, x_1) dx_1 dt_1$$

$$+12rt^2 x^6 \int_0^t \int_0^x g(t_1, x_1) dx_1 dt_1$$

$$\leq \left(3 + \frac{r}{2}\right) rA, \quad t \geq 0, \quad x \geq 0,$$

and

$$\left| \frac{\partial}{\partial x} F_2 u(t, x) \right|$$

$$\leq 4 \int_0^t \int_0^x (t-t_1)^4 (x-x_1)^3 g(t_1, x_1) \int_0^{t_1} \int_0^{x_1} (x_1 - x_2) \beta(t_2)$$

$$\times |u(t_2,x_2)| dx_2 dt_2 dx_1 dt_1$$

$$-2\int_0^t \int_0^x (t-t_1)^4(x-x_1)^3 g(t_1,x_1)\int_0^{t_1}\int_0^{x_1}\alpha(t_2)(u(t_2,x_2))^2$$

$$\times dx_2 dt_2 dx_1 dt_1$$

$$+4\int_0^t\int_0^x (t-t_1)^4(x-x_1)^3 g(t_1,x_1)\int_0^{t_1}|u(t_2,x_1)| dt_2 dx_1 dt_1$$

$$+4\int_0^t\int_0^x (t-t_1)^4(x-x_1)^3 g(t_1,x_1)\int_0^{x_1}(x_1-x_2)$$

$$\times u_0(x_2) dx_2 dx_1 dt_1$$

$$\leq 4r\int_0^t\int_0^x x_1^2(t-t_1)^4(x-x_1)^3 g(t_1,x_1)\int_0^{t_1}\beta(t_2) dt_2 dx_1 dt_1$$

$$-2r^2\int_0^t\int_0^x x_1(t-t_1)^4(x-x_1)^3 g(t_1,x_1)\int_0^{t_1}\alpha(t_2) dt_2 dx_1 dt_1$$

$$+4r\int_0^t\int_0^x t_1(t-t_1)^4(x-x_1)^3 g(t_1,x_1) dx_1 dt_1$$

$$+4r\int_0^t\int_0^x x_1^2(t-t_1)^4(x-x_1)^3 g(t_1,x_1) dx_1 dt_1$$

$$\leq 4rt^4 x^5\int_0^t\int_0^x g(t_1,x_1)\int_0^{t_1}\beta(t_2) dt_2 dx_1 dt_1$$

$$-2r^2 t^4 x^4\int_0^t\int_0^x g(t_1,x_1)\int_0^{t_1}\alpha(t_2) dt_2 dx_1 dt_1$$

$$+ 4rt^5 x^3 \int_0^t \int_0^x g(t_1, x_1) dx_1 dt_1$$

$$+ 4rt^4 x^5 \int_0^t \int_0^x g(t_1, x_1) dx_1 dt_1$$

$$\leq \left(3 + \frac{r}{2}\right) rA, \quad t \geq 0, \quad x \geq 0,$$

and

$$\left| \frac{\partial^2}{\partial x^2} F_2 u(t, x) \right|$$

$$\leq 12 \int_0^t \int_0^x (t - t_1)^4 (x - x_1)^2 g(t_1, x_1) \int_0^{t_1} \int_0^{x_1} (x_1 - x_2) \beta(t_2)$$

$$\times |u(t_2, x_2)| dx_2 dt_2 dx_1 dt_1$$

$$- 6 \int_0^t \int_0^x (t - t_1)^4 (x - x_1)^2 g(t_1, x_1) \int_0^{t_1} \int_0^{x_1} \alpha(t_2) (u(t_2, x_2))^2$$

$$\times dx_2 dt_2 dx_1 dt_1$$

$$+ 12 \int_0^t \int_0^x (t - t_1)^4 (x - x_1)^2 g(t_1, x_1) \int_0^{t_1} |u(t_2, x_1)| dt_2 dx_1 dt_1$$

$$+ 12 \int_0^t \int_0^x (t - t_1)^4 (x - x_1)^2 g(t_1, x_1) \int_0^{x_1} (x_1 - x_2)$$

$$\times u_0(x_2) dx_2 dx_1 dt_1$$

$$\leq 12r \int_0^t \int_0^x x_1^2 (t - t_1)^4 (x - x_1)^2 g(t_1, x_1) \int_0^{t_1} \beta(t_2) dt_2 dx_1 dt_1$$

$$-6r^2 \int_0^t \int_0^x x_1(t-t_1)^4(x-x_1)^2 g(t_1,x_1) \int_0^{t_1} \alpha(t_2) dt_2 dx_1 dt_1$$

$$+12r \int_0^t \int_0^x t_1(t-t_1)^4(x-x_1)^2 g(t_1,x_1) dx_1 dt_1$$

$$+12r \int_0^t \int_0^x x_1^2(t-t_1)^4(x-x_1)^2 g(t_1,x_1) dx_1 dt_1$$

$$\leq \quad 12rt^4 x^4 \int_0^t \int_0^x g(t_1,x_1) \int_0^{t_1} \beta(t_2) dt_2 dx_1 dt_1$$

$$-6r^2 t^4 x^3 \int_0^t \int_0^x g(t_1,x_1) \int_0^{t_1} \alpha(t_2) dt_2 dx_1 dt_1$$

$$+12rt^5 x^2 \int_0^t \int_0^x g(t_1,x_1) dx_1 dt_1$$

$$+12rt^4 x^4 \int_0^t \int_0^x g(t_1,x_1) dx_1 dt_1$$

$$\leq \quad \left(3+\frac{r}{2}\right) rA, \quad t \geq 0, \quad x \geq 0,$$

and

$$\left| \frac{\partial^3}{\partial x^3} F_2 u(t,x) \right| \leq 24 \int_0^t \int_0^x (t-t_1)^4(x-x_1) g(t_1,x_1) \int_0^{t_1} \int_0^{x_1} (x_1-x_2)\beta(t_2)$$

$$\times |u(t_2,x_2)| dx_2 dt_2 dx_1 dt_1$$

$$- 12 \int_0^t \int_0^x (t-t_1)^4(x-x_1) g(t_1,x_1)$$

$$\times \int_0^{t_1} \int_0^{x_1} \alpha(t_2)(u(t_2,x_2))^2 dx_2 dt_2 dx_1 dt_1$$

$$+ 24 \int_0^t \int_0^x (t-t_1)^4(x-x_1) g(t_1,x_1) \int_0^{t_1} |u(t_2,x_1)| dt_2 dx_1 dt_1$$

$$+24\int_0^t\int_0^x (t-t_1)^4(x-x_1)g(t_1,x_1)\int_0^{x_1}(x_1-x_2)$$

$$\times u_0(x_2)dx_2dx_1dt_1$$

$$\leq 24r\int_0^t\int_0^x x_1^2(t-t_1)^4(x-x_1)g(t_1,x_1)\int_0^{t_1}\beta(t_2)dt_2dx_1dt_1$$

$$-12r^2\int_0^t\int_0^x x_1(t-t_1)^4(x-x_1)g(t_1,x_1)\int_0^{t_1}\alpha(t_2)dt_2dx_1dt_1$$

$$+24r\int_0^t\int_0^x t_1(t-t_1)^4(x-x_1)g(t_1,x_1)dx_1dt_1$$

$$+24r\int_0^t\int_0^x x_1^2(t-t_1)^4(x-x_1)g(t_1,x_1)dx_1dt_1$$

$$\leq 24rt^4x^3\int_0^t\int_0^x g(t_1,x_1)\int_0^{t_1}\beta(t_2)dt_2dx_1dt_1$$

$$-12r^2t^4x^2\int_0^t\int_0^x g(t_1,x_1)\int_0^{t_1}\alpha(t_2)dt_2dx_1dt_1$$

$$+24rt^5x\int_0^t\int_0^x g(t_1,x_1)dx_1dt_1$$

$$+24rt^4x^3\int_0^t\int_0^x g(t_1,x_1)dx_1dt_1$$

$$\leq \left(3+\frac{r}{2}\right)rA, \quad t\geq 0, \quad x\geq 0.$$

Consequently

$$\|F_2u\|\leq\left(3+\frac{r}{2}\right)rA, \quad \left\|\frac{\partial^2}{\partial t^2}F_2u\right\|_{\mathscr{C}^0}\leq\left(3+\frac{r}{2}\right)rA,$$

$$\left\|\frac{\partial^3}{\partial x^3}F_2u\right\|_{\mathscr{C}^0}\leq\left(3+\frac{r}{2}\right)rA.$$

By the Ascoli-Arzelà theorem, it follows that the operator $F_2 : \{u \in E : \|u\| \leq r\} \to E$ is a completely continuous operator. Let now, $u_1, u_2 \in \{u \in E : \|u\| \leq r\}$. Then

$$|F_1 u_1(t,x) - F_1 u_2(t,x)|$$

$$\leq \left(\int_0^t \int_0^x (t - t_1)^4 (x - x_1)^4 g(t_1, x_1) \right.$$

$$\times int_0^{t_1} \int_0^{x_1} (x_1 - x_2)$$

$$\times \beta(t_2) \left(|u_1(t_2, x_2)| + |u_2(t_2, x_2)| \right) dx_2 dt_2 dx_1 dt_1$$

$$+ \int_0^t \int_0^x (t - t_1)^4 (x - x_1)^4 g(t_1, x_1) \int_0^{x_1} (x_1 - x_2) dx_2 dx_1 dt_1 \right) \|u_1 - u_2\|$$

$$\leq \left(2rx^6 t^4 \int_0^t \int_0^x g(t_1, x_1) \int_0^{t_1} \beta(t_2) dt_2 dt_1 \right.$$

$$\left. + x^6 t^4 \int_0^t \int_0^x g(t_1, x_1) dx_1 dt_1 \right) \|u_1 - u_2\|$$

$$\leq (2r + 1) A \|u_1 - u_2\|, \quad t \geq 0, \quad x \geq 0,$$

and

$$\left| \frac{\partial}{\partial t} F_1 u_1(t,x) - \frac{\partial}{\partial t} F_1 u_2(t,x) \right|$$

$$\leq \left(4 \int_0^t \int_0^x (t - t_1)^3 (x - x_1)^4 g(t_1, x_1) \int_0^{t_1} \int_0^{x_1} (x_1 - x_2) \right.$$

$$\times \beta(t_2) \left(|u_1(t_2, x_2)| + |u_2(t_2, x_2)| \right) dx_2 dt_2 dx_1 dt_1$$

$$+4\int_0^t\int_0^x (t-t_1)^3(x-x_1)^4 g(t_1,x_1)\int_0^{x_1}(x_1-x_2)$$

$$\times dx_2 dx_1 dt_1\Bigg)\|u_1-u_2\|$$

$$\leq \Bigg(8rx^6 t^3 \int_0^t\int_0^x g(t_1,x_1)\int_0^{t_1}\beta(t_2)dt_2 dt_1$$

$$+4x^6 t^3\int_0^t\int_0^x g(t_1,x_1)dx_1 dt_1\Bigg)\|u_1-u_2\|$$

$$\leq (2r+1)A\|u_1-u_2\|,\quad t\geq 0,\quad x\geq 0,$$

and

$$\left|\frac{\partial}{\partial x}F_1 u_1(t,x)-\frac{\partial}{\partial x}F_1 u_2(t,x)\right|$$

$$\leq \Bigg(4\int_0^t\int_0^x (t-t_1)^4(x-x_1)^3 g(t_1,x_1)\int_0^{t_1}\int_0^{x_1}(x_1-x_2)$$

$$\times \beta(t_2)\,(|u_1(t_2,x_2)|+|u_2(t_2,x_2)|)\,dx_2 dt_2 dx_1 dt_1$$

$$+4\int_0^t\int_0^x (t-t_1)^4(x-x_1)^3 g(t_1,x_1)\int_0^{x_1}(x_1-x_2)dx_2 dx_1 dt_1\Bigg)\|u_1-u_2\|$$

$$\leq \Bigg(8rx^5 t^4\int_0^t\int_0^x g(t_1,x_1)\int_0^{t_1}\beta(t_2)dt_2 dt_1$$

$$+4x^5 t^4\int_0^t\int_0^x g(t_1,x_1)dx_1 dt_1\Bigg)\|u_1-u_2\|$$

$$\leq (2r+1)A\|u_1-u_2\|,\quad t\geq 0,\quad x\geq 0,$$

and

$$\left| \frac{\partial^2}{\partial x^2} F_1 u_1(t,x) - \frac{\partial^2}{\partial x^2} F_1 u_2(t,x) \right|$$

$$\leq \left(12 \int_0^t \int_0^x (t-t_1)^4 (x-x_1)^2 g(t_1,x_1) \int_0^{t_1} \int_0^{x_1} (x_1 - x_2) \right.$$

$$\times \beta(t_2) \left(|u_1(t_2,x_2)| + |u_2(t_2,x_2)| \right) dx_2 dt_2 dx_1 dt_1$$

$$\left. + 12 \int_0^t \int_0^x (t-t_1)^4 (x-x_1)^2 g(t_1,x_1) \int_0^{x_1} (x_1 - x_2) dx_2 dx_1 dt_1 \right) \|u_1 - u_2\|$$

$$\leq \left(24 r x^4 t^4 \int_0^t \int_0^x g(t_1,x_1) \int_0^{t_1} \beta(t_2) dt_2 dt_1 \right.$$

$$\left. + 12 x^4 t^4 \int_0^t \int_0^x g(t_1,x_1) dx_1 dt_1 \right) \|u_1 - u_2\|$$

$$\leq (2r+1)A\|u_1 - u_2\|, \quad t \geq 0, \quad x \geq 0.$$

Therefore

$$\|F_1 u_1 - F_1 u_2\| \leq (2r+1)A\|u_1 - u_2\|.$$

This completes the proof.

Theorem 5.2.5 *Suppose* (B4) *and* (B5). *Then the IVP* (5.4)-(5.5) *has at least one non-negative solution* $u \in \mathscr{C}^1([0,\infty), \mathscr{C}^2([0,\infty)))$.

Proof 5.2.6 *Set*

$$\mathscr{P} = \{u \in E : u(t,x) \geq 0, \quad t \geq 0, \quad x \geq 0\},$$

$$\Omega = \{u \in \mathscr{P} : \|u\| \leq r_1, \quad u(t,x) \leq u_0(x), \quad t \geq 0, \quad x \geq 0\},$$

$$U = \{u \in \mathscr{P} : \|u\| \leq r_1, \quad \frac{1}{2}u_0(x) \leq u(t,x) \leq u_0(x), \quad t \geq 0, \quad x \geq 0\}.$$

For $u \in E$, define the operators

$$Tu(t,x) = -F_1 u(t,x),$$

$$Su(t,x) = F_2 u(t,x), \quad t \geq 0, \quad x \geq 0.$$

1. Let $u,v \in \Omega$. Then $(I-T)(u-v) = (I+F_1)(u-v)$ and using Lemma 5.2.3, we get

$$\|(I-T)(u-v)\| \geq \|u-v\| - \|F_1(u-v)\|$$

$$\geq (1-(1+2r_1)A)\|u-v\|.$$

Thus, $I-T : \Omega \to E$ is Lipschitz invertible with $\gamma = \dfrac{1}{1-(1+2r_1)A}$.

2. By Lemma 5.2.3, we have that $S : \overline{U} \to E$ is a completely continuous operator. Therefore $S : \overline{U} \to E$ is 0-set contraction.

3. Let $v \in \overline{U}$ be arbitrarily chosen. For $u \in \Omega$, we have

$$-F_1 u(t,x) + F_2 v(t,x)$$

$$\geq -\int_0^t \int_0^x (t-t_1)^4 (x-x_1)^4 g(t_1,x_1) \int_0^{t_1} \int_0^{x_1} (x_1-x_2)\beta(t_2)$$

$$\times (u(t_2,x_2))^2 dx_2 dt_2 dx_1 dt_1$$

$$-\int_0^t \int_0^x (t-t_1)^4 (x-x_1)^4 g(t_1,x_1) \int_0^{x_1} (x_1-x_2)$$

$$\times u(t_1,x_2) dx_2 dx_1 dt_1$$

$$+\int_0^t \int_0^x (t-t_1)^4 (x-x_1)^4 g(t_1,x_1) \int_0^{t_1} \int_0^{x_1} (x_1-x_2)\beta(t_2)$$

$$\times v(t_2,x_2) dx_2 dt_2 dx_1 dt_1$$

$$+ \int_0^t \int_0^x (t-t_1)^4 (x-x_1)^4 g(t_1,x_1) \int_0^{x_1} (x_1 - x_2)$$

$$\times u_0(x_2) dx_2 dx_1 dt_1$$

$$\geq \left(\frac{r_1}{2} - r_1^2\right) \int_0^t \int_0^x (t-t_1)^4 (x-x_1)^4 g(t_1,x_1) \int_0^{t_1} \int_0^{x_1} (x_1 - x_2)\beta(t_2)$$

$$\times dx_2 dt_2 dx_1 dt_1$$

$$+ \int_0^t \int_0^x (t-t_1)^4 (x-x_1)^4 g(t_1,x_1) \int_0^{x_1} (x_1 - x_2)$$

$$\times (u_0(x_2) - u(t_1,x_2)) dx_2 dx_1 dt_1$$

$$\geq 0, \quad t \geq 0, \quad x \geq 0,$$

and

$$-F_1 u(t,x) + F_2 v(t,x) \leq \|F_1 u\| + \|F_2 v\|$$

$$\leq (1+r_1)r_1 A + \left(3 + \frac{r_1}{2}\right) r_1 A$$

$$= \left(4 + \frac{3}{2}r_1\right) r_1 A$$

$$< \frac{r_1}{2}$$

$$\leq u_0(x), \quad t \geq 0, \quad x \geq 0.$$

For $u \in \Omega$, define the operator

$$Lu(t,x) = -F_1 u(t,x) + F_2 v(t,x), \quad t \geq 0, \quad x \geq 0.$$

Then, using Lemma 5.2.3, we get

$$\|Lu\| \leq \|F_1 u\| + \|F_2 v\|$$

$$\leq r_1(1+r_1)A + \left(3+\frac{r_1}{2}\right)r_1A$$

$$= \left(4+\frac{3}{2}r_1\right)r_1A$$

$$\leq \frac{r_1}{2}.$$

Consequently $L : \Omega \to \Omega$*. Again, applying Lemma 5.2.3, we obtain*

$$\|Lu_1 - Lu_2\| \leq (2r_1+1)A\|u_1-u_2\|.$$

Therefore, $L : \Omega \to \Omega$ *is a contraction operator and there exists a unique* $u \in \Omega$ *so that* $u = Lu$ *or* $(I-T)u = Sv$*. Then* $S(\overline{U}) \subset (I-T)(\Omega)$*.*

4. *Assume that there are an* $u \in \partial U$ *and* $\lambda \geq 1$ *so that*

$$Su = (I-T)(\lambda u) \quad and \quad \lambda u \in \Omega.$$

Then

$$Su = (I+F_1)(\lambda u)$$

and applying Lemma 5.2.3, we obtain

$$\left(3+\frac{r_1}{2}\right)r_1A \geq \|Su\|$$

$$\geq \lambda\|u\| - \|F_1(\lambda u)\|$$

$$\geq \lambda\|u\| - (1+r_1)A\|\lambda u\|$$

$$= (1-(1+r_1)A)\lambda\|u\|$$

$$\geq (1-(1+r_1)A)\|u\|$$

$$= r_1(1-(1+r_1)A),$$

whereupon

$$\left(3+\frac{r_1}{2}\right)A \geq 1-(1+r_1)A \quad or \quad \left(4+\frac{3}{2}r_1\right)A \geq 1,$$

which is a contradiction.

Hence, and Proposition 2.2.61, it follows that the operator $T+S$ has at least one fixed point in $U \cap \Omega$, which is a nontrivial non-negative solution of the IVP (5.4)-(5.5). This completes the proof.

Example 5.2.7 *Below, we will illustrate our main result. Let*

$$h(x) = \log \frac{1+s^{11}\sqrt{2}+s^{22}}{1-s^{11}\sqrt{2}+s^{22}}, \quad l(s) = \arctan \frac{s^{11}\sqrt{2}}{1-s^{22}},$$

$s \in \mathbb{R}$, $s \neq \pm 1$. Then

$$h'(s) = \frac{22\sqrt{2}s^{10}(1-s^{22})}{(1-s^{11}\sqrt{2}+s^{22})(1+s^{11}\sqrt{2}+s^{22})},$$

$$l'(s) = \frac{11\sqrt{2}s^{10}(1+s^{20})}{1+s^{40}}, \quad s \in \mathbb{R}.$$

Therefore,

$$-\infty < \lim_{s\to\pm\infty}(1+s+\cdots+s^9)h(s) < \infty,$$

$$-\infty < \lim_{s\to\pm\infty}(1+s+\cdots+s^9)l(s) < \infty.$$

Hence, there exists a positive constant C_1 so that

$$(1+s+\cdots+s^9)\left(\frac{1}{44\sqrt{2}}\log\frac{1+s^{11}\sqrt{2}+s^{22}}{1-s^{11}\sqrt{2}+s^{22}}\right.$$

$$\left.+\frac{1}{22\sqrt{2}}\arctan\frac{s^{11}\sqrt{2}}{1-s^{22}}\right) \leq C_1,$$

$$(1+s+\cdots+s^9)\left(\frac{1}{44\sqrt{2}}\log\frac{1+s^{11}\sqrt{2}+s^{22}}{1-s^{11}\sqrt{2}+s^{22}}\right.$$

$$+\frac{1}{22\sqrt{2}} \arctan \frac{s^{11}\sqrt{2}}{1-s^{22}}\Bigg) \leq C_1,$$

$s \in [0,\infty)$. *Note that by [37] (pp. 707, Integral 79), we have*

$$\int \frac{dz}{1+z^4} = \frac{1}{4\sqrt{2}} \log \frac{1+z\sqrt{2}+z^2}{1-z\sqrt{2}+z^2} + \frac{1}{2\sqrt{2}} \arctan \frac{z\sqrt{2}}{1-z^2}.$$

Let

$$Q(s) = \frac{s^{10}}{(1+s^{44})(1+(1+s+\cdots+s^9)^2)^{28}}, \quad s \in [0,\infty),$$

and

$$g_1(t,x) = Q(t)Q(x), \quad t,x \in [0,\infty).$$

Then there exists a positive constant A_1 such that

$$720(1+t+\cdots+t^6)(1+x+\cdots+x^6) \int_0^t \int_0^x g_1(t_1,x_1)dx_1 dt_1 \leq A_1,$$

$t,x \geq 0$. Take $g(t,x) = \dfrac{g_1(t,x)}{280A_1}$, $A = \dfrac{1}{50}$, $r_1 = \dfrac{1}{4}$. Consider the IVP

$$u_t - u_{xx} - uu_x = u(1-u), \quad t > 0, \quad x \geq 0,$$

$$u(0,x) = \frac{1}{8} + \frac{1}{8(1+x^2)}, \quad x \geq 0.$$

Here $\alpha = -1$, $\beta = 1$ on $[0,\infty)$,

$$\frac{r_1}{2} \leq u_0(x) = \frac{1}{8} + \frac{1}{8(1+x^2)} \leq r_1, \quad x \geq 0,$$

$$1 - (1+r_1)A = \frac{39}{40} > 0, \quad \left(4 + \frac{3}{2}r_1\right)A = \frac{7}{80} < \frac{1}{2},$$

and

$$120\left(1+t+t^2+t^3+t^4\right)\left(1+x+x^2+x^3+x^4+x^5+x^6\right)$$

$$\times \int_0^t \int_0^x g(t_1,x_1)\left(1+ \int_0^{t_1}(\beta(t_2)-\alpha(t_2))dt_2\right)dx_1 dt_1$$

$$\leq\; 240(1+t)\left(1+t+t^2+t^3+t^4\right)\left(1+x+x^2+x^3+x^4+x^5+x^6\right)$$

$$\times \int_0^t \int_0^x g(t_1,x_1)dx_1dt_1$$

$$\leq\; \frac{720}{280A_1}\left(1+t+t^2+t^3+t^4+t^5\right)\left(1+x+x^2+x^3+x^4+x^5+x^6\right)$$

$$\times \int_0^t \int_0^x g_1(t_1,x_1)dx_1dt_1$$

$$\leq\; \frac{1}{280}$$

$$\leq\; A.$$

Therefore, the considered IVP has at least one non-negative solution
$u \in \mathscr{C}^1([0,\infty),\mathscr{C}^2([0,\infty)))$.

5.3   IBVPs for Nonlinear Parabolic Equations

In this section, we will give an application for existence of non-negative solutions for some IBVPs for nonlinear parabolic equations. Consider the following IBVP

$$u_t - u_{xx} \;=\; f(t,x,u), \quad t \geq 0, \quad x \in [0,L],$$

$$u(0,x) \;=\; u_0(x), \quad x \in [0,L], \qquad\qquad (5.7)$$

$$u(t,0) \;=\; u_x(t,L) = 0, \quad t \geq 0,$$

where

(B6) Let $0 < A < \varepsilon$, $l \in \mathbb{N}$, $R > 0$ is large enough and $r \in \left(0,\dfrac{1}{8}\right)$ is small enough so that

$$4r + 2\sum_{j=1}^{l} r^{p_j} < \frac{1}{8},$$

$$\varepsilon r + \left(2r + 2\sum_{j=1}^{l} r^{p_j}\right) A \leq (\varepsilon - A)R.$$

(B7) $f \in \mathscr{C}([0,\infty) \times [0,L] \times \mathbb{R})$ is such that

$$0 \leq f(t,x,u) \leq \sum_{j=1}^{l} c_j(t,x)|u|^{p_j}, \quad (t,x,u) \in [0,\infty) \times [0,L] \times \mathbb{R},$$

$p_j > 0$, $j \in \{1,\ldots,l\}$,

(B8) $u_0 \in \mathscr{C}^2([0,L])$, $u_1 \in \mathscr{C}^1([0,L])$, $0 < u_0, u_1 < r$ on $[0,L]$,

(B9) $g \in \mathscr{C}([0,\infty) \times [0,L])$ is a non-negative function such that

$$\int_0^t \int_0^{t_1} \int_0^{t_2} \int_0^x \int_0^{x_1} \int_0^{x_2} t_3^2 g(t_3,x_3)\,dx_3 dx_2 dx_1 dt_3 dt_2 dt_1 \leq A,$$

$$\int_0^t \int_0^{t_2} \int_0^x \int_0^{x_1} \int_0^{x_2} t_3^2 g(t_3,x_3)\,dx_3 dx_2 dx_1 dt_3 dt_2 \leq A,$$

$$\int_0^t \int_0^{t_1} \int_0^{t_2} \int_0^x \int_0^{x_2} t_3^2 g(t_3,x_3)\,dx_3 dx_2 dt_3 dt_2 dt_1 \leq A,$$

$$\int_0^t \int_0^{t_1} \int_0^{t_2} \int_0^x t_3^2 g(t_3,x_3)\,dx_3 dt_3 dt_2 dt_1 \leq A,$$

$$\int_0^t \int_0^{t_1} \int_0^{t_2} \int_0^x \int_0^{x_1} \int_0^{x_2} x_3^2 g(t_3,x_3)\,dx_3 dx_2 dx_1 dt_3 dt_2 dt_1 \leq A,$$

$$\int_0^t \int_0^{t_2} \int_0^x \int_0^{x_1} \int_0^{x_2} x_3^2 g(t_3,x_3)\,dx_3 dx_2 dx_1 dt_3 dt_2 \leq A,$$

$$\int_0^t \int_0^{t_1} \int_0^{t_2} \int_0^x \int_0^{x_2} x_3^2 g(t_3,x_3)\,dx_3 dx_2 dt_3 dt_2 dt_1 \leq A,$$

$$\int_0^t \int_0^{t_1} \int_0^{t_2} \int_0^x x_3^2 g(t_3,x_3)\,dx_3 dt_3 dt_2 dt_1 \leq A,$$

$$\int_0^t \int_0^{t_1} \int_0^{t_2} \int_0^x \int_0^{x_1} \int_0^{x_2} g(t_3,x_3) \int_0^{x_3} x_4$$

$$\times \int_0^{t_3} c_j(t_5,x_4)dt_5dx_4dx_3dx_2dx_1dt_3dt_2dt_1 \leq A,$$

$$\int_0^t \int_0^{t_2} \int_0^x \int_0^{x_1} \int_0^{x_2} g(t_3,x_3) \int_0^{x_3} x_4 \int_0^{t_3}$$

$$c_j(t_5,x_4)dt_5dx_4dx_3dx_2dx_1dt_3dt_2 \leq A,$$

$$\int_0^t \int_0^{t_1} \int_0^{t_2} \int_0^x \int_0^{x_2} g(t_3,x_3) \int_0^{x_3} x_4 \int_0^{t_3}$$

$$c_j(t_5,x_4)dt_5dx_4dx_3dx_2dt_3dt_2dt_1 \leq A,$$

$$\int_0^t \int_0^{t_1} \int_0^{t_2} \int_0^x \int_0^{x_1} \int_0^{x_2} g(t_3,x_3)$$

$$\times \int_{x_3}^L \int_0^{t_3} c_j(t_5,x_4)dt_5dx_4dx_3dx_2dx_1dt_3dt_2dt_1 \leq A,$$

$$\int_0^t \int_0^{t_2} \int_0^x \int_0^{x_1} \int_0^{x_2} g(t_3,x_3) \int_{x_3}^L \int_0^{t_3}$$

$$c_j(t_5,x_4)dt_5dx_4dx_3dx_2dx_1dt_3dt_2 \leq A,$$

$$\int_0^t \int_0^{t_1} \int_0^{t_2} \int_0^x \int_0^{x_2} g(t_3,x_3) \int_{x_3}^L \int_0^{t_3}$$

$$c_j(t_5,x_4)dt_5dx_4dx_3dx_2dt_3dt_2dt_1 \leq A,$$

$$\int_0^t \int_0^{t_1} \int_0^{t_2} \int_0^x g(t_3,x_3) \int_{x_3}^L \int_0^{t_3} c_j(t_5,x_4)dt_5dx_4dx_3dt_3dt_2dt_1 \leq A,$$

$$\int_0^t \int_0^{t_1} \int_0^{t_2} \int_0^x \int_0^{x_1} \int_0^{x_2} (L-x_3)g(t_3,x_3)dx_3dx_2dx_1dt_3dt_2dt_1 \leq A,$$

$$\int_0^t \int_0^{t_2} \int_0^x \int_0^{x_1} \int_0^{x_2} (L-x_3)g(t_3,x_3)dx_3dx_2dx_1dt_3dt_2 \leq A,$$

$$\int_0^t \int_0^{t_1} \int_0^{t_2} \int_0^x \int_0^{x_2} (L-x_3)g(t_3,x_3)dx_3dx_2dt_3dt_2dt_1 \leq A,$$

$$\int_0^t \int_0^{t_1} \int_0^{t_2} \int_0^x (L-x_3)g(t_3,x_3)dx_3dt_3dt_2dt_1 \leq A,$$

$$\int_1^2 \int_1^{t_1} \int_1^{t_2} \int_0^L \int_0^{x_1} \int_0^{x_2} g(t_3,x_3)$$

$$\times \int_0^{x_3} x_4 u_0(x_4)dx_4dx_3dx_2dx_1dt_3dt_2dt_1 > \frac{A}{2},$$

$$j \in \{1,\dots,l\}, \quad (t,x) \in [0,\infty) \times [0,L].$$

Our main result is as follows.

Theorem 5.3.1 *Suppose (B6)-(B9). Then the IBVP (5.7) has at least one non-negative solution* $u \in \mathscr{C}^1([0,\infty), \mathscr{C}^2([0,L]))$.

5.3.1 Some preliminary results

Let $E = \mathscr{C}^1([0,\infty), \mathscr{C}^2([0,L]))$ be endowed with the norm

$$\|u\| = \left\{ \sup_{(t,x)\in[0,\infty)\times[0,L]} |u(t,x)|, \quad \sup_{(t,x)\in[0,\infty)\times[0,L]} \left|\frac{\partial}{\partial t}u(t,x)\right|, \right.$$

$$\left. \sup_{(t,x)\in[0,\infty)\times[0,L]} \left|\frac{\partial}{\partial x}u(t,x)\right|, \quad \sup_{(t,x)\in[0,\infty)\times[0,L]} \left|\frac{\partial^2}{\partial x^2}u(t,x)\right| \right\}$$

provided it exists.

Lemma 5.3.2 *Let $u \in E$ be a solution to the integral equation*

$$0 = -\int_0^t \int_0^{t_1} \int_0^{t_2} \int_0^x \int_0^{x_1} \int_0^{x_2} g(t_3,x_3) \int_0^{t_3}$$

$$u(t_5,x_3)dt_5dx_3dx_2dx_1dt_3dt_2dt_1$$

$$+ \int_0^t \int_0^{t_1} \int_0^{t_2} \int_0^x \int_0^{x_1} \int_0^{x_2} g(t_3,x_3) \int_0^{x_3} x_4 \left(-u(t_3,x_4) + u_0(x_4) \right)$$

$$+ \int_0^{t_3} f(t_5, x_4, u(t_5, x_4)) dt_5 \Bigg) dx_4 dx_3 dx_2 dx_1 dt_3 dt_2 dt_1$$

$$+ \int_0^t \int_0^{t_1} \int_0^{t_2} \int_0^x \int_0^{x_1} \int_0^{x_2} x_3 g(t_3, x_3) \int_{x_3}^L \Big(-u(t_3, x_4) + u_0(x_3) $$

$$+ \int_0^{t_3} f(t_5, x_4, u(t_5, x_4)) dt_5 \Bigg) dx_4 dx_3 dx_2 dx_1 dt_3 dt_2 dt_1, \qquad (5.8)$$

$(t, x) \in [0, \infty) \times [0, L]$. *Then it is a solution to the IBVP* (5.7).

Proof 5.3.3 *We differentiate trice with respect to t and then trice with respect to x the equation* (5.8) *and we get*

$$0 = -g(t, x) \int_0^t \int_0^{t_1} u(t_2, x) dt_2 dt_1$$

$$+ g(t, x) \int_0^x x_1 \Big(-u(t, x_1) + u_0(x_1)$$

$$+ \int_0^t f(t_2, x_1, u(t_2, x_1)) dt_2 \Bigg) dx_1$$

$$+ x g(t, x) \int_x^L \Big(-u(t, x_1) + u_0(x_1)$$

$$+ \int_0^t f(t_2, x_1, u(t_2, x_1)) dt_2 \Bigg) dx_1,$$

$(t, x) \in [0, \infty) \times [0, L]$, *whereupon*

$$0 = -\int_0^t \int_0^{t_1} u(t_2, x) dt_2 dt_1$$

$$+ \int_0^x x_1 \Big(-u(t, x_1) + u_0(x_1)$$

$$+ \int_0^t f(t_2, x_1, u(t_2, x_1)) dt_2 \bigg) dx_1$$

$$+ x \int_x^L \bigg(-u(t, x_1) + u_0(x_1)$$

$$+ \int_0^t f(t_2, x_1, u(t_2, x_1)) dt_2 \bigg) dx_1, \qquad (5.9)$$

$(t,x) \in [0,\infty) \times [0,L]$. *Now we differentiate the last equation with respect to t and we find*

$$0 = -u(t,x)$$

$$+ \int_0^x x_1 \bigg(-u_t(t, x_1) + f(t, x_1, u(t, x_1)) \bigg) dx_1 \qquad (5.10)$$

$$+ x \int_x^L \bigg(-u_t(t, x_1) + f(t, x_1, u(t, x_1)) \bigg) dx_1,$$

$(t,x) \in [0,\infty) \times [0,L]$. *Now we differentiate with respect to x the last equation and we find*

$$0 = -u_x(t,x)$$

$$+ x(-u_t(t,x) + f(t,x,u(t,x)))$$

$$- x(-u_t(t,x) + f(t,x,u(t,x)))$$

$$+ \int_x^L (-u_t(t, x_1) + f(t, x_1, u(t, x_1))) dx_1 \qquad (5.11)$$

$$= -u_x(t,x)$$

$$+ \int_x^L (-u_t(t, x_1) + f(t, x_1, u(t, x_1))) dx_1,$$

$(t,x) \in [0,\infty) \times [0,L]$. *Now we differentiate the last equation with respect to x and we find*

$$0 = -u_{xx}(t,x) + u_t(t,x) - f(t,x,u(t,x)), \quad (t,x) \in [0,\infty) \times [0,L].$$

We put $t=0$ in (5.9) and we find

$$0 = \int_0^x x_1\left(-u(0,x_1) + u_0(x_1)\right) dx_1$$

$$+x \int_x^L \left(-u(0,x_1) + u_0(x_1)\right) dx_1, \quad x \in [0,L],$$

which we differentiate with respect to x and we get

$$0 = x(-u(0,x) + u_0(x)) + \int_x^L (-u(0,x_1) + u_0(x_1)) dx_1$$

$$-x(-u(0,x) + u_0(x))$$

$$= \int_x^L (-u(0,x_1) + u_0(x_1)) dx_1, \quad x \in [0,L].$$

Again, we differentiate in x and we find

$$u(0,x) = u_0(x), \quad x \in [0,L].$$

Now we put $x=0$ in (5.10) and we get

$$u(t,0) = 0, \quad t \in [0,\infty).$$

We put $x=L$ in (5.11) and we find

$$u_x(t,L) = 0, \quad t \in [0,\infty).$$

This completes the proof.

For $u \in E$, define the operators

$$Gu(t,x) = -\int_0^t \int_0^{t_1} \int_0^{t_2} \int_0^x \int_0^{x_1} \int_0^{x_2} g(t_3,x_3)$$

$$\times \int_0^{t_3} u(t_5,x_3)dt_5dx_3dx_2dx_1dt_3dt_2dt_1$$

$$Fu(t,x) = \int_0^t \int_0^{t_1} \int_0^{t_2} \int_0^x \int_0^{x_1} \int_0^{x_2} g(t_3,x_3) \int_0^{x_3} x_4 \left(-u(t_3,x_4) + u_0(x_4) \right.$$

$$+ \int_0^{t_3} f(t_5,x_4,u(t_5,x_4))dt_5 \Bigg) dx_4 dx_3 dx_2 dx_1 dt_3 dt_2 dt_1$$

$$+ \int_0^t \int_0^{t_1} \int_0^{t_2} \int_0^x \int_0^{x_1} \int_0^{x_2} x_3 g(t_3,x_3) \int_{x_3}^L \left(-u(t_3,x_4) + u_0(x_3) \right.$$

$$+ \int_0^{t_3} f(t_5,x_4,u(t_5,x_4))dt_5 \Bigg) dx_4 dx_3 dx_2 dx_1 dt_3 dt_2 dt_1,$$

$(t,x) \in [0,\infty) \times [0,L]$.

Lemma 5.3.4 *Suppose* $(B6)$-$(B9)$*. For* $u \in E$ *and* $\|u\| \le r$*, we have the following estimations.*

$$\|Gu\| \le rA,$$

$$\|Fu\| \le \left(2r + 2 \sum_{j=1}^l r^{p_j} \right) A.$$

Proof 5.3.5 *Using* $(B8)$ *and* $(B9)$*, we have*

$$|Gu(t,x)|$$

$$= \left| -\int_0^t \int_0^{t_1} \int_0^{t_2} \int_0^x \int_0^{x_1} \int_0^{x_2} g(t_3,x_3) \int_0^{t_3} u(t_5,x_3)dt_5dx_3dx_2dx_1dt_3dt_2dt_1 \right|$$

$$\le \int_0^t \int_0^{t_1} \int_0^{t_2} \int_0^x \int_0^{x_1} \int_0^{x_2} g(t_3,x_3) \int_0^{t_3} |u(t_5,x_3)|dt_5dx_3dx_2dx_1dt_3dt_2dt_1$$

$$\le rA,$$

$$\left| \frac{\partial}{\partial t} Gu(t,x) \right|$$

$$= \left| - \int_0^t \int_0^{t_2} \int_0^x \int_0^{x_1} \int_0^{x_2} g(t_3,x_3) \int_0^{t_3} u(t_5,x_3) dt_5 dx_3 dx_2 dx_1 dt_3 dt_2 dt_1 \right|$$

$$\le \int_0^t \int_0^{t_1} \int_0^{t_2} \int_0^x \int_0^{x_1} \int_0^{x_2} g(t_3,x_3) \int_0^{t_3} |u(t_5,x_3)| dt_5 dx_3 dx_2 dx_1 dt_3 dt_2$$

$$\le rA,$$

$$\left| \frac{\partial}{\partial x} Gu(t,x) \right|$$

$$= \left| - \int_0^t \int_0^{t_1} \int_0^{t_2} \int_0^x \int_0^{x_2} g(t_3,x_3) \int_0^{t_3} u(t_5,x_3) dt_5 dx_3 dx_2 dx_1 dt_3 dt_2 dt_1 \right|$$

$$\le \int_0^t \int_0^{t_1} \int_0^{t_2} \int_0^x \int_0^{x_1} \int_0^{x_2} g(t_3,x_3) \int_0^{t_3} |u(t_5,x_3)| dt_5 dx_3 dx_2 dt_3 dt_2 dt_1$$

$$\le rA,$$

$$\left| \frac{\partial^2}{\partial x^2} Gu(t,x) \right|$$

$$= \left| - \int_0^t \int_0^{t_1} \int_0^{t_2} \int_0^x g(t_3,x_3) \int_0^{t_3} u(t_5,x_3) dt_5 dx_3 dx_2 dx_1 dt_3 dt_2 dt_1 \right|$$

$$\le \int_0^t \int_0^{t_1} \int_0^{t_2} \int_0^x \int_0^{x_1} \int_0^{x_2} g(t_3,x_3) \int_0^{t_3} |u(t_5,x_3)| dt_5 dx_3 dt_3 dt_2 dt_1$$

$$\le rA, \quad (t,x) \in [0,\infty) \times [0,L].$$

Thus,

$$\|Gu\| \leq rA.$$

Next,

$$|Fu(t,x)| = \left| \int_0^t \int_0^{t_1} \int_0^{t_2} \int_0^x \int_0^{x_1} \int_0^{x_2} g(t_3,x_3) \int_0^{x_3} x_4 \left(-u(t_3,x_4) + u_0(x_4) \right.\right.$$

$$\left. + \int_0^{t_3} f(t_5,x_4,u(t_5,x_4))dt_5 \right) dx_4 dx_3 dx_2 dx_1 dt_3 dt_2 dt_1$$

$$+ \int_0^t \int_0^{t_1} \int_0^{t_2} \int_0^x \int_0^{x_1} \int_0^{x_2} x_3 g(t_3,x_3) \int_{x_3}^L \left(-u(t_3,x_4) + u_0(x_3) \right.$$

$$\left.\left. + \int_0^{t_3} f(t_5,x_4,u(t_5,x_4))dt_5 \right) dx_4 dx_3 dx_2 dx_1 dt_3 dt_2 dt_1 \right|$$

$$\leq \int_0^t \int_0^{t_1} \int_0^{t_2} \int_0^x \int_0^{x_1} \int_0^{x_2} g(t_3,x_3) \int_0^{x_3} x_4 \left(|u(t_3,x_4)| + u_0(x_4) \right.$$

$$\left. + \int_0^{t_3} \sum_{j=1}^l c_j(t_5,x_4)|u(t_5,x_4)|^{p_j} dt_5 \right) dx_4 dx_3 dx_2 dx_1 dt_3 dt_2 dt_1$$

$$+ \int_0^t \int_0^{t_1} \int_0^{t_2} \int_0^x \int_0^{x_1} \int_0^{x_2} x_3 g(t_3,x_3) \int_{x_3}^L \left(|u(t_3,x_4)| + u_0(x_3) \right.$$

$$\left. + \int_0^{t_3} \sum_{j=1}^l c_j(t_5,x_4)|u(t_5,x_4)|^{p_j} dt_5 \right) dx_4 dx_3 dx_2 dx_1 dt_3 dt_2 dt_1$$

$$\leq r \int_0^t \int_0^{t_1} \int_0^{t_2} \int_0^x \int_0^{x_1} \int_0^{x_2} x_3^2 g(t_3,x_3) dx_3 dx_2 dx_1 dt_3 dt_2 dt_1$$

$$+ \sum_{j=1}^{l} r^{p_j} \int_0^t \int_0^{t_1} \int_0^{t_2} \int_0^x \int_0^{x_1} \int_0^{x_2} g(t_3,x_3) \int_0^{x_3} x_4 \int_0^{t_3} c_j(t_5,x_4)$$

$$\times \, dt_5 dx_4 dx_3 dx_2 dx_1 dt_3 dt_2 dt_1$$

$$+ r \int_0^t \int_0^{t_1} \int_0^{t_2} \int_0^x \int_0^{x_1} \int_0^{x_2} (L-x_3) g(t_3,x_3) dx_3 dx_2 dx_1 dt_3 dt_2 dt_1$$

$$+ \sum_{j=1}^{l} r^{p_j} \int_0^t \int_0^{t_1} \int_0^{t_2} \int_0^x \int_0^{x_1} \int_0^{x_2} g(t_3,x_3) \int_{x_3}^{L} x_4 \int_0^{t_3} c_j(t_5,x_4)$$

$$\times \, dt_5 dx_4 dx_3 dx_2 dx_1 dt_3 dt_2 dt_1$$

$$\leq rA + \sum_{j=1}^{l} r^{p_j} A + rA + \sum_{j=1}^{l} r^{p_j} A$$

$$= \left(2r + 2 \sum_{j=1}^{l} r^{p_j} \right) A, \quad (t,x) \in [0,\infty) \times [0,L],$$

and

$$\left| \frac{\partial}{\partial t} Fu(t,x) \right| = \left| \int_0^t \int_0^{t_2} \int_0^x \int_0^{x_1} \int_0^{x_2} g(t_3,x_3) \int_0^{x_3} x_4 \left(-u(t_3,x_4) + u_0(x_4) \right. \right.$$

$$\left. + \int_0^{t_3} f(t_5,x_4,u(t_5,x_4)) dt_5 \right) dx_4 dx_3 dx_2 dx_1 dt_3 dt_2$$

$$+ \int_0^t \int_0^{t_2} \int_0^x \int_0^{x_1} \int_0^{x_2} x_3 g(t_3,x_3) \int_{x_3}^{L} \left(-u(t_3,x_4) + u_0(x_3) \right.$$

$$\left. \left. + \int_0^{t_3} f(t_5,x_4,u(t_5,x_4)) dt_5 \right) dx_4 dx_3 dx_2 dx_1 dt_3 dt_2 \right|$$

$$\leq \int_0^t \int_0^{t_2} \int_0^x \int_0^{x_1} \int_0^{x_2} g(t_3,x_3) \int_0^{x_3} x_4 \left(|u(t_3,x_4)| + u_0(x_4) \right.$$

$$+ \int_0^{t_3} \sum_{j=1}^l c_j(t_5,x_4) |u(t_5,x_4)|^{p_j} dt_5 \bigg) dx_4 dx_3 dx_2 dx_1 dt_3 dt_2$$

$$+ \int_0^t \int_0^{t_2} \int_0^x \int_0^{x_1} \int_0^{x_2} x_3 g(t_3,x_3) \int_{x_3}^L \left(|u(t_3,x_4)| + u_0(x_3) \right.$$

$$+ \int_0^{t_3} \sum_{j=1}^l c_j(t_5,x_4) |u(t_5,x_4)|^{p_j} dt_5 \bigg) dx_4 dx_3 dx_2 dx_1 dt_3 dt_2$$

$$\leq r \int_0^t \int_0^{t_2} \int_0^x \int_0^{x_1} \int_0^{x_2} x_3^2 g(t_3,x_3) dx_3 dx_2 dx_1 dt_3 dt_2$$

$$+ \sum_{j=1}^l r^{p_j} \int_0^t \int_0^{t_2} \int_0^x \int_0^{x_1} \int_0^{x_2} g(t_3,x_3) \int_0^{x_3} x_4 \int_0^{t_3} c_j(t_5,x_4)$$

$$dt_5 dx_4 dx_3 dx_2 dx_1 dt_3 dt_2$$

$$+ r \int_0^t \int_0^{t_2} \int_0^x \int_0^{x_1} \int_0^{x_2} (L - x_3) g(t_3,x_3) dx_3 dx_2 dx_1 dt_3 dt_2$$

$$+ \sum_{j=1}^l r^{p_j} \int_0^t \int_0^{t_2} \int_0^x \int_0^{x_1} \int_0^{x_2} g(t_3,x_3) \int_{x_3}^L x_4 \int_0^{t_3} c_j(t_5,x_4)$$

$$dt_5 dx_4 dx_3 dx_2 dx_1 dt_3 dt_2$$

$$\leq rA + \sum_{j=1}^l r^{p_j} A + rA + \sum_{j=1}^l r^{p_j} A$$

$$= \left(2r + 2 \sum_{j=1}^l r^{p_j} \right) A, \quad (t,x) \in [0,\infty) \times [0,L],$$

and

$$\left|\frac{\partial}{\partial x}Fu(t,x)\right| = \left|\int_0^t \int_0^{t_1} \int_0^{t_2} \int_0^x \int_0^{x_2} g(t_3,x_3)\int_0^{x_3} x_4\left(-u(t_3,x_4)+u_0(x_4)\right.\right.$$

$$\left.+\int_0^{t_3} f(t_5,x_4,u(t_5,x_4))dt_5\right) dx_4 dx_3 dx_2 dt_3 dt_2 dt_1$$

$$+\int_0^t \int_0^{t_1} \int_0^{t_2} \int_0^x \int_0^{x_2} x_3 g(t_3,x_3)\int_{x_3}^L \left(-u(t_3,x_4)+u_0(x_3)\right.$$

$$\left.\left.+\int_0^{t_3} f(t_5,x_4,u(t_5,x_4))dt_5\right) dx_4 dx_3 dx_2 dt_3 dt_2 dt_1\right|$$

$$\leq \int_0^t \int_0^{t_1} \int_0^{t_2} \int_0^x \int_0^{x_2} g(t_3,x_3)\int_0^{x_3} x_4\left(|u(t_3,x_4)|+u_0(x_4)\right.$$

$$\left.+\int_0^{t_3} \sum_{j=1}^l c_j(t_5,x_4)|u(t_5,x_4)|^{p_j}dt_5\right) dx_4 dx_3 dx_2 dt_3 dt_2 dt_1$$

$$+\int_0^t \int_0^{t_1} \int_0^{t_2} \int_0^x \int_0^{x_2} x_3 g(t_3,x_3)\int_{x_3}^L \left(|u(t_3,x_4)|+u_0(x_3)\right.$$

$$\left.+\int_0^{t_3} \sum_{j=1}^l c_j(t_5,x_4)|u(t_5,x_4)|^{p_j}dt_5\right) dx_4 dx_3 dx_2 dt_3 dt_2 dt_1$$

$$\leq r\int_0^t \int_0^{t_1} \int_0^{t_2} \int_0^x \int_0^{x_2} x_3^2 g(t_3,x_3)dx_3 dx_2 dt_3 dt_2 dt_1$$

$$+\sum_{j=1}^l r^{p_j}\int_0^t \int_0^{t_1} \int_0^{t_2} \int_0^x \int_0^{x_2} g(t_3,x_3)\int_0^{x_3} x_4 \int_0^{t_3} c_j(t_5,x_4)$$

$$dt_5 dx_4 dx_3 dx_2 dt_3 dt_2 dt_1$$

$$+ r \int_0^t \int_0^{t_1} \int_0^{t_2} \int_0^x \int_0^{x_2} (L - x_3) g(t_3, x_3) dx_3 dx_2 dt_3 dt_2 dt_1$$

$$+ \sum_{j=1}^l r^{p_j} \int_0^t \int_0^{t_1} \int_0^{t_2} \int_0^x \int_0^{x_2} g(t_3, x_3) \int_{x_3}^L x_4 \int_0^{t_3} c_j(t_5, x_4)$$

$$dt_5 dx_4 dx_3 dx_2 dt_3 dt_2 dt_1$$

$$\leq rA + \sum_{j=1}^l r^{p_j} A + rA + \sum_{j=1}^l r^{p_j} A$$

$$= \left(2r + 2 \sum_{j=1}^l r^{p_j} \right) A, \quad (t, x) \in [0, \infty) \times [0, L],$$

and

$$\left| \frac{\partial^2}{\partial x^2} F u(t, x) \right| = \left| \int_0^t \int_0^{t_1} \int_0^{t_2} \int_0^x g(t_3, x_3) \int_0^{x_3} x_4 \left(-u(t_3, x_4) + u_0(x_4) \right. \right.$$

$$\left. + \int_0^{t_3} f(t_5, x_4, u(t_5, x_4)) dt_5 \right) dx_4 dx_3 dt_3 dt_2 dt_1$$

$$+ \int_0^t \int_0^{t_1} \int_0^{t_2} \int_0^x x_3 g(t_3, x_3) \int_{x_3}^L \left(-u(t_3, x_4) + u_0(x_3) \right.$$

$$\left. \left. + \int_0^{t_3} f(t_5, x_4, u(t_5, x_4)) dt_5 \right) dx_4 dx_3 dt_3 dt_2 dt_1 \right|$$

$$\leq \int_0^t \int_0^{t_1} \int_0^{t_2} \int_0^x g(t_3, x_3) \int_0^{x_3} x_4 \left(|u(t_3, x_4)| + u_0(x_4) \right.$$

$$+ \int_0^{t_3} \sum_{j=1}^{l} c_j(t_5,x_4) |u(t_5,x_4)|^{p_j} dt_5 \Bigg) dx_4 dx_3 dt_3 dt_2 dt_1$$

$$+ \int_0^{t} \int_0^{t_1} \int_0^{t_2} \int_0^{x} x_3 g(t_3,x_3) \int_{x_3}^{L} \Bigg(|u(t_3,x_4)| + u_0(x_3)$$

$$+ \int_0^{t_3} \sum_{j=1}^{l} c_j(t_5,x_4) |u(t_5,x_4)|^{p_j} dt_5 \Bigg) dx_4 dx_3 dt_3 dt_2 dt_1$$

$$\leq r \int_0^{t} \int_0^{t_1} \int_0^{t_2} \int_0^{x} x_3^2 g(t_3,x_3) dx_3 dt_3 dt_2 dt_1$$

$$+ \sum_{j=1}^{l} r^{p_j} \int_0^{t} \int_0^{t_1} \int_0^{t_2} \int_0^{x} g(t_3,x_3) \int_0^{x_3} x_4 \int_0^{t_3} c_j(t_5,x_4)$$

$$\times dt_5 dx_4 dx_3 dt_3 dt_2 dt_1$$

$$+ r \int_0^{t} \int_0^{t_1} \int_0^{t_2} \int_0^{x} (L-x_3) g(t_3,x_3) dx_3 dt_3 dt_2 dt_1$$

$$+ \sum_{j=1}^{l} r^{p_j} \int_0^{t} \int_0^{t_1} \int_0^{t_2} \int_0^{x} g(t_3,x_3) \int_{x_3}^{L} x_4 \int_0^{t_3} c_j(t_5,x_4)$$

$$\times dt_5 dx_4 dx_3 dt_3 dt_2 dt_1$$

$$\leq rA + \sum_{j=1}^{l} r^{p_j} A + rA + \sum_{j=1}^{l} r^{p_j} A$$

$$= \left(2r + 2 \sum_{j=1}^{l} r^{p_j} \right) A, \quad (t,x) \in [0,\infty) \times [0,L].$$

Therefore

$$\|Fu\| \le \left(r + 2 \sum_{j=1}^{l} r^{p_j} \right) A.$$

This completes the proof.

5.3.2 Proof of the main result

For $u \in E$, define the operators

$$Tu(t,x) = (1 - \varepsilon)u(t,x) + Gu(t,x),$$

$$Su(t,x) = \varepsilon u(t,x) + Fu(t,x), \quad (t,x) \in [0,\infty) \times [0,L].$$

Note that if $u \in E$ is a fixed point of the operator $T + S$, then

$$u(t,x) = Tu(t,x) + Su(t,x)$$

$$= (1 - \varepsilon)u(t,x) + Gu(t,x) + \varepsilon u(t,x) + Fu(t,x)$$

$$= u(t,x) + Fu(t,x) + Gu(t,x), \quad (t,x) \in [0,\infty) \times [0,L],$$

or

$$0 = Gu(t,x) + Fu(t,x), \quad (t,x) \in [0,\infty) \times [0,L].$$

Therefore, any fixed point $u \in E$ of the operator $T + S$ is a solution of the IBVP (5.7). Define

$$\widetilde{\mathscr{P}} = \{ u \in E : u(t,x) \ge 0, \quad (t,x) \in [0,\infty) \times [0,L] \},$$

$\mathscr{P}$ be the set of all equi-continuous families in $\widetilde{\mathscr{P}}$(an example for an equi-continuous family is the family $\{ (3 + \sin(t+n))e^{x-n} : t \in [0,\infty), x \in [0,L] \}_{n \in \mathbb{N}}$),

$$\Omega = \{ u \in \mathscr{P} : \|u\| \le R \},$$

$$U = \Big\{ u \in \mathscr{P} : \frac{1}{2}u_0(x) \le u(t,x) \le u_0(x), \quad (t,x) \in [0,\infty) \times [0,L],$$

$$u(t,x) < \frac{3}{4}u_0(x), \quad (t,x) \in [1,2] \times [0,L], \quad \|u\| < r \Big\}.$$

1. For $u \in \Omega$, we have

    $$(I-T)u(t,x) = \varepsilon u(t,x) - Gu(t,x), \quad (t,x) \in [0,\infty) \times [0,L].$$

    Then, for $u,v \in \Omega$, using Lemma 5.3.4, we find

    $$\|(I-T)(u-v)\| \geq \varepsilon\|u-v\| - \|G(u-v)\|$$

    $$\geq (\varepsilon - A)\|u-v\|.$$

    Thus, $I - T : \Omega \to E$ is Lipschitz invertible with a constant $\gamma = \dfrac{1}{\varepsilon - A}$.

2. Let $u \in \overline{U}$. By Lemma 5.3.4, we have

    $$\|Su\| \leq \varepsilon\|u\| + \|Fu\|$$

    $$\leq \varepsilon r + \left(2r + 2\sum_{j=1}^{l} r^{p_j}\right) A.$$

    Therefore $S : \overline{U} \to E$ is uniformly bounded. Since $S : \overline{U} \to E$ is continuous, we obtain that $S(\overline{U})$ is equi-continuous and then $S : \overline{U} \to E$ is relatively compact. Consequently $S : \overline{U} \to E$ is a 0-set contraction.

3. Let $u \in \overline{U}$. For $z \in \Omega$, define the operator

    $$Lz(t,x) = Tz(t,s) + Su(t,s), \quad (t,x) \in [0,\infty) \times [0,L].$$

    For $z \in \Omega$, we get

    $$\|Lz\| = \|Tz + Su\|$$

    $$\leq \|Tz\| + \|Su\|$$

    $$\leq (1 - \varepsilon + A)R + \varepsilon r + \left(2r + 2\sum_{j=1}^{l} r^{p_j}\right) A$$

$$\leq \quad (1-\varepsilon+A)R+(\varepsilon-A)R$$

$$= \quad R,$$

i.e., $L:\Omega \rightarrow \Omega$. Next, for $z_1, z_2 \in \Omega$, we have

$$\|Lz_1 - Lz_2\| \quad = \quad \|Tz_1 - Tz_2\|$$

$$= \quad \|T(z_1 - z_2)\|$$

$$\leq \quad (1-\varepsilon+A)\|z_1 - z_2\|.$$

Therefore, $L:\Omega \rightarrow \Omega$ is a contraction mapping. Hence, there exists a unique $z \in \Omega$ such that

$$z = Lz$$

or

$$(I - T)z = Su.$$

Consequently $S(\overline{U}) \subset (I - T)(\Omega)$.

4. Assume that there are $u \in \partial U$ and $\lambda \geq 1$ such that

$$Su = (I - T)(\lambda u), \quad \lambda u \in \Omega.$$

We have

$$\varepsilon u + Fu = \varepsilon \lambda u - G(\lambda u)$$

or

$$\varepsilon(\lambda - 1)u = Fu + G(\lambda u).$$

Suppose that $\lambda = 1$. Then, for $t \in (2, \infty)$ and $x = L$, we have

$$\frac{1}{8}A \quad > \quad rA$$

$$\geq \quad -Gu$$

$$= \quad Fu$$

$$> \frac{1}{4} \int_1^2 \int_1^{t_1} \int_1^{t_2} \int_0^L \int_0^{x_1} \int_0^{x_2} g(t_3, x_3)$$

$$\times \int_0^{x_3} x_4 u_0(x_4) dx_4 dx_3 dx_2 dx_1 dt_3 dt_2 dt_1$$

$$> \frac{A}{8},$$

which is a contradiction. Therefore $\lambda > 1$. Next, for $u \in \partial U$, we have

$$\varepsilon(\lambda - 1)r = \varepsilon(\lambda - 1)\|u\|$$

$$= \|Fu - G(\lambda u)\|$$

$$\leq \|Fu\| + \|G(\lambda u)\|$$

$$\leq \left(3r + 2\sum_{j=1}^{l} r^{p_j}\right) A.$$

Hence, for $\lambda u \in \Omega, u \in \partial U, t \in (2, \infty)$ and $x = L$, we get

$$\left(3r + 2\sum_{j=1}^{l} r^{p_j}\right) A \geq \varepsilon(\lambda - 1)|u|$$

$$= |Fu + G(\lambda u)|$$

$$\geq Fu - |G(\lambda u)|$$

$$\geq \frac{1}{4} \int_1^2 \int_1^{t_1} \int_1^{t_2} \int_0^L \int_0^{x_1} \int_0^{x_2} g(t_3, x_3)$$

$$\times \int_0^{x_3} x_4 u_0(x_4) dx_4 dx_3 dx_2 dx_1 dt_3 dt_2 dt_1 - Ar$$

$$\geq \frac{A}{8} - Ar,$$

whereupon

$$\left(4r + 2\sum_{j=1}^{l} r^{p_j}\right) A \geq \frac{A}{8}$$

or

$$4r + 2\sum_{j=1}^{l} r^{p_j} \geq \frac{1}{8},$$

which is a contradiction.

By (1), (2), (3), and (4) and Proposition 2.2.61, we conclude that the operator $T + S$ has a fixed point in $U \cap \Omega$. This completes the proof.

6

Applications to Hyperbolic Equations

In this chapter, we give some applications of some fixed point theorems proved in Chapters 2 and 3 for existence of classical solutions for some classes IVPs for one, two, and n dimensional wave equations. Some of the results in this chapter can be found in [29].

6.1 Applications to One-Dimensional Hyperbolic Equations

Consider the following IVP

$$u_{tt} - u_{xx} = f(t,x,u_t), \quad x \geq 2t \geq 0, \tag{6.1}$$

$$u(0,x) = u_0(x), \quad u_t(0,x) = u_1(x), \quad x \geq 0. \tag{6.2}$$

Let

$$E = \left\{ u \in \mathscr{C}^2([0,\infty) \times [0,\infty)) : \|u\| = \max \left\{ \sup_{x \geq 2t \geq 0} |u(t,x)|, \right. \right.$$

$$\sup_{x \geq 2t \geq 0} \left| \frac{\partial}{\partial t} u(t,x) \right|,$$

$$\sup_{x \geq 2t \geq 0} \left| \frac{\partial^2}{\partial t^2} u(t,x) \right|, \sup_{x \geq 2t \geq 0} \left| \frac{\partial}{\partial x} u(t,x) \right|,$$

$$\left. \left. \sup_{x \geq 2t \geq 0} \left| \frac{\partial^2}{\partial x^2} u(t,x) \right| \right\} < \infty \right\}.$$

Suppose that

(C1) $f \in \mathscr{C}([0,\infty) \times [0,\infty) \times \mathbb{R}), 0 \leq f(t,x,v) \leq a(t,x)|v|^p, x \geq 2t \geq 0,$
$v \in \mathbb{R}$, where $p > 0$, $p \neq 1$, $a \in \mathscr{C}([0,\infty) \times [0,\infty)), a > 0$ on $[0,\infty) \times$

$[0, \infty)$, and there exist $g \in \mathscr{C}([0, \infty) \times [0, \infty))$ a positive function on $(0, \infty) \times (0, \infty)$ and constants $A > B > 0$ such that

$$100\left(1 + t + \cdots + t^6\right)\left(1 + x + \cdots + x^6\right)\int_0^t \int_0^x g(t_1, x_1)$$

$$\times \left(1 + \int_0^{t_1} \int_0^{x_1} a(t_2, x_2)dx_2dt_2\right)dx_1dt_1 \leq A, \quad x \geq 2t \geq 0,$$

$$\int_2^3 \int_2^3 g(t_1, x_1)dx_1dt_1 \geq B.$$

(C2) $u_0, u_1 \in \mathscr{C}^2([0, \infty))$, $\|u_0\|, \|u_1\| \leq r_1$, where $r_1 > 0$ is chosen so that

$$r_1 \geq \left(4\left(\frac{Aq}{2B}\right)^p\right)^{\frac{1}{1-p}} \quad \text{if} \quad p \in (0, 1)$$

and

$$0 < r_1 \leq \left(\frac{1}{4}\left(\frac{2B}{Aq}\right)^p\right)^{\frac{1}{p-1}} \quad \text{if} \quad p \in (1, \infty),$$

for some fixed constant $q > 2$.

In the next subsection, it will be given examples for the constants A, B, q, r_1, p and for the functions f, g, u_0, u_1 that satisfy the conditions $(C1)$ and $(C2)$.

For $u \in E$, define the operators

$$Tu(t, x) = \int_0^t \int_0^x (t - t_1)^4 (x - x_1)^4 g(t_1, x_1)\int_0^{x_1} (x_1 - x_2)$$

$$\times \left(u(t_1, x_2) - u_0(x_2) - t_1u_1(x_2)\right)dx_2dx_1dt_1$$

$$- \int_0^t \int_0^x (t - t_1)^4 (x - x_1)^4 g(t_1, x_1)$$

$$\times \int_0^{t_1} (t_1 - t_2)u(t_2, x_1)dt_2dx_1dt_1,$$

$$F_1 u(t,x) = -\int_0^t \int_0^x (t-t_1)^4 (x-x_1)^4 g(t_1,x_1) \int_0^{t_1} \int_0^{x_1} (t_1-t_2)(x_1-x_2)$$

$$\times f(t_2,x_2,u_t(t_2,x_2)) dx_2 dt_2 dx_1 dt_1,$$

$x \geq 2t \geq 0.$

Lemma 6.1.1 *Suppose* (C1) *and* (C2). *If a function* $u \in E$ *is a solution of the integral equation*

$$Tu(t,x) + F_1 u(t,x) = 0, \quad x \geq 2t \geq 0, \tag{6.3}$$

then it is a solution of the IVP (6.1)–(6.2).

Proof 6.1.2 *We differentiate five times respect to* t *and five times with respect to* x *the equation* (6.3). *Then*

$$0 = g(t,x) \int_0^x \int_0^{x_1} (-u(t,x_2) + u_0(x_2) + t u_1(x_2)) dx_2 dx_1$$

$$+ g(t,x) \int_0^t \int_0^{t_1} u(t_2,x) dt_2 dt_1$$

$$+ g(t,x) \int_0^t \int_0^{t_1} \int_0^x \int_0^{x_1} f(t_2,x_2,u_t(t_2,x_2)) dx_2 dx_1 dt_2 dt_1,$$

$x \geq 2t \geq 0,$ *whereupon*

$$0 = \int_0^x \int_0^{x_1} (-u(t,x_2) + u_0(x_2) + t u_1(x_2)) dx_2 dx_1$$

$$+ \int_0^t \int_0^{t_1} u(t_2,x) dt_2 dt_1 \tag{6.4}$$

$$+ \int_0^t \int_0^{t_1} \int_0^x \int_0^{x_1} f(t_2,x_2,u_t(t_2,x_2)) dx_2 dx_1 dt_2 dt_1,$$

$x \geq 2t \geq 0$. *We differentiate the equation (6.4) with respect to t and we get*

$$0 = \int_0^x \int_0^{x_1} (-u_t(t,x_2) + u_1(x_2)) \, dx_2 dx_1$$

$$+ \int_0^t u(t_1,x) \, dt_1 \tag{6.5}$$

$$+ \int_0^t \int_0^x \int_0^{x_1} f(t_1,x_2,u_t(t_1,x_2)) \, dx_2 dx_1 dt_1,$$

$x \geq 2t \geq 0$, *which we differentiate in t and we get*

$$0 = -\int_0^x \int_0^{x_1} u_{tt}(t,x_2) \, dx_2 dx_1 + u(t,x)$$

$$+ \int_0^x \int_0^{x_1} f(t,x_2,u_t(t,x_2)) \, dx_2 dx_1, \quad x \geq 2t \geq 0.$$

The last equation we differentiate twice in x and we get

$$0 = -u_{tt}(t,x) + u_{xx}(t,x) + f(t,x,u_t(t,x)), \quad (t,x) \in (0,\infty) \times (0,\infty).$$

Now, we put $t = 0$ in (6.5) and we find

$$0 = \int_0^x \int_0^{x_1} (-u_t(0,x_2) + u_1(x_2)) \, dx_2 dx_1, \quad x \in [0,\infty),$$

which we differentiate twice in x and we obtain

$$0 = -u_t(0,x) + u_1(x), \quad x \in [0,\infty).$$

Now, we put $t = 0$ in (6.4) and we arrive to

$$0 = \int_0^x \int_0^{x_1} (-u(0,x_2) + u_0(x_2)) \, dx_2 dx_1, \quad x \in [0,\infty),$$

which we differentiate twice in x and we find

$$0 = -u(0,x) + u_0(x), \quad x \in [0,\infty).$$

This completes the proof.

Lemma 6.1.3 *Suppose* (C1), (C2), *and* $r \geq r_1$. *For* $u \in \{v \in E : \|v\| \leq r\}$ *we have*

$$\|Tu\| \leq 2Ar, \quad \|F_1 u\| \leq A\|u\|^p \leq Ar^p,$$

$$\left\|\frac{\partial^3}{\partial t^3} F_1 u\right\|_{\mathscr{C}^0} \leq Ar^p, \quad \left\|\frac{\partial^3}{\partial x^3} F_1 u\right\|_{\mathscr{C}^0} \leq Ar^p$$

and $F_1 : \{v \in E : \|v\| \leq r\} \to E$ *is a completely continuous operator.*

Proof 6.1.4 *Let* $u \in \{v \in E : \|v\| \leq r\}$ *be arbitrarily chosen and fixed.*
Then

$$
\begin{aligned}
|Tu(t,x)| &\leq \int_0^t \int_0^x (t-t_1)^4 (x-x_1)^4 g(t_1,x_1) \int_0^{x_1} (x_1-x_2) \\
&\quad \times \left(|u(t_1,x_2)| + u_0(x_2) + t_1 u_1(x_2) \right) dx_2 dx_1 dt_1 \\
&\quad + \int_0^t \int_0^x (t-t_1)^4 (x-x_1)^4 g(t_1,x_1) \\
&\quad \times \int_0^{t_1} (t_1-t_2)|u(t_2,x_1)| dt_2 dx_1 dt_1 \\
&\leq t^4 x^4 ((2+t)x^2 + t^2) r \int_0^t \int_0^x g(t_1,x_1) dx_1 dt_2 \\
&\leq Ar, \quad x \geq 2t \geq 0,
\end{aligned}
$$

and

$$
\begin{aligned}
\left|\frac{\partial}{\partial t} Tu(t,x)\right| &\leq 4 \int_0^t \int_0^x (t-t_1)^3 (x-x_1)^4 g(t_1,x_1) \\
&\quad \times \int_0^{x_1} (x_1-x_2) \left(|u(t_1,x_2)| + u_0(x_2) \right. \\
&\quad \left. + t_1 u_1(x_2) \right) dx_2 dx_1 dt_1 \\
&\quad + 4 \int_0^t \int_0^x (t-t_1)^3 (x-x_1)^4 g(t_1,x_1)
\end{aligned}
$$

$$\times \int_0^{t_1} (t_1 - t_2)|u(t_2,x_1)|dt_2dx_1dt_1$$

$$\leq 4t^3x^4((2+t)x^2 + t^2)r$$

$$\times \int_0^t \int_0^x g(t_1,x_1)dx_1dt_2$$

$$\leq Ar, \quad x \geq 2t \geq 0,$$

and

$$\left|\frac{\partial^2}{\partial t^2}Tu(t,x)\right| \leq 12\int_0^t\int_0^x (t-t_1)^2(x-x_1)^4g(t_1,x_1)$$

$$\times \int_0^{x_1}(x_1-x_2)\left(|u(t_1,x_2)| + u_0(x_2)\right.$$

$$\left. + t_1u_1(x_2)\right)dx_2dx_1dt_1$$

$$+12\int_0^t\int_0^x (t-t_1)^2(x-x_1)^4g(t_1,x_1)$$

$$\times \int_0^{t_1}(t_1-t_2)|u(t_2,x_1)|dt_2dx_1dt_1$$

$$\leq 12t^2x^4((2+t)x^2 + t^2)r\int_0^t\int_0^x g(t_1,x_1)dx_1dt_2$$

$$\leq Ar, \quad x \geq 2t \geq 0,$$

and

$$\left|\frac{\partial}{\partial x}Tu(t,x)\right| \leq 4\int_0^t\int_0^x (t-t_1)^4(x-x_1)^3g(t_1,x_1)$$

$$\times \int_0^{x_1}(x_1-x_2)\left(|u(t_1,x_2)| + u_0(x_2)\right.$$

$$\left. + t_1u_1(x_2)\right)dx_2dx_1dt_1$$

$$+4 \int_0^t \int_0^x (t-t_1)^4 (x-x_1)^3 g(t_1,x_1)$$

$$\times \int_0^{t_1} (t_1-t_2)|u(t_2,x_1)|dt_2 dx_1 dt_1$$

$$\leq 4t^4 x^3 ((2+t)x^2 + t^2)r \int_0^t \int_0^x g(t_1,x_1)dx_1 dt_2$$

$$\leq Ar, \quad x \geq 2t \geq 0,$$

and

$$\left| \frac{\partial^2}{\partial x^2} Tu(t,x) \right| \leq 12 \int_0^t \int_0^x (t-t_1)^4 (x-x_1)^2 g(t_1,x_1)$$

$$\times \int_0^{x_1} (x_1-x_2)\left(|u(t_1,x_2)| + u_0(x_2) \right.$$

$$\left. +t_1 u_1(x_2) \right) dx_2 dx_1 dt_1$$

$$+12 \int_0^t \int_0^x (t-t_1)^4 (x-x_1)^2 g(t_1,x_1)$$

$$\times \int_0^{t_1} (t_1-t_2)|u(t_2,x_1)|dt_2 dx_1 dt_1$$

$$\leq 12t^4 x^2 ((2+t)x^2 + t^2)r \int_0^t \int_0^x g(t_1,x_1)dx_1 dt_2$$

$$\leq Ar, \quad x \geq 2t \geq 0.$$

Consequently $\|Tu\| \leq Ar$. *Next,*

$$|F_1 u(t,x)| \leq \int_0^t \int_0^x (t-t_1)^4 (x-x_1)^4 g(t_1,x_1) \int_0^{t_1} \int_0^{x_1} (t_1-t_2)(x_1-x_2)$$

$$\times a(t_2,x_2)|u_t(t_2,x_2)|^p dx_2 dt_2 dx_1 dt_1$$

$$\leq t^6 x^6 \|u\|^p \int_0^t \int_0^x g(t_1,x_1) \int_0^{t_1} \int_0^{x_1} a(t_2,x_2)dx_2dt_2dx_1dt_1$$

$$\leq A\|u\|^p$$

$$\leq Ar^p, \quad x \geq 2t \geq 0,$$

and

$$\left| \frac{\partial}{\partial t} F_1 u(t,x) \right| \leq 4 \int_0^t \int_0^x (t-t_1)^3 (x-x_1)^4 g(t_1,x_1)$$

$$\times \int_0^{t_1} \int_0^{x_1} (t_1-t_2)(x_1-x_2)$$

$$\times a(t_2,x_2)|u_t(t_2,x_2)|^p dx_2dt_2dx_1dt_1$$

$$\leq 4t^5 x^6 \|u\|^p \int_0^t \int_0^x g(t_1,x_1)$$

$$\times \int_0^{t_1} \int_0^{x_1} a(t_2,x_2)dx_2dt_2dx_1dt_1$$

$$\leq A\|u\|^p$$

$$\leq Ar^p, \quad x \geq 2t \geq 0,$$

and

$$\left| \frac{\partial^2}{\partial t^2} F_1 u(t,x) \right| \leq 12 \int_0^t \int_0^x (t-t_1)^2 (x-x_1)^4 g(t_1,x_1)$$

$$\times \int_0^{t_1} \int_0^{x_1} (t_1-t_2)(x_1-x_2)$$

$$\times a(t_2,x_2)|u_t(t_2,x_2)|^p dx_2dt_2dx_1dt_1$$

$$\leq\ 12t^4x^6\|u\|^P \int_0^t \int_0^x g(t_1,x_1)$$

$$\times \int_0^{t_1} \int_0^{x_1} a(t_2,x_2)dx_2dt_2dx_1dt_1$$

$$\leq\ A\|u\|^P$$

$$\leq\ Ar^P,\quad x \geq 2t \geq 0,$$

and

$$\left|\frac{\partial}{\partial x}F_1u(t,x)\right|\ \leq\ 4\int_0^t \int_0^x (t-t_1)^4(x-x_1)^3 g(t_1,x_1)$$

$$\times \int_0^{t_1} \int_0^{x_1} (t_1-t_2)(x_1-x_2)$$

$$\times a(t_2,x_2)|u_t(t_2,x_2)|^P dx_2dt_2dx_1dt_1$$

$$\leq\ 4t^6x^5\|u\|^P \int_0^t \int_0^x g(t_1,x_1)$$

$$\times \int_0^{t_1} \int_0^{x_1} a(t_2,x_2)dx_2dt_2dx_1dt_1$$

$$\leq\ A\|u\|^P$$

$$\leq\ Ar^P,\quad x \geq 2t \geq 0,$$

and

$$\left|\frac{\partial^2}{\partial x^2}F_1u(t,x)\right|\ \leq\ 12\int_0^t \int_0^x (t-t_1)^4(x-x_1)^2 g(t_1,x_1)$$

$$\times \int_0^{t_1} \int_0^{x_1} (t_1-t_2)(x_1-x_2)$$

$$\times a(t_2,x_2)|u_t(t_2,x_2)|^P dx_2dt_2dx_1dt_1$$

$$\leq\ 12t^6x^4\|u\|^p \int_0^t \int_0^x g(t_1,x_1)$$

$$\times \int_0^{t_1} \int_0^{x_1} a(t_2,x_2)dx_2dt_2dx_1dt_1$$

$$\leq\ A\|u\|^p$$

$$\leq\ Ar^p, \quad x \geq 2t \geq 0.$$

Consequently $\|F_1u\| \leq Ar^p$. *Moreover,*

$$\left|\frac{\partial^3}{\partial t^3}F_1u(t,x)\right| \leq\ 24 \int_0^t \int_0^x (t-t_1)(x-x_1)^4 g(t_1,x_1)$$

$$\times \int_0^{t_1} \int_0^{x_1} (t_1-t_2)(x_1-x_2)$$

$$\times a(t_2,x_2)|u_t(t_2,x_2)|^p dx_2dt_2dx_1dt_1$$

$$\leq\ 24t^3x^6\|u\|^p \int_0^t \int_0^x g(t_1,x_1)$$

$$\times \int_0^{t_1} \int_0^{x_1} a(t_2,x_2)dx_2dt_2dx_1dt_1$$

$$\leq\ A\|u\|^p$$

$$\leq\ Ar^p, \quad x \geq 2t \geq 0,$$

and

$$\left|\frac{\partial^3}{\partial x^3}F_1u(t,x)\right| \leq\ 24 \int_0^t \int_0^x (t-t_1)^4(x-x_1)g(t_1,x_1)$$

$$\times \int_0^{t_1} \int_0^{x_1} (t_1-t_2)(x_1-x_2)$$

$$\times a(t_2,x_2)|u_t(t_2,x_2)|^p dx_2dt_2dx_1dt_1$$

$$\leq 24t^6x^3\|u\|^p \int_0^t \int_0^x g(t_1,x_1)$$

$$\times \int_0^{t_1} \int_0^{x_1} a(t_2,x_2)dx_2dt_2dx_1dt_1$$

$$\leq A\|u\|^p$$

$$\leq Ar^p, \quad x \geq 2t \geq 0.$$

Therefore, $F_1 : \{v \in E : \|v\| \leq r\} \to E$ is a completely continuous operator. This completes the proof.

Theorem 6.1.5 *Suppose (C1) and (C2). Then the IVP (6.1)-(6.2) has at least one non-negative solution $u \in \mathscr{C}^2([0,\infty) \times [0,\infty))$.*

Proof 6.1.6 *For $u \in E$, define the operators*

$$Fu = (I + F_1)u.$$

Note that any fixed point $u \in E$ of the operator $T + F$ is a solution of the IVP (6.1), (6.2). Denote

$$\mathscr{P} = \{u \in E : u(t,x) \geq 0, \quad x \geq 2t \geq 0\},$$

$$K = \left\{ u \in \mathscr{P} : u(t,x) \leq u_0(x) + tu_1(x), \right.$$

$$x \geq 2t \geq 0 \quad and \quad if \quad u,v \in K \quad one \quad has$$

$$(u(t,x) - v(t,x))\left(\int_0^x (x - x_1)(u(t,x_1) - v(t,x_1))dx_1 \right.$$

$$\left. -2\int_0^t (t - t_1)(u(t_1,x) - v(t_1,x))dt_1 \right) \geq 0, \quad x \geq 2t \geq 0,$$

$$\inf_{x \geq 2t \geq 0} |u(t,x) - v(t,x)| \geq \frac{1}{q}\|u - v\| \right\}$$

$$\cup \left\{ \frac{u_0(x+t) + u_0(x-t)}{2} + \frac{1}{2}\int_{x-t}^{x+t} u_1(s)ds, \quad x \geq 2t \geq 0 \right\}.$$

Example 6.1.7 *Let* $u_0(x) = u_1(x) = e^{-x} + 2$, $x \geq 0$,

$$u(t,x) \ = \ 1 + (1+t)e^{-x} + 2t, \quad v(t,x) = (1+t)e^{-x} + 2t,$$

$x \geq 2t \geq 0$. *Then*

$$u_0(x) + tu_1(x) \ = \ e^{-x} + 2 + t(e^{-x} + 2)$$

$$= \ 2 + (1+t)e^{-x} + 2t$$

$$\geq \ u(t,x)$$

$$\geq \ v(t,x), \quad x \geq 2t \geq 0,$$

and

$$\int_0^x (x - x_1)(u(t,x_1) - v(t,x_1))dx_1$$

$$-2\int_0^t (t - t_1)(u(t_1,x) - v(t_1,x))dt_1$$

$$= \ \int_0^x (x - x_1)dx_1 - 2\int_0^t (t - t_1)dt_1$$

$$= \ \frac{x^2}{2} - t^2$$

$$\geq \ 2t^2 - t^2$$

$$= \ t^2$$

$$\geq \ 0, \quad x \geq 2t \geq 0.$$

Note that $0 \in K$ *and if* $u \in K$, *we have*

$$\inf_{x \geq 2t \geq 0} u(t,x) \geq \frac{1}{q}\|u\|.$$

Observe that if for some $u \in K$ we have $Tu = 0$, then

$$u_{tt} - u_{xx} = 0, \quad x \geq 2t \geq 0,$$

$$u(0,x) = u_0(x), \quad u_t(0,x) = u_1(x), \quad x \geq 0,$$

and if $u \in \mathscr{C}^2([0,\infty) \times [0,\infty))$ satisfies the last IVP, then $Tu = 0$. Therefore $0 \in T(K)$. By the definitions of the operators F and F_1, we have $I - F = -F_1$ and the operator $I - F : K \to \mathscr{P}$ is a completely continuous operator and then it is a 0-set contraction. Let $u \in K$. Then

$$\|Tu\| \geq \left| \int_0^4 \int_0^4 (4 - t_1)^4 (4 - x_1)^4 g(t_1, x_1) \right.$$

$$\times \int_0^{x_1} (x_1 - x_2) \left(u(t_1, x_2) - u_0(x_2) \right.$$

$$\left. - t_1 u_1(x_2) \right) dx_2 dx_1 dt_1$$

$$- \int_0^4 \int_0^4 (4 - t_1)^4 (4 - x_1)^4 g(t_1, x_1)$$

$$\left. \times \int_0^{t_1} (t_1 - t_2) u(t_2, x_1) dt_2 dx_1 dt_1 \right|$$

$$\geq \int_0^4 \int_0^4 (4 - t_1)^4 (4 - x_1)^4 g(t_1, x_1)$$

$$\times \int_0^{x_1} (x_1 - x_2) \left(-u(t_1, x_2) + u_0(x_2) + t_1 u_1(x_2) \right) dx_2 dx_1 dt_1$$

$$+ \int_0^4 \int_0^4 (4 - t_1)^4 (4 - x_1)^4 g(t_1, x_1)$$

$$\times \int_0^{t_1} (t_1 - t_2)u(t_2, x_1)dt_2 dx_1 dt_1$$

$$\geq \int_0^4 \int_0^4 (4 - t_1)^4 (4 - x_1)^4 g(t_1, x_1) \int_0^{t_1} (t_1 - t_2)u(t_2, x_1)dt_2 dx_1 dt_1$$

$$\geq \frac{1}{q}\|u\| \int_0^4 \int_0^4 (4 - t_1)^4 (4 - x_1)^4 g(t_1, x_1) \int_0^{t_1} (t_1 - t_2)dt_2 dx_1 dt_1$$

$$\geq \frac{1}{q}\|u\| \int_2^4 \int_2^4 (4 - t_1)^4 (4 - x_1)^4 g(t_1, x_1) \int_0^{t_1} (t_1 - t_2)dt_2 dx_1 dt_1$$

$$\geq \frac{1}{q}\|u\| \int_2^4 \int_2^4 (4 - t_1)^4 (4 - x_1)^4 g(t_1, x_1) \int_0^2 (2 - t_2)dt_2 dx_1 dt_1$$

$$\geq \frac{1}{q}\|u\| \int_2^4 \int_2^4 (4 - t_1)^4 (4 - x_1)^4 g(t_1, x_1) \int_0^1 (2 - t_2)dt_2 dx_1 dt_1$$

$$\geq \frac{1}{q}\|u\| \int_2^4 \int_2^4 (4 - t_1)^4 (4 - x_1)^4 g(t_1, x_1)dx_1 dt_1$$

$$\geq \frac{1}{q}\|u\| \int_2^3 \int_2^3 (4 - t_1)^4 (4 - x_1)^4 g(t_1, x_1)dx_1 dt_1$$

$$\geq \frac{1}{q}\|u\| \int_2^3 \int_2^3 g(t_1, x_1)dx_1 dt_1$$

$$\geq \frac{B}{q}\|u\|.$$

Thus, $T^{-1}: T(K) \to K$ exists and it is bounded. Now, suppose that $u, v \in K$ be arbitrarily chosen. Without loss of generality, suppose that

$$\int_0^x (x - x_1)(u(t, x_1) - v(t, x_1))dx_1 \geq 2 \int_0^t (t - t_1)(u(t_1, x) - v(t_1, x))dt_1,$$

$$u(t,x) \geq v(t,x), \quad x \geq 2t \geq 0.$$

Then

$$Tu(t,x) - Tv(t,x)$$

$$= \int_0^t \int_0^x (t-t_1)^4 (x-x_1)^4 g(t_1,x_1)$$

$$\times \int_0^{x_1} (x_1 - x_2)(u(t_1,x_2) - v(t_1,x_2))dx_2 dx_1 dt_1$$

$$- \int_0^t \int_0^x (t-t_1)^4 (x-x_1)^4 g(t_1,x_1)$$

$$\times \int_0^{t_1} (t_1 - t_2)(u(t_2,x_2) - v(t_2,x_2))dt_2 dx_1 dt_1$$

$$\geq \int_0^t \int_0^x (t-t_1)^4 (x-x_1)^4 g(t_1,x_1)$$

$$\times \int_0^{t_1} (t_1 - t_2)(u(t_2,x_2) - v(t_2,x_2))dt_2 dx_1 dt_1$$

$$\geq 0, \quad x \geq 2t \geq 0.$$

Hence, and above computations, we obtain

$$\|Tu - Tv\| \geq \int_0^4 \int_0^4 (4-t_1)^4 (4-x_1)^4 g(t_1,x_1)$$

$$\times \int_0^{t_1} (t_1 - t_2)(u(t_2,x_2) - v(t_2,x_2))dt_2 dx_1 dt_1$$

$$\geq \frac{B}{q}\|u-v\|.$$

*Thus, T^{-1} a Lipschitz operator with some constant $\delta = \frac{q}{B}$. Therefore,
(D1) and (D2) hold. Let*

$$\overline{U} = \{u \in T(K) : \frac{Ar_1}{4} \leq \|u\| \leq \frac{Ar_1}{2}\} \cup \{0\}.$$

Assume that there are $y \in T^{-1}(U)$ and $\lambda > 0$ so that

$$y = \lambda Ty + Fy \quad or \quad -F_1 y = \lambda Ty.$$

Denote $v = Ty \in \overline{U}$. Then

$$\|y\| \le \|T^{-1}v\| \le \frac{q}{B}\|v\| \le \frac{qAr_1}{2B}.$$

Note that $\dfrac{qAr_1}{2B} > r_1$. Hence, we obtain

$$\|F_1 y\| \le A\left(\frac{qAr_1}{2B}\right)^p.$$

We have

$$\frac{Ar_1}{4} \le \|v\|$$

$$= \|Ty\|$$

$$< \lambda\|Ty\|$$

$$= \|F_1 y\| \le A\left(\frac{qAr_1}{2B}\right)^p$$

or

$$0 < r_1 < \left(4\left(\frac{Aq}{2B}\right)^p\right)^{\frac{1}{1-p}} \quad if \quad p \in (0,1)$$

and

$$r_1 > \left(\frac{1}{4}\left(\frac{2B}{Aq}\right)^p\right)^{\frac{1}{p-1}} \quad if \quad p \in (1,\infty),$$

This is a contradiction. Hence, and Proposition 3.3.12, it follows that the IVP (6.1)-(6.2) has at least one non-negative solution $u \in \mathscr{C}^2([0,\infty) \times [0,\infty))$. This completes the proof.

Theorem 6.1.8 *Suppose (C1), (C2), and $p \in (0,1)$. Let $u \in \mathscr{C}^2([0,\infty) \times [0,\infty))$ be a solution of the IVP (6.1)-(6.2) such that*

$$u_t(t,x) \ge 0, \quad u_{xx}(t,x) \le 0, \quad x \ge 2t \ge 0.$$

Then

$$u_t(t,x) \le (1-p)^{\frac{1}{1-p}} \left(\frac{1}{1-p}(u_1(x))^{1-p} + \int_0^t a(t_1,x)dt_1 \right)^{\frac{1}{1-p}}, \quad t \in [0,\infty).$$

(6.6)

Proof 6.1.9 *We have*

$$u_t(t,x) - u_1(x) = \int_0^t u_{xx}(t_1,x)dt_1 + \int_0^t f(t_1,x,u_t(t_1,x))dt_1,$$

$x \ge 2t \ge 0$. *Hence, using that* $u_{xx}(t,x) \le 0$, $x \ge 2t \ge 0$ *and* (A2), *we obtain*

$$u_t(t,x) \le u_1(x) + \int_0^t a(t_1,x)(u_t(t_1,x))^p dt_1,$$

$x \ge 2t \ge 0$. *Fix* $x \in [0,\infty)$. *We apply the Gronwall-Bellman-Bihari inequality for* $w(s) = s^p$, $s \in [0,\infty)$, *and*

$$G(s) = \int_0^s \frac{dr}{w(r)} = \frac{1}{1-p}s^{1-p},$$

$$G^{-1}(s) = (1-p)^{\frac{1}{1-p}}s^{\frac{1}{1-p}}, \quad s \in [0,\infty),$$

and we find (6.6). *Because* $x \in [0,\infty)$ *was arbitrarily chosen, the inequality* (6.6) *is valid for any* $x \ge 2t \ge 0$. *This completes the proof.*

Example 6.1.10 *Below, we will illustrate our main result. Let*

$$h(x) = \log \frac{1 + s^{11}\sqrt{2} + s^{22}}{1 - s^{11}\sqrt{2} + s^{22}}, \quad s \in \mathbb{R}, \qquad l(s) = \arctan \frac{s^{11}\sqrt{2}}{1 - s^{22}},$$

$s \in \mathbb{R}$, $s \ne \pm 1$. *Then*

$$h'(s) = \frac{22\sqrt{2}s^{10}(1 - s^{22})}{(1 - s^{11}\sqrt{2} + s^{22})(1 + s^{11}\sqrt{2} + s^{22})}, \quad s \in \mathbb{R},$$

$$l'(s) = \frac{11\sqrt{2}s^{10}(1 + s^{20})}{1 + s^{40}}, \quad s \in \mathbb{R}, \quad s \ne \pm 1.$$

Therefore

$$-\infty < \lim_{s \to \pm\infty} \left(1 + |s| + s^2 + \cdots + |s|^9\right) h(s) < \infty,$$

$$-\infty \ < \ \lim_{s \to \pm\infty} \left(1 + |s| + s^2 + \cdots + |s|^9\right) l(s) < \infty.$$

Note that $\lim_{s \to \pm 1} |l(s)| = \dfrac{\pi}{2}$ *and by [37] (pp. 707, Integral 79), we have*

$$\int \frac{dz}{1 + z^4} = \frac{1}{4\sqrt{2}} \log \frac{1 + z\sqrt{2} + z^2}{1 - z\sqrt{2} + z^2} + \frac{1}{2\sqrt{2}} \arctan \frac{z\sqrt{2}}{1 - z^2}.$$

Hence, there exists a positive constant C_1 *so that*

$$\left(1 + t + t^2 \cdots + t^9\right) \int_0^t \frac{y^{10}}{1 + y^{44}} dy \ \leq \ C_1,$$

$$\left(1 + |s| + s^2 + \cdots + |s|^9\right) \int_0^s sign(s) \frac{y^{10}}{1 + y^{44}} dy \ \leq \ C_1,$$

$t \in [0, \infty)$, $s \in \mathbb{R}$. *Let*

$$Q(s) = \frac{s^{10}}{(1 + s^{44})(1 + s + s^2)^{28}}, \quad s \in \mathbb{R},$$

and

$$g(t, x) = Q(t)Q(x), \quad x \geq 2t \geq 0.$$

Then there exist constants $A > B > 0$ *such that*

$$100 \left(1 + t + \cdots + t^9\right) \left(1 + x + \cdots + x^9\right)$$

$$\int_0^t \int_0^x g(t_1, x_1) dx_1 dt_1 \leq A, \quad x \geq 2t \geq 0,$$

$$\int_2^3 \int_2^3 g(t_1, x_1) dx_1 dt_1 \ \geq \ B.$$

Fix $q > 2$ *and for* $p > 0$, $p \neq 1$, *fix* $r_1 > 0$ *so that*

$$r_1 > \left(4 \left(\frac{Aq}{2B}\right)^p\right)^{\frac{1}{1-p}} \quad if \quad p \in (0, 1)$$

and

$$0 < r_1 < \left(\frac{1}{4} \left(\frac{2B}{Aq} \right)^p \right)^{\frac{1}{p-1}} \quad \text{if} \quad p \in (1, \infty).$$

Note that $a(t,x) = 1$, $x \geq 2t \geq 0$. *Then, by Theorem 6.1.5, it follows that the IVP*

$$u_{tt} - u_{xx} = |u_t|^p, \quad x \geq 2t \geq 0,$$

$$u(0,x) = u_t(0,x) = \frac{r_1}{100(1+x)^9}, \quad x \geq 0,$$

has a non-negative solution $u \in \mathscr{C}^2([0,\infty) \times [0,\infty))$.

6.2 Applications to IVPs for a Class Two-Dimensional Nonlinear Wave Equations

In this section, we will investigate the following IVP

$$u_{tt} - u_{xx} - u_{yy} = f(u), \quad -\infty < x, y < \infty, \quad 0 < t < \infty, \tag{6.7}$$

$$u(x,y,0) = u_t(x,y,0) = 0, \quad -\infty < x, y < \infty, \tag{6.8}$$

where

(C3) $f : \mathbb{R} \to (-\infty, 0]$ is a continuous function such that

$$-f(v) \leq \sum_{j=1}^{l} a_j |v|^{l_j}, \quad v \in \mathbb{R},$$

for some nonnegative constants a_j, $j \in \{1, \ldots, l\}$, and for some positive constants l_j, $j \in \{1, \ldots, l\}$, $l \in \mathbb{N}$.

Our main result is as follows.

Theorem 6.2.1 *Suppose* (C3). *Then the IVP* (6.7)-(6.8) *has at least one solution* $u \in \mathscr{C}^2(\mathbb{R}^2 \times [0,\infty))$ *such that* $u \not\equiv 0$ *on* $\mathbb{R}^2 \times [0,\infty)$.

Proof 6.2.2 *Firstly, we will note that in [21] is shown that the function*

$$G(x,y,t,\xi,\eta,\tau) = -\frac{1}{2\pi} \frac{H\left(t-\tau-\sqrt{(x-\xi)^2+(y-\eta)^2}\right)}{\sqrt{(t-\tau)^2-(x-\xi)^2-(y-\eta)^2}},$$

$-\infty < x,y,\xi,\eta < \infty,\ 0 < t,\tau < \infty$, *where* $H(\cdot)$ *denotes the Heaviside function, is the Green function for the considered IVP (6.7)-(6.8), i.e., we have*

$$\frac{\partial^2}{\partial t^2}G(x,y,t,\xi,\eta,\tau) - \frac{\partial^2}{\partial x^2}G(x,y,t,\xi,\eta,\tau) - \frac{\partial^2}{\partial y^2}G(x,y,t,\xi,\eta,\tau)$$

$$= \delta(x-\xi)\delta(y-\eta)\delta(t-\tau),$$

$-\infty < x,y,\xi,\eta < \infty,\ 0 < t,\tau < \infty,$

$$G(x,y,0,\xi,\eta,\tau) = \frac{\partial}{\partial t}G(x,y,0,\xi,\eta,\tau) = 0,$$

$-\infty < x,y,\xi,\eta < \infty.$ *Here* $\delta(\cdot)$ *is the Dirac delta function. Thus, if*

$$u(x,y,t) = \int_{-\infty}^{\infty}\int_{-\infty}^{\infty}\int_{0}^{\infty} G(x,y,t,\xi,\eta,\tau)f(u(\xi,\eta,\tau))d\tau d\eta d\xi, \quad (6.9)$$

$-\infty < x,y < \infty,\ 0 < t < \infty$, *we have*

$$\frac{\partial^2}{\partial t^2}u(x,y,t) - \frac{\partial^2}{\partial x^2}u(x,y,t) - \frac{\partial^2}{\partial y^2}u(x,y,t)$$

$$= \int_{-\infty}^{\infty}\int_{-\infty}^{\infty}\int_{0}^{\infty} \left(\frac{\partial^2}{\partial t^2}G(x,y,t,\xi,\eta,\tau) - \frac{\partial^2}{\partial x^2}G(x,y,t,\xi,\eta,\tau)\right.$$

$$\left. - \frac{\partial^2}{\partial y^2}G(x,y,t,\xi,\eta,\tau)\right)f(u(\xi,\eta,\tau))d\tau d\eta d\xi$$

$$= \int_{-\infty}^{\infty}\int_{-\infty}^{\infty}\int_{0}^{\infty} \delta(x-\xi)\delta(y-\eta)\delta(t-\tau)f(u(\xi,\eta,\tau))d\tau d\eta d\xi$$

$$= f(u(x,y,t)), \quad -\infty < x,y < \infty, \quad 0 < t < \infty,$$

and

$$u(x,y,0) = \frac{\partial}{\partial t}u(x,y,0) = 0, \quad -\infty < x,y < \infty,$$

i.e., u, defined by (6.9), is a solution of the IVP (6.7), (6.8). Note that

$$G(x,y,t,\xi,\eta,\tau)\le 0,\quad -\infty<x,y,\xi,\eta<\infty,\quad 0<t,\tau<\infty.$$

Below we suppose

(C4) *a, b, c, a_1, b_1, A, and ε are constants such that*

$$c\ge b>a>0,\quad A>0,\quad \varepsilon\in(0,1),$$

$$\left(1+\frac{\varepsilon}{2}\right)\frac{c}{4}\le b,$$

$$0<a_1<b_1,$$

(C5) $g\in\mathscr{C}^2(\mathbb{R}^2\times[0,\infty))$ *be a non-negative function such that*

$$\int_{-\infty}^{x}\int_{-\infty}^{x_1}\int_{-\infty}^{x_2}\int_{-\infty}^{y}\int_{-\infty}^{y_1}\int_{-\infty}^{y_2}\int_{0}^{t}\int_{0}^{t_1}\int_{0}^{t_2}g(x_3,y_3,t_3)$$

$$dt_3dt_2dt_1dy_3dy_2dy_1dx_3dx_2dx_1\le A,$$

$$\int_{-\infty}^{x}\int_{-\infty}^{x_2}\int_{-\infty}^{y}\int_{-\infty}^{y_1}\int_{-\infty}^{y_2}\int_{0}^{t}\int_{0}^{t_1}\int_{0}^{t_2}g(x_3,y_3,t_3)$$

$$dt_3dt_2dt_1dy_3dy_2dy_1dx_3dx_2\le A,$$

$$\int_{-\infty}^{x}\int_{-\infty}^{y}\int_{-\infty}^{y_1}\int_{-\infty}^{y_2}\int_{0}^{t}\int_{0}^{t_1}\int_{0}^{t_2}g(x_3,y_3,t_3)$$

$$dt_3dt_2dt_1dy_3dy_2dy_1dx_3\le A,$$

$$\int_{-\infty}^{x}\int_{-\infty}^{x_1}\int_{-\infty}^{x_2}\int_{-\infty}^{y}\int_{-\infty}^{y_2}\int_{0}^{t}\int_{0}^{t_1}\int_{0}^{t_2}g(x_3,y_3,t_3)$$

$$dt_3dt_2dt_1dy_3dy_2dx_3dx_2dx_1\le A,$$

$$\int_{-\infty}^{x}\int_{-\infty}^{x_1}\int_{-\infty}^{x_2}\int_{-\infty}^{y}\int_{0}^{t}\int_{0}^{t_1}\int_{0}^{t_2}g(x_3,y_3,t_3)$$

$$dt_3dt_2dt_1dy_3dx_3dx_2dx_1\le A,$$

$$\int_{-\infty}^{x}\int_{-\infty}^{x_1}\int_{-\infty}^{x_2}\int_{-\infty}^{y}\int_{-\infty}^{y_1}\int_{-\infty}^{y_2}\int_{0}^{t}\int_{0}^{t_2} g(x_3,y_3,t_3)$$

$$dt_3 dt_2 dy_3 dy_2 dy_1 dx_3 dx_2 dx_1 \leq A,$$

$$\int_{-\infty}^{x}\int_{-\infty}^{x_1}\int_{-\infty}^{x_2}\int_{-\infty}^{y}\int_{-\infty}^{y_1}\int_{-\infty}^{y_2}\int_{0}^{t} g(x_3,y_3,t_3)$$

$$dt_3 dy_3 dy_2 dy_1 dx_3 dx_2 dx_1 \leq A,$$

$$-\int_{-\infty}^{x}\int_{-\infty}^{x_1}\int_{-\infty}^{x_2}\int_{-\infty}^{y}\int_{-\infty}^{y_1}\int_{-\infty}^{y_2}\int_{0}^{t}\int_{0}^{t_1}\int_{0}^{t_2} g(x_3,y_3,t_3)$$

$$\int_{-\infty}^{\infty}\int_{-\infty}^{\infty}\int_{0}^{\infty} G(x_3,y_3,t_3,\xi,\eta,\tau)d\tau d\eta d\xi$$

$$dt_3 dt_2 dt_1 dy_3 dy_2 dy_1 dx_3 dx_2 dx_1 \leq A,$$

$$-\int_{-\infty}^{x}\int_{-\infty}^{x_2}\int_{-\infty}^{y}\int_{-\infty}^{y_1}\int_{-\infty}^{y_2}\int_{0}^{t}\int_{0}^{t_1}\int_{0}^{t_2} g(x_3,y_3,t_3)$$

$$\int_{-\infty}^{\infty}\int_{-\infty}^{\infty}\int_{0}^{\infty} G(x_3,y_3,t_3,\xi,\eta,\tau)d\tau d\eta d\xi dt_3 dt_2 dt_1$$

$$dy_3 dy_2 dy_1 dx_3 dx_2 \leq A,$$

$$-\int_{-\infty}^{x}\int_{-\infty}^{y}\int_{-\infty}^{y_1}\int_{-\infty}^{y_2}\int_{0}^{t}\int_{0}^{t_1}\int_{0}^{t_2} g(x_3,y_3,t_3)$$

$$\int_{-\infty}^{\infty}\int_{-\infty}^{\infty}\int_{0}^{\infty} G(x_3,y_3,t_3,\xi,\eta,\tau)$$

$$d\tau d\eta d\xi dt_3 dt_2 dt_1 dy_3 dy_2 dy_1 dx_3 \leq A,$$

$$-\int_{-\infty}^{x}\int_{-\infty}^{x_1}\int_{-\infty}^{x_2}\int_{-\infty}^{y}\int_{-\infty}^{y_2}\int_{0}^{t}\int_{0}^{t_1}\int_{0}^{t_2} g(x_3,y_3,t_3)$$

$$\int_{-\infty}^{\infty}\int_{-\infty}^{\infty}\int_{0}^{\infty} G(x_3,y_3,t_3,\xi,\eta,\tau)$$

$$d\tau d\eta d\xi dt_3 dt_2 dt_1 dy_3 dy_2 dx_3 dx_2 dx_1 \leq A,$$

$$-\int_{-\infty}^{x}\int_{-\infty}^{x_1}\int_{-\infty}^{x_2}\int_{-\infty}^{y}\int_{0}^{t}\int_{0}^{t_1}\int_{0}^{t_2} g(x_3,y_3,t_3)$$

$$\int_{-\infty}^{\infty}\int_{-\infty}^{\infty}\int_{0}^{\infty} G(x_3,y_3,t_3,\xi,\eta,\tau)d\tau d\eta d\xi$$

$$dt_3 dt_2 dt_1 dy_3 dx_3 dx_2 dx_1 \leq A,$$

$$-\int_{-\infty}^{x}\int_{-\infty}^{x_1}\int_{-\infty}^{x_2}\int_{-\infty}^{y}\int_{-\infty}^{y_1}\int_{-\infty}^{y_2}\int_{0}^{t}\int_{0}^{t_2} g(x_3,y_3,t_3)$$

$$\int_{-\infty}^{\infty}\int_{-\infty}^{\infty}\int_{0}^{\infty} G(x_3,y_3,t_3,\xi,\eta,\tau)$$

$$\leq d\tau d\eta d\xi dt_3 dt_2 dy_3 dy_2 dy_1 dx_3 dx_2 dx_1 \leq A,$$

$$-\int_{-\infty}^{x}\int_{-\infty}^{x_1}\int_{-\infty}^{x_2}\int_{-\infty}^{y}\int_{-\infty}^{y_1}\int_{-\infty}^{y_2}\int_{0}^{t} g(x_3,y_3,t_3)$$

$$\int_{-\infty}^{\infty}\int_{-\infty}^{\infty}\int_{0}^{\infty} G(x_3,y_3,t_3,\xi,\eta,\tau)$$

$$\times d\tau d\eta d\xi dt_3 dy_3 dy_2 dy_1 dx_3 dx_2 dx_1 \leq A,$$

$$-\infty < x,y < \infty,\ 0 < t < \infty.$$

Let $E = \mathscr{C}^2(\mathbb{R}^2 \times [0,\infty))$ be endowed with the norm

$$\|u\| = \max\Bigg\{ \sup_{\mathbb{R}^2 \times [0,\infty)} |u(x,y,t)|,\quad \sup_{\mathbb{R}^2 \times [0,\infty)} |u_t(x,y,t)|$$

$$\sup_{\mathbb{R}^2 \times [0,\infty)} |u_{tt}(x,y,t)|,\quad \sup_{\mathbb{R}^2 \times [0,\infty)} |u_x(x,y,t)|,$$

$$\sup_{\mathbb{R}^2 \times [0,\infty)} |u_{xx}(x,y,t)|,\quad \sup_{\mathbb{R}^2 \times [0,\infty)} |u_y(x,y,t)|,$$

$$\sup_{\mathbb{R}^2 \times [0,\infty)} |u_{yy}(x,y,t)| \Bigg\},$$

provided it exists,

$$\widetilde{\mathscr{P}} = \{u \in E : u \geq 0 \quad on \quad \mathbb{R}^2 \times [0,\infty)\},$$

$\widetilde{\widetilde{\mathscr{P}}}$ *be the set of all equi-continuous families in* $\widetilde{\mathscr{P}}$ *(an example for an equi-continuous family is the family* $\{(3+\sin(t+n))(3+\sin(x+n))(3+\sin(y+n)) : t \in [0,\infty), \quad x,y \in \mathbb{R}\}_{n\in\mathbb{N}}$*),*

$$\alpha(u) = \min_{(x,y,t)\in[a_1,b_1]^3} |u(x,y,t)|, \quad u \in E,$$

$$\mathscr{P} = \left\{ u \in \widetilde{\widetilde{\mathscr{P}}} \quad \text{on} \quad \mathbb{R}^2 \times [0,\infty), \right.$$

$$\|u\| \leq \frac{c}{4} \quad \text{if} \quad \alpha(u) \leq a,$$

$$\left. \alpha(u) > a \quad \text{if} \quad c \geq \|u\| > b \right\},$$

$$\mathscr{P}_c = \{u \in \mathscr{P} : \|u\| \leq c\},$$

$$S(\alpha,a,c) = \{u \in \mathscr{P} : \alpha(u) \geq a, \quad \|u\| \leq c\},$$

$$\Omega = \left\{ u \in \mathscr{P} : \|u\| \leq c + \frac{3A\left(c + \sum_{j=1}^{l} a_j c^{l_j}\right)}{2+\varepsilon} \right\}.$$

Let $R = 2A\left(c + \sum_{j=1}^{l} a_j c^{l_j}\right)$. *For* $u \in E$, *define the operators*

$$Q_1 u(x,y,t) = -\int_{-\infty}^{x}\int_{-\infty}^{x_1}\int_{-\infty}^{x_2}\int_{-\infty}^{y}\int_{-\infty}^{y_1}\int_{-\infty}^{y_2}\int_{0}^{t}\int_{0}^{t_1}\int_{0}^{t_2} g(x_3,y_3,t_3)$$

$$\times u(x_3,y_3,t_3) dt_3 dt_2 dt_1 dy_3 dy_2 dy_1 dx_3 dx_2 dx_1$$

$$+\int_{-\infty}^{x}\int_{-\infty}^{x_1}\int_{-\infty}^{x_2}\int_{-\infty}^{y}\int_{-\infty}^{y_1}\int_{-\infty}^{y_2}\int_{0}^{t}\int_{0}^{t_1}\int_{0}^{t_2} g(x_3,y_3,t_3)$$

$$\int_{-\infty}^{\infty}\int_{-\infty}^{\infty}\int_{0}^{\infty} G(x_3,y_3,t_3,\xi,\eta,\tau)f(u(\xi,\eta,\tau))$$

$$d\tau d\eta d\xi dt_3 dt_2 dt_1 dy_3 dy_2 dy_1 dx_3 dx_2 dx_1,$$

$$Qu(x,y,t) = R + Q_1u(x,y,t),$$

$-\infty < x,y < \infty,\ 0 < t < \infty.$

Lemma 6.2.3 *Suppose* $(C3) - (C5)$. *For* $u \in E$ *and* $\|u\| \leq c$, *we have*

$$\|Q_1u\| \leq A\left(c + \sum_{j=1}^{l} a_jc^{l_j}\right) \quad and \quad \|Qu\| \leq 3A\left(c + \sum_{j=1}^{l} a_jc^{l_j}\right).$$

For $u \in \mathscr{P}_c$, *we have* $Qu \geq 0$ *on* $\mathbb{R}^2 \times [0,\infty)$.

Proof 6.2.4 *Let* $u \in E$ *and* $\|u\| \leq c$. *Then*

$$|Q_1u(x,y,t)| = \left| -\int_{-\infty}^{x}\int_{-\infty}^{x_1}\int_{-\infty}^{x_2}\int_{-\infty}^{y}\int_{-\infty}^{y_1}\int_{-\infty}^{y_2}\int_{0}^{t}\int_{0}^{t_1}\int_{0}^{t_2} g(x_3,y_3,t_3)\right.$$

$$\times u(x_3,y_3,t_3)dt_3dt_2dt_1dy_3dy_2dy_1dx_3dx_2dx_1$$

$$+\int_{-\infty}^{x}\int_{-\infty}^{x_1}\int_{-\infty}^{x_2}\int_{-\infty}^{y}\int_{-\infty}^{y_1}\int_{-\infty}^{y_2}\int_{0}^{t}\int_{0}^{t_1}\int_{0}^{t_2} g(x_3,y_3,t_3)$$

$$\int_{-\infty}^{\infty}\int_{-\infty}^{\infty}\int_{0}^{\infty} G(x_3,y_3,t_3,\xi,\eta,\tau)f(u(\xi,\eta,\tau))$$

$$\left. d\tau d\eta d\xi dt_3dt_2dt_1dy_3dy_2dy_1dx_3dx_2dx_1 \right|$$

$$\leq \int_{-\infty}^{x}\int_{-\infty}^{x_1}\int_{-\infty}^{x_2}\int_{-\infty}^{y}\int_{-\infty}^{y_1}\int_{-\infty}^{y_2}\int_{0}^{t}\int_{0}^{t_1}\int_{0}^{t_2} g(x_3,y_3,t_3)$$

$$\times |u(x_3,y_3,t_3)|dt_3dt_2dt_1dy_3dy_2dy_1dx_3dx_2dx_1$$

$$+\int_{-\infty}^{x}\int_{-\infty}^{x_1}\int_{-\infty}^{x_2}\int_{-\infty}^{y}\int_{-\infty}^{y_1}\int_{-\infty}^{y_2}\int_{0}^{t}\int_{0}^{t_1}\int_{0}^{t_2} g(x_3,y_3,t_3)$$

$$\times \int_{-\infty}^{\infty}\int_{-\infty}^{\infty}\int_{0}^{\infty} G(x_3,y_3,t_3,\xi,\eta,\tau)f(u(\xi,\eta,\tau))$$

$$\times d\tau d\eta d\xi dt_3dt_2dt_1dy_3dy_2dy_1dx_3dx_2dx_1$$

$$\leq \int_{-\infty}^{x} \int_{-\infty}^{x_1} \int_{-\infty}^{x_2} \int_{-\infty}^{y} \int_{-\infty}^{y_1} \int_{-\infty}^{y_2} \int_{0}^{t} \int_{0}^{t_1} \int_{0}^{t_2} g(x_3, y_3, t_3)$$

$$\times |u(x_3, y_3, t_3)| dt_3 dt_2 dt_1 dy_3 dy_2 dy_1 dx_3 dx_2 dx_1$$

$$- \sum_{j=1}^{l} a_j \int_{-\infty}^{x} \int_{-\infty}^{x_1} \int_{-\infty}^{x_2} \int_{-\infty}^{y} \int_{-\infty}^{y_1} \int_{-\infty}^{y_2} \int_{0}^{t} \int_{0}^{t_1} \int_{0}^{t_2} g(x_3, y_3, t_3)$$

$$\times \int_{-\infty}^{\infty} \int_{-\infty}^{\infty} \int_{0}^{\infty} G(x_3, y_3, t_3, \xi, \eta, \tau) |u(\xi, \eta, \tau)|^{l_j}$$

$$\times d\tau d\eta d\xi dt_3 dt_2 dt_1 dy_3 dy_2 dy_1 dx_3 dx_2 dx_1$$

$$\leq A \left(c + \sum_{j=1}^{l} a_j c^{l_j} \right),$$

$-\infty < x, y < \infty$, $0 < t < \infty$, *and*

$$\left| \frac{\partial}{\partial t} Q_1 u(x, y, t) \right| = \left| - \int_{-\infty}^{x} \int_{-\infty}^{x_1} \int_{-\infty}^{x_2} \int_{-\infty}^{y} \int_{-\infty}^{y_1} \int_{-\infty}^{y_2} \int_{0}^{t} \int_{0}^{t_2} g(x_3, y_3, t_3) \right.$$

$$\times u(x_3, y_3, t_3) dt_3 dt_2 dy_3 dy_2 dy_1 dx_3 dx_2 dx_1$$

$$+ \int_{-\infty}^{x} \int_{-\infty}^{x_1} \int_{-\infty}^{x_2} \int_{-\infty}^{y} \int_{-\infty}^{y_1} \int_{-\infty}^{y_2} \int_{0}^{t} \int_{0}^{t_2} g(x_3, y_3, t_3)$$

$$\int_{-\infty}^{\infty} \int_{-\infty}^{\infty} \int_{0}^{\infty} G(x_3, y_3, t_3, \xi, \eta, \tau) f(u(\xi, \eta, \tau))$$

$$\left. d\tau d\eta d\xi dt_3 dt_2 dy_3 dy_2 dy_1 dx_3 dx_2 dx_1 \right|$$

$$\leq \int_{-\infty}^{x} \int_{-\infty}^{x_1} \int_{-\infty}^{x_2} \int_{-\infty}^{y} \int_{-\infty}^{y_1} \int_{-\infty}^{y_2} \int_{0}^{t} \int_{0}^{t_2} g(x_3, y_3, t_3)$$

$$\times |u(x_3, y_3, t_3)| dt_3 dt_2 dy_3 dy_2 dy_1 dx_3 dx_2 dx_1$$

$$+ \int_{-\infty}^{x} \int_{-\infty}^{x_1} \int_{-\infty}^{x_2} \int_{-\infty}^{y} \int_{-\infty}^{y_1} \int_{-\infty}^{y_2} \int_{0}^{t} \int_{0}^{t_2} g(x_3, y_3, t_3)$$

$$\times \int_{-\infty}^{\infty} \int_{-\infty}^{\infty} \int_0^{\infty} G(x_3, y_3, t_3, \xi, \eta, \tau) f(u(\xi, \eta, \tau))$$

$$\times d\tau d\eta d\xi dt_3 dt_2 dy_3 dy_2 dy_1 dx_3 dx_2 dx_1$$

$$\leq \int_{-\infty}^{x} \int_{-\infty}^{x_1} \int_{-\infty}^{x_2} \int_{-\infty}^{y} \int_{-\infty}^{y_1} \int_{-\infty}^{y_2} \int_0^t \int_0^{t_2} g(x_3, y_3, t_3) |u(x_3, y_3, t_3)|$$

$$\times dt_3 dt_2 dy_3 dy_2 dy_1 dx_3 dx_2 dx_1$$

$$-\sum_{j=1}^{l} a_j \int_{-\infty}^{x} \int_{-\infty}^{x_1} \int_{-\infty}^{x_2} \int_{-\infty}^{y} \int_{-\infty}^{y_1} \int_{-\infty}^{y_2} \int_0^t \int_0^{t_2} g(x_3, y_3, t_3)$$

$$\times \int_{-\infty}^{\infty} \int_{-\infty}^{\infty} \int_0^{\infty} G(x_3, y_3, t_3, \xi, \eta, \tau) |u(\xi, \eta, \tau)|^{l_j}$$

$$\times d\tau d\eta d\xi dt_3 dt_2 dy_3 dy_2 dy_1 dx_3 dx_2 dx_1$$

$$\leq A\left(c + \sum_{j=1}^{l} a_j c^{l_j}\right),$$

$-\infty < x, y < \infty, \ 0 < t < \infty$, and

$$\left| \frac{\partial^2}{\partial t^2} Q_1 u(x, y, t) \right| = \left| -\int_{-\infty}^{x} \int_{-\infty}^{x_1} \int_{-\infty}^{x_2} \int_{-\infty}^{y} \int_{-\infty}^{y_1} \int_{-\infty}^{y_2} \int_0^t g(x_3, y_3, t_3) \right.$$

$$\times u(x_3, y_3, t_3) dt_3 dy_3 dy_2 dy_1 dx_3 dx_2 dx_1$$

$$+ \int_{-\infty}^{x} \int_{-\infty}^{x_1} \int_{-\infty}^{x_2} \int_{-\infty}^{y} \int_{-\infty}^{y_1} \int_{-\infty}^{y_2} \int_0^t g(x_3, y_3, t_3)$$

$$\times \int_{-\infty}^{\infty} \int_{-\infty}^{\infty} \int_0^{\infty} G(x_3, y_3, t_3, \xi, \eta, \tau) f(u(\xi, \eta, \tau))$$

$$\left. \times d\tau d\eta d\xi dt_3 dy_3 dy_2 dy_1 dx_3 dx_2 dx_1 \right|$$

$$\leq \int_{-\infty}^{x} \int_{-\infty}^{x_1} \int_{-\infty}^{x_2} \int_{-\infty}^{y} \int_{-\infty}^{y_1} \int_{-\infty}^{y_2} \int_0^t g(x_3, y_3, t_3)$$

$$\times |u(x_3, y_3, t_3)| dt_3 dy_3 dy_2 dy_1 dx_3 dx_2 dx_1$$

$$+ \int_{-\infty}^{x} \int_{-\infty}^{x_1} \int_{-\infty}^{x_2} \int_{-\infty}^{y} \int_{-\infty}^{y_1} \int_{-\infty}^{y_2} \int_{0}^{t} g(x_3, y_3, t_3)$$

$$\int_{-\infty}^{\infty} \int_{-\infty}^{\infty} \int_{0}^{\infty} G(x_3, y_3, t_3, \xi, \eta, \tau) f(u(\xi, \eta, \tau))$$

$$d\tau d\eta d\xi dt_3 dy_3 dy_2 dy_1 dx_3 dx_2 dx_1$$

$$\leq \int_{-\infty}^{x} \int_{-\infty}^{x_1} \int_{-\infty}^{x_2} \int_{-\infty}^{y} \int_{-\infty}^{y_1} \int_{-\infty}^{y_2} \int_{0}^{t} g(x_3, y_3, t_3) |u(x_3, y_3, t_3)|$$

$$dt_3 dt_2 dt_1 dy_3 dy_2 dy_1 dx_3 dx_2 dx_1$$

$$- \sum_{j=1}^{l} a_j \int_{-\infty}^{x} \int_{-\infty}^{x_1} \int_{-\infty}^{x_2} \int_{-\infty}^{y} \int_{-\infty}^{y_1} \int_{-\infty}^{y_2} \int_{0}^{t} g(x_3, y_3, t_3)$$

$$\int_{-\infty}^{\infty} \int_{-\infty}^{\infty} \int_{0}^{\infty} G(x_3, y_3, t_3, \xi, \eta, \tau) |u(\xi, \eta, \tau)|^{l_j}$$

$$d\tau d\eta d\xi dt_3 dy_3 dy_2 dy_1 dx_3 dx_2 dx_1$$

$$\leq A \left(c + \sum_{j=1}^{l} a_j c^{l_j} \right),$$

$-\infty < x, y < \infty$, $0 < t < \infty$, *and*

$$\left| \frac{\partial}{\partial x} Q_1 u(x, y, t) \right| = \left| - \int_{-\infty}^{x} \int_{-\infty}^{x_2} \int_{-\infty}^{y} \int_{-\infty}^{y_1} \int_{-\infty}^{y_2} \int_{0}^{t} \int_{0}^{t_1} \int_{0}^{t_2} g(x_3, y_3, t_3) \right.$$

$$\times u(x_3, y_3, t_3) dt_3 dt_2 dt_1 dy_3 dy_2 dy_1 dx_3 dx_2$$

$$+ \int_{-\infty}^{x} \int_{-\infty}^{x_2} \int_{-\infty}^{y} \int_{-\infty}^{y_1} \int_{-\infty}^{y_2} \int_{0}^{t} \int_{0}^{t_1} \int_{0}^{t_2} g(x_3, y_3, t_3)$$

$$\times \int_{-\infty}^{\infty} \int_{-\infty}^{\infty} \int_{0}^{\infty} G(x_3, y_3, t_3, \xi, \eta, \tau) f(u(\xi, \eta, \tau))$$

$$\left. \times d\tau d\eta d\xi dt_3 dt_2 dt_1 dy_3 dy_2 dy_1 dx_3 dx_2 \right|$$

$$\leq \int_{-\infty}^{x} \int_{-\infty}^{x_2} \int_{-\infty}^{y} \int_{-\infty}^{y_1} \int_{-\infty}^{y_2} \int_{0}^{t} \int_{0}^{t_1} \int_{0}^{t_2} g(x_3, y_3, t_3)$$

$$\times |u(x_3, y_3, t_3)| dt_3 dt_2 dt_1 dy_3 dy_2 dy_1 dx_3 dx_2$$

$$+ \int_{-\infty}^{x} \int_{-\infty}^{x_2} \int_{-\infty}^{y} \int_{-\infty}^{y_1} \int_{-\infty}^{y_2} \int_{0}^{t} \int_{0}^{t_1} \int_{0}^{t_2} g(x_3, y_3, t_3)$$

$$\times \int_{-\infty}^{\infty} \int_{-\infty}^{\infty} \int_{0}^{\infty} G(x_3, y_3, t_3, \xi, \eta, \tau) f(u(\xi, \eta, \tau))$$

$$\times d\tau d\eta d\xi dt_3 dt_2 dt_1 dy_3 dy_2 dy_1 dx_3 dx_2$$

$$\leq \int_{-\infty}^{x} \int_{-\infty}^{x_2} \int_{-\infty}^{y} \int_{-\infty}^{y_1} \int_{-\infty}^{y_2} \int_{0}^{t} \int_{0}^{t_1} \int_{0}^{t_2} g(x_3, y_3, t_3)$$

$$\times |u(x_3, y_3, t_3)| dt_3 dt_2 dt_1 dy_3 dy_2 dy_1 dx_3 dx_2$$

$$- \sum_{j=1}^{l} a_j \int_{-\infty}^{x} \int_{-\infty}^{x_2} \int_{-\infty}^{y} \int_{-\infty}^{y_1} \int_{-\infty}^{y_2} \int_{0}^{t} \int_{0}^{t_1} \int_{0}^{t_2} g(x_3, y_3, t_3)$$

$$\times \int_{-\infty}^{\infty} \int_{-\infty}^{\infty} \int_{0}^{\infty} G(x_3, y_3, t_3, \xi, \eta, \tau) |u(\xi, \eta, \tau)|^{l_j}$$

$$\times d\tau d\eta d\xi dt_3 dt_2 dt_1 dy_3 dy_2 dy_1 dx_3 dx_2$$

$$\leq A \left(c + \sum_{j=1}^{l} a_j c^{l_j} \right),$$

$-\infty < x, y < \infty$, $0 < t < \infty$, *and*

$$\left| \frac{\partial^2}{\partial x^2} Q_1 u(x,y,t) \right| = \left| -\int_{-\infty}^{x}\int_{-\infty}^{y}\int_{-\infty}^{y_1}\int_{-\infty}^{y_2}\int_{0}^{t}\int_{0}^{t_1}\int_{0}^{t_2} g(x_3,y_3,t_3) \right.$$

$$\times u(x_3,y_3,t_3)\,dt_3\,dt_2\,dt_1\,dy_3\,dy_2\,dy_1\,dx_3$$

$$+ \int_{-\infty}^{x}\int_{-\infty}^{y}\int_{-\infty}^{y_1}\int_{-\infty}^{y_2}\int_{0}^{t}\int_{0}^{t_1}\int_{0}^{t_2} g(x_3,y_3,t_3)$$

$$\times \int_{-\infty}^{\infty}\int_{-\infty}^{\infty}\int_{0}^{\infty} G(x_3,y_3,t_3,\xi,\eta,\tau)f(u(\xi,\eta,\tau))$$

$$\left. \times d\tau\,d\eta\,d\xi\,dt_3\,dt_2\,dt_1\,dy_3\,dy_2\,dy_1\,dx_3 \right|$$

$$\leq \int_{-\infty}^{x}\int_{-\infty}^{y}\int_{-\infty}^{y_1}\int_{-\infty}^{y_2}\int_{0}^{t}\int_{0}^{t_1}\int_{0}^{t_2} g(x_3,y_3,t_3)$$

$$\times |u(x_3,y_3,t_3)|\,dt_3\,dt_2\,dt_1\,dy_3\,dy_2\,dy_1\,dx_3$$

$$+ \int_{-\infty}^{x}\int_{-\infty}^{y}\int_{-\infty}^{y_1}\int_{-\infty}^{y_2}\int_{0}^{t}\int_{0}^{t_1}\int_{0}^{t_2} g(x_3,y_3,t_3)$$

$$\times \int_{-\infty}^{\infty}\int_{-\infty}^{\infty}\int_{0}^{\infty} G(x_3,y_3,t_3,\xi,\eta,\tau)f(u(\xi,\eta,\tau))$$

$$\times d\tau\,d\eta\,d\xi\,dt_3\,dt_2\,dt_1\,dy_3\,dy_2\,dy_1\,dx_3$$

$$\leq \int_{-\infty}^{x}\int_{-\infty}^{y}\int_{-\infty}^{y_1}\int_{-\infty}^{y_2}\int_{0}^{t}\int_{0}^{t_1}\int_{0}^{t_2} g(x_3,y_3,t_3)|u(x_3,y_3,t_3)|$$

$$\times dt_3\,dt_2\,dt_1\,dy_3\,dy_2\,dy_1\,dx_3$$

$$-\sum_{j=1}^{l} a_j \int_{-\infty}^{x} \int_{-\infty}^{y} \int_{-\infty}^{y_1} \int_{-\infty}^{y_2} \int_{0}^{t} \int_{0}^{t_1} \int_{0}^{t_2} g(x_3, y_3, t_3)$$

$$\times \int_{-\infty}^{\infty} \int_{-\infty}^{\infty} \int_{0}^{\infty} G(x_3, y_3, t_3, \xi, \eta, \tau) |u(\xi, \eta, \tau)|^{l_j}$$

$$\times d\tau d\eta d\xi dt_3 dt_2 dt_1 dy_3 dy_2 dy_1 dx_3$$

$$\leq A \left(c + \sum_{j=1}^{l} a_j c^{l_j} \right),$$

$-\infty < x, y < \infty, \ 0 < t < \infty,$ *and*

$$\left| \frac{\partial}{\partial y} Q_1 u(x, y, t) \right| = \left| - \int_{-\infty}^{x} \int_{-\infty}^{x_1} \int_{-\infty}^{x_2} \int_{-\infty}^{y} \int_{-\infty}^{y_2} \int_{0}^{t} \int_{0}^{t_1} \int_{0}^{t_2} g(x_3, y_3, t_3) \right.$$

$$\times u(x_3, y_3, t_3) dt_3 dt_2 dt_1 dy_3 dy_2 dx_3 dx_2 dx_1$$

$$+ \int_{-\infty}^{x} \int_{-\infty}^{x_1} \int_{-\infty}^{x_2} \int_{-\infty}^{y} \int_{-\infty}^{y_2} \int_{0}^{t} \int_{0}^{t_1} \int_{0}^{t_2} g(x_3, y_3, t_3)$$

$$\times \int_{-\infty}^{\infty} \int_{-\infty}^{\infty} \int_{0}^{\infty} G(x_3, y_3, t_3, \xi, \eta, \tau) f(u(\xi, \eta, \tau))$$

$$\left. \times d\tau d\eta d\xi dt_3 dt_2 dt_1 dy_3 dy_2 dx_3 dx_2 dx_1 \right|$$

$$\leq \int_{-\infty}^{x} \int_{-\infty}^{x_1} \int_{-\infty}^{x_2} \int_{-\infty}^{y} \int_{-\infty}^{y_2} \int_{0}^{t} \int_{0}^{t_1} \int_{0}^{t_2} g(x_3, y_3, t_3)$$

$$\times |u(x_3, y_3, t_3)| dt_3 dt_2 dt_1 dy_3 dy_2 dx_3 dx_2 dx_1$$

$$+ \int_{-\infty}^{x} \int_{-\infty}^{x_1} \int_{-\infty}^{x_2} \int_{-\infty}^{y} \int_{-\infty}^{y_2} \int_{0}^{t} \int_{0}^{t_1} \int_{0}^{t_2} g(x_3, y_3, t_3)$$

$$\times \int_{-\infty}^{\infty} \int_{-\infty}^{\infty} \int_{0}^{\infty} G(x_3, y_3, t_3, \xi, \eta, \tau) f(u(\xi, \eta, \tau))$$

$$\times d\tau d\eta d\xi dt_3 dt_2 dt_1 dy_3 dy_2 dx_3 dx_2 dx_1$$

$$\leq \int_{-\infty}^{x} \int_{-\infty}^{x_1} \int_{-\infty}^{x_2} \int_{-\infty}^{y} \int_{-\infty}^{y_2} \int_{0}^{t} \int_{0}^{t_1} \int_{0}^{t_2} g(x_3, y_3, t_3)$$

$$\times |u(x_3, y_3, t_3)| dt_3 dt_2 dt_1 dy_3 dy_2 dx_3 dx_2 dx_1$$

$$- \sum_{j=1}^{l} a_j \int_{-\infty}^{x} \int_{-\infty}^{x_1} \int_{-\infty}^{x_2} \int_{-\infty}^{y} \int_{-\infty}^{y_2} \int_{0}^{t} \int_{0}^{t_1} \int_{0}^{t_2} g(x_3, y_3, t_3)$$

$$\times \int_{-\infty}^{\infty} \int_{-\infty}^{\infty} \int_{0}^{\infty} G(x_3, y_3, t_3, \xi, \eta, \tau) |u(\xi, \eta, \tau)|^{l_j}$$

$$\times d\tau d\eta d\xi dt_3 dt_2 dt_1 dy_3 dy_2 dx_3 dx_2 dx_1$$

$$\leq A\left(c + \sum_{j=1}^{l} a_j c^{l_j}\right),$$

$-\infty < x, y < \infty,\ 0 < t < \infty,$ *and*

$$\left| \frac{\partial^2}{\partial y^2} Q_1 u(x, y, t) \right| = \left| - \int_{-\infty}^{x} \int_{-\infty}^{x_1} \int_{-\infty}^{x_2} \int_{-\infty}^{y} \int_{0}^{t} \int_{0}^{t_1} \int_{0}^{t_2} g(x_3, y_3, t_3) \right.$$

$$\times u(x_3, y_3, t_3) dt_3 dt_2 dt_1 dy_3 dx_3 dx_2 dx_1$$

$$+ \int_{-\infty}^{x} \int_{-\infty}^{x_1} \int_{-\infty}^{x_2} \int_{-\infty}^{y} \int_{0}^{t} \int_{0}^{t_1} \int_{0}^{t_2} g(x_3, y_3, t_3)$$

$$\times \int_{-\infty}^{\infty} \int_{-\infty}^{\infty} \int_{0}^{\infty} G(x_3, y_3, t_3, \xi, \eta, \tau) f(u(\xi, \eta, \tau))$$

$$\times d\tau d\eta d\xi dt_3 dt_2 dt_1 dy_3 dx_3 dx_2 dx_1 \Bigg|$$

$$\leq \int_{-\infty}^{x} \int_{-\infty}^{x_1} \int_{-\infty}^{x_2} \int_{-\infty}^{y} \int_{0}^{t} \int_{0}^{t_1} \int_{0}^{t_2} g(x_3,y_3,t_3)$$

$$\times |u(x_3,y_3,t_3)| dt_3 dt_2 dt_1 dy_3 dx_3 dx_2 dx_1$$

$$+ \int_{-\infty}^{x} \int_{-\infty}^{x_1} \int_{-\infty}^{x_2} \int_{-\infty}^{y} \int_{0}^{t} \int_{0}^{t_1} \int_{0}^{t_2} g(x_3,y_3,t_3)$$

$$\times \int_{-\infty}^{\infty} \int_{-\infty}^{\infty} \int_{0}^{\infty} G(x_3,y_3,t_3,\xi,\eta,\tau) f(u(\xi,\eta,\tau))$$

$$\times d\tau d\eta d\xi dt_3 dt_2 dt_1 dy_3 dx_3 dx_2 dx_1$$

$$\leq \int_{-\infty}^{x} \int_{-\infty}^{x_1} \int_{-\infty}^{x_2} \int_{-\infty}^{y} \int_{0}^{t} \int_{0}^{t_1} \int_{0}^{t_2} g(x_3,y_3,t_3) |u(x_3,y_3,t_3)|$$

$$\times dt_3 dt_2 dt_1 dy_3 dx_3 dx_2 dx_1$$

$$- \sum_{j=1}^{l} a_j \int_{-\infty}^{x} \int_{-\infty}^{x_1} \int_{-\infty}^{x_2} \int_{-\infty}^{y} \int_{0}^{t} \int_{0}^{t_1} \int_{0}^{t_2} g(x_3,y_3,t_3)$$

$$\times \int_{-\infty}^{\infty} \int_{-\infty}^{\infty} \int_{0}^{\infty} G(x_3,y_3,t_3,\xi,\eta,\tau) |u(\xi,\eta,\tau)|^{l_j}$$

$$\times d\tau d\eta d\xi dt_3 dt_2 dt_1 dy_3 dx_3 dx_2 dx_1$$

$$\leq A \left(c + \sum_{j=1}^{l} a_j c^{l_j} \right),$$

$-\infty < x, y < \infty$, $0 < t < \infty$. *Let* $u \in \mathscr{P}_c$. *Then*

$$Qu(x,y,t) \quad = \quad R + Q_1 u(x,y,t)$$

$$\geq \quad R - |Q_1 u(x,y,t)|$$

$$\geq \quad A\left(c + \sum_{j=1}^{l} a_j c^{l_j}\right) > 0,$$

$-\infty < x, y < \infty$, $0 < t < \infty$. *This completes the proof.*

For $u \in E$, *define the operators*

$$Tu \quad = \quad -(1+\varepsilon)u,$$

$$Fu \quad = \quad (2+\varepsilon)u + Qu.$$

Lemma 6.2.5 *Suppose* $(C3) - (C5)$. *If* $u \in E$ *is a fixed point of the operator* $T + F$, *then it is a solution of the IVP* (6.7)-(6.8).

Proof 6.2.6 *We have*

$$u \quad = \quad Tu + Fu$$

$$= \quad -(1+\varepsilon)u + (2+\varepsilon)u + Qu$$

$$= \quad u + Qu,$$

whereupon $Qu = 0$. *We differentiate trice in* x, *trice in* y *and trice in* t, *the last equation and we get*

$$0 \quad = \quad -g(x,y,t)u(x,y,t)$$

$$+ g(x,y,t)\int_{-\infty}^{\infty}\int_{-\infty}^{\infty}\int_{0}^{\infty} G(x,y,t,\xi,\eta,\tau)f(u(\xi,\eta,\tau))d\tau d\eta d\xi,$$

$-\infty < x, y < \infty$, $0 < t < \infty$. *Hence, we arrive at* (6.9) *and therefore* u *is a solution of the IVP* (6.7), (6.8). *This completes the proof.*

1. For $u \in \mathscr{P}$, we have

$$\|Tu\| = (1+\varepsilon)\|u\|,$$

i.e., $T : \mathscr{P} \to E$ is an expansive operator with $h = 1 + \varepsilon$.

2. For $u \in \mathscr{P}_c$, we get

$$\|Fu\| \leq (2+\varepsilon)\|u\| + \|Qu\|$$

$$\leq (2+\varepsilon)c + 3A\left(c + \sum_{j=1}^{l} a_j c^{l_j}\right).$$

Therefore, $F : \mathscr{P}_c \to E$ is uniformly bounded. Since $F : \mathscr{P}_c \to E$ is continuous, we obtain that $F(\mathscr{P}_c)$ is equi-continuous and then $F : \mathscr{P}_c \to E$ is relatively compact. Consequently, $F : \mathscr{P}_c \to E$ is 0-set contraction.

3. Let $\lambda \in [0,1]$, $z_0 \in B(0, \varepsilon c) \cap \mathscr{P}_c$ and $u \in \mathscr{P}_c$ be arbitrarily chosen. Take

$$z = \frac{\lambda(2+\varepsilon)u + \lambda Qu + (1-\lambda)z_0}{2+\varepsilon}.$$

We have

$$z = \frac{\lambda Fu + (1-\lambda)z_0}{2+\varepsilon}$$

or

$$\lambda Fu + (1-\lambda)z_0 = (I-T)z.$$

Next, $z \in \mathscr{P}$ and

$$\|z\| \leq \frac{\lambda(2+\varepsilon)\|u\| + \lambda\|Qu\| + (1-\lambda)\|z_0\|}{2+\varepsilon}$$

$$\leq \frac{\lambda(2+\varepsilon)c + 3A\left(c + \sum_{j=1}^{l} a_j c^{l_j}\right) + (1-\lambda)\varepsilon c}{2+\varepsilon}$$

$$\leq \frac{\lambda(2+\varepsilon)c + 3A\left(c + \sum_{j=1}^{l} a_j c^{l_j}\right) + (1-\lambda)(2+\varepsilon)c}{2+\varepsilon}$$

$$= \frac{(2+\varepsilon)c + 3A\left(c + \sum_{j=1}^{l} a_j c^{l_j}\right)}{2+\varepsilon}.$$

Thus, $z \in \Omega$ and

$$\lambda F(\mathscr{P}_c) + (1-\lambda)z_0 \subset (I-T)(\Omega), \quad \lambda \in [0,1].$$

4. *Note that*

$$b \in \{u \in S(\alpha, a, b) : \alpha(u) > a\}, \ i.e. \ \{u \in S(\alpha, a, b) : \alpha(u) > a\} \neq \emptyset.$$

Moreover, if $u \in S(\alpha, a, b)$, then $\alpha(u) \geq a$ and

$$\alpha(Tu + Fu) = \alpha(u + Qu)$$

$$> \alpha(u)$$

$$\geq a.$$

5. *Let $u \in S(\alpha, a, c)$ and $\|Tu + Fu\| > b$. Then, by the definition of $\mathscr{P}$, we have $\alpha(Tu + Fu) > a$. Let $z_0 = \dfrac{\varepsilon}{2}u$. Then*

$$\|z_0\| = \frac{\varepsilon}{2}\|u\|$$

$$\leq \frac{\varepsilon}{2}c$$

and $z_0 \in B(0, \varepsilon c) \cap \mathscr{P}_c$. Hence,

$$\alpha(Tu + z_0) = \alpha\left(-(1+\varepsilon)u + \frac{\varepsilon}{2}u\right)$$

$$= \alpha\left(-\left(1+\frac{\varepsilon}{2}\right)u\right)$$

$$= \left(1+\frac{\varepsilon}{2}\right)\alpha(u)$$

$$\geq \left(1+\frac{\varepsilon}{2}\right)a$$

$$\geq a.$$

6. *Let $u \in S(\alpha,a,c)$ and $\alpha(u) = a$. Then $\|u\| \leq \dfrac{c}{4}$. Take $z_0 = \dfrac{\varepsilon}{2}u$.
We have $z_0 \in B(0,\varepsilon c) \cap \mathscr{P}_c$ and*

$$
\begin{aligned}
\|Tu + z_0\| &= \left\| -(1+\varepsilon)u + \frac{\varepsilon}{2}u \right\| \\
&= \left(1 + \frac{\varepsilon}{2}\right)\|u\| \\
&\leq \left(1 + \frac{\varepsilon}{2}\right)\frac{c}{4} \\
&\leq b.
\end{aligned}
$$

By 1, 2, 3, 4, 5, 6 and Theorem 3.2.1, we conclude that the operator $T + F$ has a fixed point $u \in S(\alpha,a,c) \cap \Omega$ with $\alpha(u) > a$. This completes the proof.

Example 6.2.7 *Consider the IVP*

$$
u_{tt} - u_{xx} - u_{yy} = -u^2 - 3\frac{u^4}{1+u^8} - 4u^{10}, \quad -\infty < x,y < \infty, \quad 0 < t < \infty,
$$
$$
\tag{6.10}
$$
$$
u(x,y,0) = u_t(x,y,0) = 0, \quad -\infty < x,y < \infty. \tag{6.11}
$$

Here

$$
f(u) = -u^2 - 3\frac{u^4}{1+u^8} - 4u^{10}, \quad u \in \mathbb{R}.
$$

Then

$$
-f(u) \leq u^2 + 3u^4 + 4u^{10}, \quad u \in \mathbb{R}.
$$

For $\varepsilon = \dfrac{1}{2}$, the conditions (C4) take the form

$$
c \geq b > a > 0, \quad A > 0,
$$
$$
\frac{5}{2}c + 3A\left(c + c^2 + 3c^4 + 4c^{10}\right) < \frac{1}{8},
$$
$$
\frac{5c}{16} \leq b, \quad 0 < a_1 < b_1.
$$

Then, we take

$$a_1 = \frac{1}{4}, \quad b_1 = \frac{1}{2}, \quad a = \frac{1}{128}, \quad b = \frac{1}{64}, \quad c = \frac{1}{32}, \quad A = \frac{1}{10000000000}$$

and we get that (C4) *holds. Observe that*

$$-\frac{1}{2\pi} \int_{-\infty}^{\infty} \int_{-\infty}^{\infty} \int_0^{\infty} G(x,y,t,\xi,\eta,\tau) n d\tau d\eta d\xi$$

$$= \frac{1}{2\pi} \int_0^{\infty} \int_{\sqrt{(x-\xi)^2+(y-\eta)^2} \le t-\tau} \frac{1}{\sqrt{(t-\tau)^2-(x-\xi)^2-(y-\eta)^2}} d\xi d\eta d\tau$$

$$= \int_0^t \int_0^{t-\tau} \frac{\rho}{\sqrt{(t-\tau)^2-\rho^2}} d\rho d\tau$$

$$= \int_0^t (t-\tau) d\tau$$

$$= \frac{t^2}{2}.$$

Now, we will construct a function g so that (C5) *holds. Let*

$$h(t) = \log \frac{1+t^4\sqrt{2}+t^8}{1-t^4\sqrt{2}+t^8}, \quad l(t) = \arctan \frac{t^4\sqrt{2}}{1-t^8}, \quad t \ge 0.$$

We have

$$h'(t) = \frac{1}{(1+t^4\sqrt{2}+t^8)(1-t^4\sqrt{2}+t^8)} \left((4\sqrt{2}t^3+8t^7)(1-t^4\sqrt{2}+t^8) \right.$$

$$\left. - (1+t^4\sqrt{2}+t^8)(-4\sqrt{2}t^3+8t^7) \right)$$

$$= \frac{1}{(1+t^4\sqrt{2}+t^8)(1-t^4\sqrt{2}+t^8)} \left(4\sqrt{2}t^3-8t^7+4\sqrt{2}t^{11}+8t^7 \right)$$

$$-8\sqrt{2}t^{11}+8t^{15}+4\sqrt{2}t^3-8t^7+8t^7-8\sqrt{2}t^{11}+4\sqrt{2}t^{11}-8t^{15}\Bigg)$$

$$=-\frac{8\sqrt{2}t^3(t^8-1)}{(1+t^4\sqrt{2}+t^8)(1-t^4\sqrt{2}+t^8)}, \quad t\geq 0.$$

Thus,

$$\sup_{t\geq 0} h(t) = h(1) = \log\frac{2+\sqrt[4]{2}}{2-\sqrt[4]{2}},$$

h is an increasing function on $[0,1]$ *and it is a decreasing function on* $[1,\infty)$. *Next,*

$$
\begin{aligned}
l'(t) &= \frac{1}{1+\frac{2t^8}{(1-t^8)^2}}\frac{4\sqrt{2}t^3(1-t^8)+8t^7t^4\sqrt{2}}{(1-t^8)^2}\\[2mm]
&= \frac{4\sqrt{2}t^3-4\sqrt{2}t^{11}+8\sqrt{2}t^{11}}{1+t^{16}}\\[2mm]
&= \frac{4\sqrt{2}t^3(1+t^8)}{1+t^{16}}, \quad t\geq 0.
\end{aligned}
$$

Therefore, l is an increasing function on $[0,\infty)$. *Note that*

$$
\begin{aligned}
\lim_{t\to\infty} th(t) &= \lim_{t\to\infty}\frac{h(t)}{\frac{1}{t}}\\[2mm]
&= \lim_{t\to\infty}\frac{h'(t)}{-\frac{1}{t^2}}\\[2mm]
&= \lim_{t\to\infty}\frac{8\sqrt{2}t^5(t^8-1)}{(t^8+\sqrt{2}t^4+1)(t^8-\sqrt{2}t^4+1)}\\[2mm]
&= 0,
\end{aligned}
$$

and

$$\lim_{t\to\infty} t^2 h(t) = \lim_{t\to\infty}\frac{h(t)}{\frac{1}{t^2}}$$

$$= \lim_{t \to \infty} \frac{h'(t)}{-\frac{2}{t^3}}$$

$$= \lim_{t \to \infty} \frac{4\sqrt{2}t^6(t^8 - 1)}{(t^8 + \sqrt{2}t^4 + 1)(t^8 - \sqrt{2}t^4 + 1)}$$

$$= 0,$$

and

$$\lim_{t \to \infty} t l(t) = \lim_{t \to \infty} \frac{l(t)}{\frac{1}{t}}$$

$$= \lim_{t \to \infty} \frac{l'(t)}{-\frac{1}{t^2}}$$

$$= -\lim_{t \to \infty} \frac{4\sqrt{2}t^5(1 + t^8)}{1 + t^{16}}$$

$$= 0,$$

and

$$\lim_{t \to \infty} t^2 l(t) = \lim_{t \to \infty} \frac{l(t)}{\frac{1}{t^2}}$$

$$= \lim_{t \to \infty} \frac{l'(t)}{-\frac{2}{t^3}}$$

$$= -2\sqrt{2} \lim_{t \to \infty} \frac{t^6(t^8 + 1)}{1 + t^{16}}$$

$$= 0.$$

Consequently, there exists a constant $B > 1$ such that

$$\frac{1}{16\sqrt{2}} \log \frac{1 + t^4\sqrt{2} + t^8}{1 - t^4\sqrt{2} + t^8} + \frac{1}{8\sqrt{2}} \arctan \frac{t^4\sqrt{2}}{1 - t^8} \le B,$$

$$t\left(\frac{1}{16\sqrt{2}} \log \frac{1 + t^4\sqrt{2} + t^8}{1 - t^4\sqrt{2} + t^8} + \frac{1}{8\sqrt{2}} \arctan \frac{t^4\sqrt{2}}{1 - t^8} \right) \le B,$$

$$t^2\left(\frac{1}{16\sqrt{2}}\log\frac{1+t^4\sqrt{2}+t^8}{1-t^4\sqrt{2}+t^8}+\frac{1}{8\sqrt{2}}\arctan\frac{t^4\sqrt{2}}{1-t^8}\right)\leq B,\quad t\geq 0.$$

Let

$$f_1(t)=\frac{t^3}{B(1+t^{16})},\quad t\geq 0,$$

$$f_2(x)=\frac{|x|^3}{2B(3+\pi)(1+x^2)^4(1+x^{16})(1+x^6)},\quad -\infty<x<\infty.$$

Then, we can take

$$g(x,y,t)=\frac{1}{100000000000000000000}\frac{1}{t^2+1}f_1(t)f_2(x)f_2(y),$$

$-\infty<x,y<\infty,\quad 0<t<\infty.$ *To check that* (C5) *holds, we have a need to estimate the following integrals*

$$I_1(t)=\int_0^t f_1(s)ds,$$

$$I_2(t)=\int_0^t\int_0^{t_1}f_1(s)dsdt_1=\int_0^t(t-s)f_1(s)ds,$$

$$I_3(t)=\int_0^t\int_0^{t_1}\int_0^{t_2}f_1(s)dsdt_2dt_1=\frac{1}{2}\int_0^t(t-s)^2f_1(s)ds,\quad t\geq 0,$$

and

$$J_1(x)=\int_{-\infty}^x f_2(s)ds,$$

$$J_2(x)=\int_{-\infty}^x\int_{-\infty}^{x_1}f_2(s)dsdx_1=\int_{-\infty}^x(x-s)f_2(s)ds,$$

$$J_3(x)=\int_{-\infty}^x\int_{-\infty}^{x_1}\int_{-\infty}^{x_2}f_2(s)dsdx_2dx_1=\frac{1}{2}\int_{-\infty}^x(x-s)^2f_2(s)ds,$$
$$-\infty<x<\infty.$$

1. Firstly, we will estimate $I_1(t)$, $I_2(t)$, $I_3(t)$ for $t \geq 0$. We have

$$I_1(t) = \frac{1}{4B} \int_0^t \frac{ds^4}{1+s^{16}}$$

$$= \frac{1}{4B} \int_0^{t^4} \frac{dz}{1+z^4}$$

$$= \frac{1}{4B} \left(\frac{1}{4\sqrt{2}} \log \frac{1+z\sqrt{2}+z^2}{1-z\sqrt{2}+z^2} + \frac{1}{2\sqrt{2}} \arctan \frac{z\sqrt{2}}{1-z^2} \right) \Big|_{z=0}^{z=t^4}$$

$$= \frac{1}{4B} \left(\frac{1}{4\sqrt{2}} \log \frac{1+t\sqrt{2}+t^2}{1-t\sqrt{2}+t^2} + \frac{1}{2\sqrt{2}} \arctan \frac{t\sqrt{2}}{1-t^2} \right)$$

$$\leq 1, \quad t \geq 0,$$

and

$$I_2(t) = \int_0^t (t-s)f_1(s)ds$$

$$\leq t I_1(t)$$

$$= \frac{t}{4B} \left(\frac{1}{4\sqrt{2}} \log \frac{1+t\sqrt{2}+t^2}{1-t\sqrt{2}+t^2} + \frac{1}{2\sqrt{2}} \arctan \frac{t\sqrt{2}}{1-t^2} \right)$$

$$\leq 1, \quad t \geq 0,$$

and

$$I_3(t) = \frac{1}{2} \int_0^t (t-s)^2 f_1(s)ds$$

$$\leq \frac{t^2}{2} I_1(t)$$

$$= \frac{t^2}{8B} \left(\frac{1}{4\sqrt{2}} \log \frac{1+t\sqrt{2}+t^2}{1-t\sqrt{2}+t^2} + \frac{1}{2\sqrt{2}} \arctan \frac{t\sqrt{2}}{1-t^2} \right)$$

$$\leq 1, \quad t \geq 0.$$

2. *Now, we will estimate* $J_1(x)$, $J_2(x)$, *and* $J_3(x)$ *for* $-\infty < x < \infty$.

(a) *Let* $x \leq 0$. *Note that, for* $s \in (-\infty, x]$, *we have*

$$x^2 \leq s^2, \quad x \leq -s, \quad -sx \leq s^2,$$

and

$$(x-s)^2 = x^2 - 2xs + s^2 \leq 4s^2, \quad x - s \leq -2s.$$

Then

$$J_1(x) = \frac{1}{2B(3+\pi)} \int_{-\infty}^{x} \frac{|s|^3}{(1+s^2)^4(1+s^{16})(1+s^6)} ds$$

$$\leq \frac{1}{2B(3+\pi)} \int_{-\infty}^{x} \frac{ds}{1+s^2}$$

$$= \frac{1}{2B(3+\pi)} \arctan s \Big|_{s=-\infty}^{s=x}$$

$$\leq \frac{\pi}{2B(3+\pi)}$$

$$\leq 1, \quad -\infty < x < \infty,$$

$$J_2(x) = \frac{1}{2B(3+\pi)} \int_{-\infty}^{x} \frac{(x-s)|s|^3}{(1+s^2)^4(1+s^{16})(1+s^6)} ds$$

$$\leq \frac{1}{2B(3+\pi)} \int_{-\infty}^{x} \frac{x-s}{(1+s^2)^3} ds$$

$$\leq -\frac{1}{B(3+\pi)} \int_{-\infty}^{x} \frac{s}{(1+s^2)^3} ds$$

$$= -\frac{1}{2B(3+\pi)} \int_{-\infty}^{x} \frac{d(1+s^2)}{(1+s^2)^3}$$

$$= \frac{1}{4B(3+\pi)(1+s^2)^2} \Big|_{s=-\infty}^{s=x}$$

$$= \frac{1}{4B(3+\pi)(1+x^2)^2}$$

$$\leq 1, \quad -\infty < x < \infty,$$

$$J_3(x) = \frac{1}{4B(3+\pi)} \int_{-\infty}^{x} \frac{(x-s)^2 |s|^3}{(1+s^2)^4(1+s^{16})(1+s^6)} ds$$

$$\leq \frac{1}{B(3+\pi)} \int_{-\infty}^{x} \frac{s^2}{(1+s^2)^3} ds$$

$$= \frac{1}{B(3+\pi)} \left(-\frac{s}{4(1+s^2)^2} + \frac{s}{8(1+s^2)} + \frac{1}{8} \arctan s \right) \Big|_{s=-\infty}^{s=x}$$

$$= \frac{1}{B(3+\pi)} \left(-\frac{x}{4(1+x^2)^2} + \frac{x}{8(1+x^2)} \right.$$

$$\left. + \frac{1}{8} \arctan x + \frac{\pi}{16} \right)$$

$$\leq \frac{1}{B(3+\pi)} \left(2 + \frac{\pi}{8} \right)$$

$$\leq \frac{1}{2B}$$

$$\leq 1, \quad -\infty < x < \infty.$$

(b) *Let $x \geq 0$. Then, using the previous case, f_2 is an even function and the computations for I_1, I_2, and I_3, we find*

$$J_1(x) = \int_{-\infty}^{x} f_2(s)ds$$

$$= \int_{-\infty}^{-x} f_2(s)ds + \int_{-x}^{x} f_2(s)ds$$

$$= J_1(-x) + 2\int_{0}^{x} f_2(s)ds$$

$$\leq \frac{\pi}{2B(3+\pi)}$$

$$+ \frac{1}{B(3+\pi)} \int_{0}^{x} \frac{s^3}{(1+s^2)^3(1+s^{16})(1+s^6)(1+s^2)} ds$$

$$\leq \frac{\pi}{2B(3+\pi)} + \frac{1}{B(3+\pi)} \int_{0}^{x} \frac{s^3}{1+s^{16}} ds$$

$$= \frac{\pi}{2B(3+\pi)} + \frac{1}{3+\pi} I_1(x)$$

$$\leq \frac{\pi}{2B(3+\pi)} + \frac{1}{3+\pi}$$

$$\leq \frac{1}{2} + \frac{1}{2}$$

$$= 1, \quad -\infty < x < \infty.$$

Now, using that $sf_2(s)$ is an odd function on $(-\infty,\infty)$, the computations for J_2 in the previous case and using that $x - s \leq -2s$ for $s \in (-\infty,-x)$, we find

$$J_2(x) = \int_{-\infty}^{x} (x-s)f_2(s)ds$$

$$= \int_{-\infty}^{-x} (x-s)f_2(s)ds + \int_{-x}^{x} (x-s)f_2(s)ds$$

$$= \int_{-\infty}^{-x} (x-s)f_2(s)ds + x\int_{-x}^{x} f_2(s)ds$$

$$= \int_{-\infty}^{-x} (x-s)f_2(s)ds + 2x\int_{0}^{x} f_2(s)ds$$

$$= \frac{x}{B(3+\pi)} \int_{0}^{x} \frac{s^3}{(1+s^2)^4(1+s^{16})(1+s^6)}ds$$

$$-2\int_{-\infty}^{-x} sf_2(s)ds$$

$$\leq \frac{x}{B(3+\pi)} \int_{0}^{x} \frac{s^3}{1+s^{16}}ds - \frac{1}{B(3+\pi)} \int_{-\infty}^{-x} \frac{s}{(1+s^2)^3}ds$$

$$= \frac{x}{3+\pi}I_1(x) + \frac{1}{2B(3+\pi)(1+x^2)^2}$$

$$\leq \frac{1}{3+\pi} + \frac{1}{2}$$

$$\leq 1, \quad -\infty < x < \infty,$$

and since $(x-s)^2 \le 4s^2$, $s \in (-\infty, -x)$,

$$J_3(x) = \frac{1}{2}\int_{-\infty}^{x}(x-s)^2 f_2(s)ds$$

$$= \frac{1}{2}\int_{-\infty}^{-x}(x-s)^2 f_2(s)ds + \frac{1}{2}\int_{-x}^{x}(x^2-2xs+s^2)f_2(s)ds$$

$$= \frac{1}{2}\int_{-\infty}^{-x}(x-s)^2 f_2(s)ds + \frac{1}{2}\int_{-x}^{x}(x^2+s^2)f_2(s)ds$$

$$= \frac{1}{2}\int_{-\infty}^{-x}(x-s)^2 f_2(s)ds + \int_{0}^{x}(x^2+s^2)f_2(s)ds$$

$$= \frac{1}{2}\int_{-\infty}^{-x}(x-s)^2 f_2(s)ds$$

$$+x^2\frac{1}{2B(3+\pi)}\int_{0}^{x}\frac{s^3}{(1+s^2)^4(1+s^{16})(1+s^6)}ds$$

$$+\frac{1}{2B(3+\pi)}\int_{0}^{x}\frac{s^5}{(1+s^2)^4(1+s^{16})(1+s^6)}ds$$

$$\le \frac{1}{2}\int_{-\infty}^{-x}(x-s)^2 f_2(s)ds + \frac{x^2}{2B(3+\pi)}\int_{0}^{x}\frac{s^3}{1+s^{16}}ds$$

$$+\frac{1}{2B(3+\pi)}\int_{0}^{x}\frac{s^3}{1+s^{16}}ds$$

$$= \frac{1}{2}\int_{-\infty}^{-x}(x-s)^2 f_2(s)ds + \frac{1}{2(3+\pi)}x^2 I_1(x)$$

$$+\frac{1}{2(3+\pi)}I_1(x)$$

$$\leq \frac{1}{2} \int_{-\infty}^{-x} (x-s)^2 f_2(s) ds + \frac{1}{3+\pi}$$

$$\leq 2 \int_{-\infty}^{-x} s^2 f_2(s) ds + \frac{1}{3+\pi}$$

$$= \frac{1}{B(3+\pi)} \int_{-\infty}^{-x} \frac{s^2}{(1+s^2)^3} ds + \frac{1}{3+\pi}$$

$$\leq \frac{1}{2B} + \frac{1}{2}$$

$$\leq 1, \quad -\infty < x < \infty.$$

By Theorem 6.2.1, it follows that the IVP (6.10), (6.11) *has at least one solution* $u \in \mathscr{C}^2(\mathbb{R}^2 \times [0,\infty))$ *such that* $u \not\equiv 0$ *on* $\mathbb{R}^2 \times [0,\infty)$.

6.3 An IVP for Nonlinear Wave Equations in any Spaces Dimension

The main aim of this section is to investigate the following IVP

$$u_{tt} - \Delta u = f(t,x,u), \quad t \geq 0, \quad x \in \mathbb{R}^n,$$

$$u(0,x) = u_0(x), \quad x \in \mathbb{R}^n, \tag{6.12}$$

$$u_t(0,x) = u_1(x), \quad x \in \mathbb{R}^n,$$

where $u_0 \in \mathscr{C}^2(\mathbb{R}^n, \mathbb{R}_+)$, $u_1 \in \mathscr{C}^1(\mathbb{R}^n, \mathbb{R}_-)$ and $f \in \mathscr{C}(\mathbb{R}_+ \times \mathbb{R}^n \times \mathbb{R})$ satisfying a general growth condition.

We will start with the following useful lemma.

Lemma 6.3.1 *If* $u \in \mathscr{C}^2(\mathbb{R}_+ \times \mathbb{R}^n)$ *is a solution of the integral equation*

$$0 = \frac{1}{2^{n+1} n(n-2)\alpha(n)} \int_0^t \prod_{j=1}^n \int_0^{x_j} (x_j - s_j)^2 (t-t_1)^2 g(t_1,s) \int_{\mathbb{R}^n} \frac{1}{|s-y|^{n-2}}$$

$$\times \left(-\int_0^{t_1} (t_1 - t_2) f(t_2, y, u(t_2, y)) dt_2 + u(t_1, y) - u_0(y) - t_1 u_1(y) \right)$$

$dy ds dt_1$

$$+ \frac{1}{2^{n+1}} \int_0^t \prod_{j=1}^n \int_0^{x_j} (x_j - s_j)^2 (t - t_1)^2 g(t_1, s) \int_0^{t_1} (t_1 - t_2) u(t_2, s) dt_2 ds dt_1,$$

$$(6.13)$$

where

$$s = (s_1, \ldots, s_n), \quad ds = ds_n \ldots ds_1,$$

$$\prod_{j=1}^n \int_0^{x_j} (x_j - s_j)^2 (\cdot) ds = \int_0^{x_1} \ldots \int_0^{x_n} (x_1 - s_1)^2 \ldots (x_n - s_n)^2 (\cdot) ds_n \ldots ds_1,$$

$\alpha(n)$ *is the volume of the unit ball in* $\mathbb{R}^n$ *and* $g \in \mathscr{C}(\mathbb{R}_+ \times \mathbb{R}^n)$ *be a non-negative function, then it is a solution of the IVP* (6.12).

Proof 6.3.2 *Let* $u \in \mathscr{C}^2(\mathbb{R}_+ \times \mathbb{R}^n)$ *is a solution of the integral equation* (6.13). *We differentiate the integral equation* (6.13) *three times in* t, *three times in* x_1 *and so on three times in* x_n, *and we find*

$$0 = \frac{1}{n(n-2)\alpha(n)} g(t,x) \int_{\mathbb{R}^n} \frac{1}{|x-y|^{n-2}} \left(-\int_0^t (t-t_1) f(t_1, y, u(t_1, y)) dt_1 \right.$$

$$\left. + u(t,y) - u_0(y) - t u_1(y) \right) dy$$

$$+ g(t,x) \int_0^t (t - t_1) u(t_1, x) dt_1, \quad (t,x) \in \mathbb{R}_+ \times \mathbb{R}^n,$$

whereupon

$$0 = \frac{1}{n(n-2)\alpha(n)} \int_{\mathbb{R}^n} \frac{1}{|x-y|^{n-2}} \left(-\int_0^t (t-t_1) f(t_1, y, u(t_1, y)) dt_1 \right.$$

$$\left. + u(t,y) - u_0(y) - t u_1(y) \right) dy$$

$$+ \int_0^t (t - t_1) u(t_1, x) dt_1, \quad (t,x) \in \mathbb{R}_+ \times \mathbb{R}^n.$$

$$(6.14)$$

Now, we differentiate twice in t the last equation and we obtain

$$0 = \frac{1}{n(n-2)\alpha(n)} \int_{\mathbb{R}^n} \frac{1}{|x-y|^{n-2}} (-f(t,y,u(t,y)) + u_{tt}(t,y)) \, dy$$

$$+ u(t,x), \quad (t,x) \in \mathbb{R}_+ \times \mathbb{R}^n,$$

or

$$u(t,x) = -\frac{1}{n(n-2)\alpha(n)} \int_{\mathbb{R}^n} \frac{1}{|x-y|^{n-2}} (-f(t,y,u(t,y)) + u_{tt}(t,y)) \, dy,$$

$(t,x) \in \mathbb{R}_+ \times \mathbb{R}^n$. *Hence and the Poisson equation, we arrive at*

$$-\Delta u(t,x) = f(t,x,u(t,x)) - u_{tt}(t,x), \quad (t,x) \in \mathbb{R}_+ \times \mathbb{R}^n.$$

Now, we put $t = 0$ in the equation (6.14) and we find

$$0 = \frac{1}{n(n-2)\alpha(n)} \int_{\mathbb{R}^n} \frac{1}{|x-y|^{n-2}} (u(0,y) - u_0(y)) \, dy, \quad x \in \mathbb{R}^n,$$

and by the Poisson equation, we obtain

$$u(0,x) = u_0(x), \quad x \in \mathbb{R}^n.$$

We differentiate once in t the equation (6.14) and we get

$$0 = \frac{1}{n(n-2)\alpha(n)} \int_{\mathbb{R}^n} \frac{1}{|x-y|^{n-2}}$$

$$\times \left(-\int_0^t f(t_1,y,u(t_1,y)) dt_1 + u_t(t,y) - u_1(y) \right) dy$$

$$+ \int_0^t u(t_1,x) dt_1, \quad (t,x) \in \mathbb{R}_+ \times \mathbb{R}^n.$$

We put $t = 0$ in the last equation and we find

$$0 = \frac{1}{n(n-2)\alpha(n)} \int_{\mathbb{R}^n} (u_t(0,y) - u_1(y)) \, dy, \quad x \in \mathbb{R}^n.$$

From here and from the Poisson equation, we obtain

$$u_t(0,x) = u_1(x), \quad x \in \mathbb{R}^n.$$

This completes the proof.

Now, we will list the assumptions we will use in this section:

(C6) $f \in \mathscr{C}(\mathbb{R}_+ \times \mathbb{R}^n \times \mathbb{R})$ is such that

$$0 \le -f(t,x,u) \le \sum_{j=1}^{l} c_j(t,x)|u|^{p_j}, \quad (t,x,u) \in \mathbb{R}_+ \times \mathbb{R} \times \mathbb{R},$$

$p_j > 0$ are given constants, $c_j \in \mathscr{C}(\mathbb{R}_+ \times \mathbb{R}^n)$ are given functions, $j \in \{1,\ldots,l\}$, $l \in \mathbb{N}$.

(C7) There exists an $r > 0$ such that $0 \le u_0, -u_1 \le \dfrac{r}{2}$ on $\mathbb{R}_+ \times \mathbb{R}^n$, and

$$(3+2n)\left(\sum_{j=1}^{l} r^{p_j} + 2r\right) l < 1,$$

for some given constant $l > 1$.

Let

$$g_k(t,x) = \int_{\mathbb{R}^n} \frac{1}{|x-y|^{n-2}} \int_0^t c_k(t_1,y)dt_1 dy, \quad k \in \{1,\ldots,l\},$$

$(t,x) \in \mathbb{R}_+ \times \mathbb{R}^n$. Below, we suppose

(C8) there exists a non-negative function $g \in \mathscr{C}(\mathbb{R}_+ \times \mathbb{R}^n)$ and a positive constant B so that

$$t^2(2+t)(1+t+t^2)\int_0^t \prod_{j=1,j\neq m}^{n} x_j^2(1+x_m+x_m^2)$$

$$\times \int_0^{x_j} \int_0^{x_m} g(t_1,s)(1+g_k(t_1,s)+\log|s|)dsdt_1 \le A,$$

$k \in \{1,\ldots,l\}$, $m \in \{1,\ldots,n\}$, $(t,x) \in \mathbb{R}_+ \times \mathbb{R}^n$,

$$\frac{B}{2^{n+1}n(n-2)\alpha(n)}\int_1^{\frac{3}{2}} \prod_{j=1}^{n} \int_1^{\frac{3}{2}} (2-s_j)^2(2-t_1)^2 g(t_1,s)$$

$$\times \int_{n \le |y| \le 2n} \frac{1}{|s-y|^{n-2}}dydsdt_1 \ge \frac{A}{l}.$$

Our main result is as follows.

Theorem 6.3.3 *Suppose (C6)-(C8). Then the IVP (6.12) has at least one non trivial non-negative solution* $u \in \mathscr{C}^2(\mathbb{R}_+ \times \mathbb{R}^n)$ *satisfying* $\|u\| < \min(r,R)$ *and*

$$u(t,x) \geq u_0(x),\ (t,x) \in \mathbb{R}_+ \times \mathbb{R}^n,\ u(t,x) \geq u_0(x) + B,\ t \in [1,2],\quad x_j \in [1,2],$$

$$j \in \{1,\dots,n\},\quad x = (x_1,\dots,x_n),$$

Let $E = \mathscr{C}^2(\mathbb{R}_+ \times \mathbb{R}^n)$ be endowed with the norm

$$\|u\| \quad = \quad \|u\|_\infty + \|u_t\|_\infty + \|u_{tt}\|_\infty$$

$$+ \sum_{j=1}^n \|u_{x_j}\|_\infty + \sum_{j=1}^n \|u_{x_j x_j}\|_\infty,$$

provided it exists, where

$$\|v\|_\infty = \sup_{(t,x)\in\mathbb{R}_+\times\mathbb{R}^n} |v(t,x)|.$$

For $u \in E$, define the operator

$$Fu(t,x) \quad = \quad \frac{1}{2^{n+1}n(n-2)\alpha(n)}$$

$$\times \int_0^t \prod_{j=1}^n \int_0^{x_j} (x_j - s_j)^2 (t - t_1)^2 g(t_1,s) \int_{\mathbb{R}^n} \frac{1}{|s-y|^{n-2}}$$

$$\times \left(- \int_0^{t_1} (t_1 - t_2) f(t_2,y,u(t_2,y)) dt_2 \right.$$

$$\left. + u(t_1,y) - u_0(y) - t_1 u_1(y) \right) dy\, ds\, dt_1$$

$$+ \frac{1}{2^{n+1}} \int_0^t \prod_{j=1}^n \int_0^{x_j} (x_j - s_j)^2 (t - t_1)^2 g(t_1,s)$$

$$\times \int_0^{t_1} (t_1 - t_2) u(t_2,s) dt_2\, ds\, dt_1,$$

$(t,x) \in \mathbb{R}_+ \times \mathbb{R}^n$.

Lemma 6.3.4 *Suppose (C6)-(C8). For $u \in E$, $\|u\| \leq r$, we have*

$$\|Fu\| \leq (3+2n) \left(\sum_{j=1}^{l} r^{p_j} + 2r \right) A.$$

Proof 6.3.5 *We have*

$$
\begin{aligned}
|Fu(t,x)| \leq \; & \frac{1}{2^{n+1}n(n-2)\alpha(n)} \\
& \times \int_0^t \prod_{j=1}^{n} \int_0^{x_j} (x_j - s_j)^2 (t-t_1)^2 g(t_1,s) \int_{\mathbb{R}^n} \frac{1}{|s-y|^{n-2}} \\
& \times \left(\int_0^{t_1} (t_1 - t_2) \sum_{k=1}^{l} c_k(t_2,y) |u(t_2,y)|^{p_k} dt_2 \right. \\
& \left. + u(t_1,y) + u_0(y) - t_1 u_1(y) \right) dy\, ds\, dt_1 \\
& + \frac{1}{2^{n+1}} \int_0^t \prod_{j=1}^{n} \int_0^{x_j} (x_j - s_j)^2 (t-t_1)^2 g(t_1,s) \\
& \times \int_0^{t_1} (t_1 - t_2) |u(t_2,s)| dt_2\, ds\, dt_1 \\
\leq \; & \frac{1}{2^{n+1}n(n-2)\alpha(n)} \int_0^t \prod_{j=1}^{n} \int_0^{x_j} (x_j - s_j)^2 (t-t_1)^2 g(t_1,s) \\
& \times \int_{\mathbb{R}^n} \frac{1}{|s-y|^{n-2}} \\
& \times \left(\sum_{k=1}^{l} r^{p_k} \int_0^{t_1} (t_1 - t_2) c_k(t_2,y) dt_2 + (2+t_1)r \right) dy\, ds\, dt_1 \\
& + \frac{1}{2^{n+1}} r \int_0^t \prod_{j=1}^{n} t_1^2 \int_0^{x_j} (x_j - s_j)^2 (t-t_1)^2 g(t_1,s) ds\, dt_1 \\
\leq \; & \sum_{k=1}^{l} r^{p_k} \frac{t^4}{2^{n+1}n(n-2)\alpha(n)} \int_0^t \prod_{j=1}^{n} x_j^2 \int_0^{x_j} g(t_1,s) g_k(t_1,s) ds\, dt_1
\end{aligned}
$$

$$+r\frac{t^3(2+t)}{2^{n+1}n(n-2)\alpha(n)}\int_0^t\prod_{j=1}^n x_j^2\int_0^{x_j}g(t_1,s)\int_{\mathbb{R}^n}\frac{1}{|s-y|^{n-2}}ds$$

$$+r\frac{1}{2^{n+1}}t^4\int_0^t\prod_{j=1}^n x_j^2\int_0^{x_j}g(t_1,s)dsdt_1$$

$$\leq\left(\sum_{k=1}^l r^{p_k}+2r\right)A,\quad(t,x)\in\mathbb{R}_+\times\mathbb{R}^n,$$

and

$$\left|\frac{\partial}{\partial t}Fu(t,x)\right|$$

$$\leq\frac{1}{2^n n(n-2)\alpha(n)}\int_0^t\prod_{j=1}^n\int_0^{x_j}(x_j-s_j)^2(t-t_1)g(t_1,s)$$

$$\times\int_{\mathbb{R}^n}\frac{1}{|s-y|^{n-2}}\left(\int_0^{t_1}(t_1-t_2)\sum_{k=1}^l c_k(t_2,y)|u(t_2,y)|^{p_k}dt_2\right.$$

$$\left.+u(t_1,y)+u_0(y)-t_1u_1(y)\right)dydsdt_1$$

$$+\frac{1}{2^n}\int_0^t\prod_{j=1}^n\int_0^{x_j}(x_j-s_j)^2(t-t_1)g(t_1,s)$$

$$\times\int_0^{t_1}(t_1-t_2)|u(t_2,s)|dt_2dsdt_1$$

$$\leq\frac{1}{2^n n(n-2)\alpha(n)}\int_0^t\prod_{j=1}^n\int_0^{x_j}(x_j-s_j)^2(t-t_1)g(t_1,s)$$

$$\times\int_{\mathbb{R}^n}\frac{1}{|s-y|^{n-2}}$$

$$\times\left(\sum_{k=1}^l r^{p_k}\int_0^{t_1}(t_1-t_2)c_k(t_2,y)dt_2+(2+t_1)r\right)dydsdt_1$$

$$+\frac{1}{2^n}r\int_0^t\prod_{j=1}^n t_1^2\int_0^{x_j}(x_j-s_j)^2(t-t_1)g(t_1,s)dsdt_1$$

$$\leq \sum_{k=1}^{l} r^{p_k} \frac{t^3}{2^n n(n-2)\alpha(n)} \int_0^t \prod_{j=1}^n x_j^2 \int_0^{x_j} g(t_1,s)g_k(t_1,s)dsdt_1$$

$$+r\frac{t^2(2+t)}{2^n n(n-2)\alpha(n)} \int_0^t \prod_{j=1}^n x_j^2 \int_0^{x_j} g(t_1,s) \int_{\mathbb{R}^n} \frac{1}{|s-y|^{n-2}}ds$$

$$+r\frac{1}{2^n}t^3 \int_0^t \prod_{j=1}^n x_j^2 \int_0^{x_j} g(t_1,s)dsdt_1$$

$$\leq \left(\sum_{k=1}^{l} r^{p_k} + 2r\right)A, \quad (t,x) \in \mathbb{R}_+ \times \mathbb{R}^n,$$

and

$$\left|\frac{\partial^2}{\partial t^2}Fu(t,x)\right|$$

$$\leq \frac{1}{2^n n(n-2)\alpha(n)} \int_0^t \prod_{j=1}^n \int_0^{x_j} (x_j - s_j)^2 g(t_1,s) \int_{\mathbb{R}^n} \frac{1}{|s-y|^{n-2}}$$

$$\times \left(\int_0^{t_1} (t_1 - t_2) \sum_{k=1}^{l} c_k(t_2,y)|u(t_2,y)|^{p_k}dt_2\right.$$

$$\left. + u(t_1,y) + u_0(y) - t_1 u_1(y)\right)dydsdt_1$$

$$+\frac{1}{2^n} \int_0^t \prod_{j=1}^n \int_0^{x_j} (x_j - s_j)^2 g(t_1,s) \int_0^{t_1} (t_1 - t_2)|u(t_2,s)|dt_2 dsdt_1$$

$$\leq \frac{1}{2^n n(n-2)\alpha(n)} \int_0^t \prod_{j=1}^n \int_0^{x_j} (x_j - s_j)^2 g(t_1,s) \int_{\mathbb{R}^n} \frac{1}{|s-y|^{n-2}}$$

$$\times \left(\sum_{k=1}^{l} r^{p_k} \int_0^{t_1} (t_1 - t_2)c_k(t_2,y)dt_2 + (2+t_1)r\right)dydsdt_1$$

$$+\frac{1}{2^n}r\int_0^t\prod_{j=1}^n t_1^2\int_0^{x_j}(x_j-s_j)^2 g(t_1,s)dsdt_1$$

$$\leq\sum_{k=1}^{'l} r^{p_k}\frac{t^2}{2^n n(n-2)\alpha(n)}\int_0^t\prod_{j=1}^n x_j^2\int_0^{x_j} g(t_1,s)g_k(t_1,s)dsdt_1$$

$$+r\frac{t(2+t)}{2^n n(n-2)\alpha(n)}\int_0^t\prod_{j=1}^n x_j^2\int_0^{x_j} g(t_1,s)\int_{\mathbb{R}^n}\frac{1}{|s-y|^{n-2}}ds$$

$$+r\frac{1}{2^n}t^2\int_0^t\prod_{j=1}^n x_j^2\int_0^{x_j} g(t_1,s)dsdt_1$$

$$\leq\left(\sum_{k=1}^l r^{p_k}+2r\right)A,\quad (t,x)\in\mathbb{R}_+\times\mathbb{R}^n,$$

and

$$\left|\frac{\partial}{\partial x_m}Fu(t,x)\right|$$

$$\leq\frac{1}{2^n n(n-2)\alpha(n)}\int_0^t\prod_{j=1,j\neq m}^n\int_0^{x_j}(x_j-s_j)^2$$

$$\times\int_0^{x_m}(x_m-s_m)(t-t_1)^2 g(t_1,s)$$

$$\times\int_{\mathbb{R}^n}\frac{1}{|s-y|^{n-2}}\left(\int_0^{t_1}(t_1-t_2)\sum_{k=1}^l c_k(t_2,y)|u(t_2,y)|^{p_k}dt_2\right.$$

$$\left.+u(t_1,y)+u_0(y)-t_1 u_1(y)\right)dydsdt_1$$

$$+\frac{1}{2^n}\int_0^t \prod_{j=1,j\neq m}^n \int_0^{x_j}(x_j-s_j)^2$$

$$\times \int_0^{x_m}(x_m-s_m)(t-t_1)^2 g(t_1,s)\int_0^{t_1}(t_1-t_2)|u(t_2,s)|dt_2 ds dt_1$$

$$\leq \frac{1}{2^n n(n-2)\alpha(n)}\int_0^t \prod_{j=1,j\neq m}^n$$

$$\times \int_0^{x_j}(x_j-s_j)^2 \int_0^{x_m}(x_m-s_m)(t-t_1)^2 g(t_1,s)$$

$$\times \int_{\mathbb{R}^n}\frac{1}{|s-y|^{n-2}}$$

$$\left(\sum_{k=1}^l r^{p_k}\int_0^{t_1}(t_1-t_2)c_k(t_2,y)dt_2+(2+t_1)r\right)dyds dt_1$$

$$+\frac{1}{2^n}r\int_0^t \prod_{j=1,j\neq m}^n t_1^2 \int_0^{x_j}(x_j-s_j)^2$$

$$\times \int_0^{x_m}(x_m-s_m)(t-t_1)^2 g(t_1,s)ds dt_1$$

$$\leq \sum_{k=1}^l r^{p_k}\frac{t^4}{2^n n(n-2)\alpha(n)}\int_0^t \prod_{j=1,j\neq m}^n x_j^2 x_m$$

$$\times \int_0^{x_j}\int_0^{x_m}g(t_1,s)g_k(t_1,s)ds dt_1$$

$$+ r\frac{t^3(2+t)}{2^n n(n-2)\alpha(n)} \int_0^t \prod_{j=1,j\neq m}^n x_j^2 x_m$$

$$\times \int_0^{x_j} \int_0^{x_m} g(t_1,s) \int_{\mathbb{R}^n} \frac{1}{|s-y|^{n-2}} ds$$

$$+ r\frac{1}{2^n} t^4 \int_0^t \prod_{j=1,j\neq m}^n x_j^2 x_m \int_0^{x_j} \int_0^{x_m} g(t_1,s) ds dt_1$$

$$\leq \left(\sum_{k=1}^l r^{p_k} + 2r \right) A, \quad (t,x) \in \mathbb{R}_+ \times \mathbb{R}^n,$$

$m \in \{1,\ldots,n\}$, *and*

$$\left| \frac{\partial^2}{\partial x_m^2} Fu(t,x) \right| \leq \frac{1}{2^n n(n-2)\alpha(n)} \int_0^t \prod_{j=1,j\neq m}^n \int_0^{x_j} (x_j - s_j)^2$$

$$\int_0^{x_m} (t-t_1)^2 g(t_1,s) \int_{\mathbb{R}^n} \frac{1}{|s-y|^{n-2}}$$

$$\times \left(\int_0^{t_1} (t_1 - t_2) \sum_{k=1}^l c_k(t_2,y) |u(t_2,y)|^{p_k} dt_2 \right.$$

$$\left. + u(t_1,y) + u_0(y) - t_1 u_1(y) \right) dy ds dt_1$$

$$+ \frac{1}{2^n} \int_0^t \prod_{j=1,j\neq m}^n \times \int_0^{x_j} (x_j - s_j)^2 \int_0^{x_m} (t-t_1)^2 g(t_1,s)$$

$$\times \int_0^{t_1} (t_1 - t_2) |u(t_2,s)| dt_2 ds dt_1$$

$$\leq \frac{1}{2^n n(n-2)\alpha(n)} \int_0^t \prod_{j=1,j\neq m}^n \times \int_0^{x_j} (x_j - s_j)^2$$

$$\times \int_0^{x_m} (t-t_1)^2 g(t_1,s) \int_{\mathbb{R}^n} \frac{1}{|s-y|^{n-2}}$$

$$\times \left(\sum_{k=1}^l r^{p_k} \int_0^{t_1} (t_1 - t_2) c_k(t_2,y) dt_2 + (2+t_1)r \right) dy ds dt_1$$

$$+ \frac{1}{2^n} r \int_0^t \prod_{j=1,j\neq m}^n t_1^2 \int_0^{x_j} (x_j - s_j)^2 \int_0^{x_m} (t-t_1)^2 g(t_1,s) ds dt_1$$

$$\leq \sum_{k=1}^l r^{p_k} \frac{t^4}{2^n n(n-2)\alpha(n)} \int_0^t \prod_{j=1,j\neq m}^n x_j^2$$

$$\times \int_0^{x_j} \int_0^{x_m} g(t_1,s) g_k(t_1,s) ds dt_1$$

$$+ r \frac{t^3(2+t)}{2^n n(n-2)\alpha(n)} \int_0^t \prod_{j=1,j\neq m}^n x_j^2 x_m \int_0^{x_j} \int_0^{x_m} g(t_1,s)$$

$$\times \int_{\mathbb{R}^n} \frac{1}{|s-y|^{n-2}} ds$$

$$+ r \frac{1}{2^n} t^4 \int_0^t \prod_{j=1,j\neq m}^n x_j^2 \int_0^{x_j} \int_0^{x_m} g(t_1,s) ds dt_1$$

$$\leq \left(\sum_{k=1}^l r^{p_k} + 2r \right) A, \quad (t,x) \in \mathbb{R}_+ \times \mathbb{R}^n, \quad m \in \{1,\ldots,n\}.$$

Hence,

$$\|Fu\| \le (3+2n)\left(\sum_{j=1}^{l} r^{p_j} + 2r\right)A.$$

This completes the proof.

Proof 6.3.6 *Now, we will prove the main result in this section. For $u \in E$, define the operators*

$$Tu(t,x) = (1-\varepsilon)u(t,x),$$

$$Su(t,x) = \varepsilon u(t,x) + Fu(t,x) \quad (t,x) \in \mathbb{R}_+ \times \mathbb{R}^n.$$

Note that if $u \in E$ is a fixed point of the operator $T + S$, then it is a solution of the IVP (6.12). Really, let $u \in E$ be a fixed point of $T + S$. Then

$$u(t,x) = (1-\varepsilon)u(t,x) + \varepsilon u(t,x) + Fu(t,x)$$

$$= u(t,x) + Fu(t,x), \quad (t,x) \in \mathbb{R}_+ \times \mathbb{R}^n$$

or

$$Fu(t,x) = 0, \quad (t,x) \in \mathbb{R}_+ \times \mathbb{R}^n.$$

Hence, and Lemma 6.3.1, we conclude that u is a solution of the IVP (6.12). Define, for $R > 1$,

$$\widetilde{\mathscr{P}} = \{u \in E : u(t,x) \ge 0, \quad (t,x) \in \mathbb{R}_+ \times \mathbb{R}^n\},$$

$\mathscr{P}$ is the set of all equi-continuous families in $\widetilde{\mathscr{P}}$(an example for an equi-continuous family is the family $\{(3+\sin(t+m))\prod_{i=1}^{n}(4+\sin(x_i+m)) : t \in \mathbb{R}_+, \quad x_i \in \mathbb{R}, i = 1,\ldots,n\}_{m\in\mathbb{N}}$),

$$\Omega = \{u \in \mathscr{P} : \|u\| \le R\},$$

$$U = \{u \in \mathscr{P} : u(t,x) \ge u_0(x), \quad (t,x) \in \mathbb{R}_+ \times \mathbb{R}^n,$$

$$u(t,x) \geq u_0(x) + B, \quad t \in [1,2], \quad x_j \in [1,2],$$

$$j \in \{1,\ldots,n\}, \quad x = (x_1,\ldots,x_n), \quad \|u\| < r\}.$$

For $u \in U$, we have that $Fu(t,x) \geq 0$.

1. For $u \in \Omega$, we have

$$(I - T)u(t,x) = \varepsilon u(t,x), \quad (t,x) \in \mathbb{R}_+ \times \mathbb{R}^n,$$

and

$$\|(I - T)u\| = \varepsilon\|u\| \geq \frac{\varepsilon}{l}\|u\|.$$

Thus, $I - T : \Omega \to E$ is Lipschitz invertible with a constant $\gamma \leq \dfrac{l}{\varepsilon}$.

2. By Lemma 6.3.4, for $u \in \overline{U}$, we find

$$\begin{aligned}
\|Su\| &\leq \varepsilon\|u\| + \|Fu\| \\
&\leq \varepsilon r + (3 + 2n)\left(\sum_{j=1}^{l} r^{p_j} + 2r\right)A.
\end{aligned}$$

Therefore, $S : \overline{U} \to E$ is uniformly bounded. Since $S : \overline{U} \to E$ is continuous, we obtain that $S(\overline{U})$ is equi-continuous and then $S : \overline{U} \to E$ is relatively compact. Consequently, $S : \overline{U} \to E$ is a 0-set contraction.

3. Let $u \in \overline{U}$. For $z \in \Omega$, define the operator

$$Lz = Tz + Su.$$

For $z \in \Omega$, we get

$$\begin{aligned}
\|Lz\| &= \|Tz + Su\| \\
&\leq \|Tz\| + \|Su\| \\
&\leq (1 - \varepsilon)R + \varepsilon r + (3 + 2n)\left(\sum_{j=1}^{l} r^{p_j} + 2r\right)A
\end{aligned}$$

$$\leq \ (1-\varepsilon)R + \frac{\varepsilon}{2l}$$

$$< \ (1-\varepsilon)R + \varepsilon R$$

$$= \ R.$$

Therefore $L : \Omega \to \Omega$. Next, for $z_1, z_2 \in \Omega$, we have

$$\|Lz_1 - Lz_2\| \ = \ \|Tz_1 - Tz_2\|$$

$$= \ (1-\varepsilon)\|z_1 - z_2\|,$$

i.e., $L : \Omega \to \Omega$ is a contraction mapping. Hence, there exists a unique $z \in \Omega$ so that

$$z \ = \ Lz$$

$$= \ Tz + Su$$

or

$$(I - T)z = Su.$$

Consequently $S(\overline{U}) \subset (I - T)(\Omega)$.

4. Assume that there are $u \in \partial U$ and $\lambda \geq 1$ such that

$$Su = (I - T)(\lambda u), \quad \lambda u \in \Omega.$$

We have
$$\varepsilon u + Fu = \varepsilon \lambda u$$

or
$$\varepsilon(\lambda - 1)u = Fu.$$

Hence,

$$\varepsilon(\lambda - 1)r \ = \ \varepsilon(\lambda - 1)\|u\|$$

$$= \ \|Fu\|$$

$$\leq \ (3+2n)\left(\sum_{j=1}^{l} r^{p_j} + 2r\right) A.$$

Thus,

$$(3+2n)\left(\sum_{j=1}^{l} r^{p_j} + 2r\right) A \geq \varepsilon(\lambda-1)\|u\|$$

$$= \|Fu\|$$

$$\geq \frac{B}{2^{n+1}n(n-2)\alpha(n)}$$

$$\times \int_{1}^{\frac{3}{2}} \prod_{j=1}^{n} \int_{1}^{\frac{3}{2}} (2-s_j)^2 (2-t_1)^2 g(t_1,s)$$

$$\times \int_{n\leq|y|\leq 2n} \frac{1}{|s-y|^{n-2}} dy\,ds\,dt_1$$

$$\geq \frac{A}{l},$$

i.e.,

$$(3+2n)\left(\sum_{j=1}^{l} r^{p_j} + 2r\right) \geq \frac{1}{l},$$

which is a contradiction.

By (1), (2), (3), (4) and Proposition 2.2.61, we conclude that the operator $T+S$ has a fixed point in $U \cap \Omega$. This completes the proof.

Example 6.3.7 *Let*

$$h(x) = \log \frac{1+s^{11}\sqrt{2}+s^{22}}{1-s^{11}\sqrt{2}+s^{22}}, \quad l(s) = \arctan \frac{s^{11}\sqrt{2}}{1-s^{22}}, \quad s \in \mathbb{R}, \quad s \neq \pm 1.$$

Then

$$h'(s) = \frac{22\sqrt{2}s^{10}(1-s^{22})}{(1-s^{11}\sqrt{2}+s^{22})(1+s^{11}\sqrt{2}+s^{22})},$$

$$l'(s) = \frac{11\sqrt{2}s^{10}(1+s^{20})}{1+s^{40}}, \quad s \in \mathbb{R}.$$

Therefore

$$-\infty < \lim_{t\to\infty} t^2(2+t)(1+t+t^2)h(t) < \infty,$$

$$-\infty < \lim_{t\to\infty} t^2(2+t)(1+t+t^2)l(t) < \infty,$$

$$-\infty < \lim_{s\to\pm\infty} s^2(1+s+s^2)h(s) < \infty,$$

$$-\infty < \lim_{s\to\pm\infty} s^2(1+s+s^2)l(s) < \infty.$$

Hence, there exists a positive constant C_1 so that

$$t^2(2+t)(1+t+t^2)$$

$$\left(\frac{1}{44\sqrt{2}}\log\frac{1+t^{11}\sqrt{2}+t^{22}}{1-t^{11}\sqrt{2}+t^{22}} + \frac{1}{22\sqrt{2}}\arctan\frac{t^{11}\sqrt{2}}{1-t^{22}}\right)$$

$$\leq C_1,$$

$$s^2(1+s+s^2)\left(\frac{1}{44\sqrt{2}}\log\frac{1+s^{11}\sqrt{2}+s^{22}}{1-s^{11}\sqrt{2}+s^{22}} + \frac{1}{22\sqrt{2}}\arctan\frac{s^{11}\sqrt{2}}{1-s^{22}}\right)$$

$$\leq C_1,$$

$t \in \mathbb{R}_+, s \in \mathbb{R}$. *Let*

$$g_1(t,x) = \frac{t^{10}}{1+t^{44}}\prod_{j=1}^{n}\frac{x_j^{10}}{1+x_j^{44}}\frac{1}{(1+g_k(t,s)+\log|s|)^4},$$

where

$$g_k(t,s) = \int_{\mathbb{R}^n} \frac{1}{|s-y|^{n-2}} \int_0^t \frac{1}{(1+|y|)^{3n}(1+t_1^2)^{10}} dt_1 dy,$$

$(t,s) \in \mathbb{R}_+ \times \mathbb{R}^n$, $k \in \{1,\ldots,n\}$, $n \geq 3$. *Moreover, there exists a positive constant* C_2 *so that*

$$\frac{1}{2^{n+1}n(n-2)\alpha(n)} \int_1^{\frac{3}{2}} \prod_{j=1}^n \int_1^{\frac{3}{2}} (2-s_j)^2 (2-t_1)^2 g_1(t,s)$$

$$\times \int_{n \leq |y| \leq 2n} \frac{1}{|s-y|^{n-2}} dy ds dt_1 \geq C_2.$$

Take

$$g(t,s) = \frac{g_1(t,x)}{C_1^{n+1}}, \quad A = 2, \quad l = 1, \quad r = \frac{1}{10000(3+2n)},$$

$$\varepsilon = \frac{1}{2}, \quad B = \frac{3}{C_2}.$$

Hence,

$$2\left(\varepsilon r + (3+2n)\left(\sum_{j=1}^l r^{p_j} + 2r\right)A\right) < \frac{\varepsilon}{l},$$

i.e., (C8) holds. Next,

$$t^2(2+t)(1+t+t^2) \int_0^t \prod_{j=1,j\neq m}^n x_j^2(1+x_m+x_m^2)$$

$$\times \int_0^{x_j} \int_0^{x_m} g(t_1,s)(1+g_k(t_1,s)+\log|s|) ds dt_1 \leq 1 \leq A,$$

$k \in \{1,\ldots,l\}$, $m \in \{1,\ldots,n\}$, $(t,x) \in \mathbb{R}_+ \times \mathbb{R}^n$, *and*

$$\frac{B}{2^{n+1}n(n-2)\alpha(n)} \int_1^{\frac{3}{2}} \prod_{j=1}^n \int_1^{\frac{3}{2}} (2-s_j)^2 (2-t_1)^2 g(t_1,s)$$

$$\times \int_{n \leq |y| \leq 2n} \frac{1}{|s-y|^{n-2}} dy ds dt_1 \geq 3 \geq \frac{A}{l}.$$

Consider the IVP

$$u_{tt} - \Delta u = -\frac{1}{(1+|x|)^{30}(1+t^2)^{10}}|u|^p, \quad (t,x) \in \mathbb{R}_+ \times \mathbb{R}^n,$$

$$u(0,x) = \frac{1}{30000(3+2n)(1+|x|^2)^{4n}},$$

$$u_t(0,x) = -\frac{1}{30000(3+2n)(1+|x|^2)^{5n}}, \quad x \in \mathbb{R}^n,$$

$$(6.15)$$

where $n \geq 3$, $p > 1$. *Here*

$$u_0(x) = \frac{1}{30000(3+2n)(1+|x|^2)^{4n}},$$

$$u_1(x) = -\frac{1}{30000(3+2n)(1+|x|^2)^{5n}}, \quad x \in \mathbb{R}^n.$$

We have $0 \leq u_0, -u_1 < \dfrac{r}{2}$ *and*

$$(3+2n)(r^p + 2r)l < 1.$$

Hence, and Theorem 6.3.3, it follows that the IVP (6.15) has at least one non trivial non-negative solution $u \in \mathscr{C}^2(\mathbb{R}_+ \times \mathbb{R}^n)$.

Bibliography

[1] R. Akhmerov, M. Kamenskii, A. Potapov, A. Rodkina and N. Sadovskii. Measures of noncompactness and condensing operators. Translated from the 1986 Russian original by A. Iacob. Operator Theory: Advances and Applications, 55. Birkhäuser Verlag, Basel, 1992. viii+249 pp.

[2] H. Amann. On the number of solutions of nonlinear equations in ordered Banach spaces, J. Functional Analysis **11** (1972), no. 4, 346–384.

[3] H. Amann. Fixed point equations and nonlinear eigenvalue problems in ordered Banach spaces, SIAM Rev. **18** (1976), no. 4, 620–709.

[4] C. Avramescu. Sur l'existence des solutions convergentes des systèmes d'équations différentielles non linéaires, Ann. Mat. Pura., **81** (1969), no. 4, 147–168.

[5] J. Ayerbe Toledano, T. Domínguez Benavides and G. López Acedo, Measures of noncompactness in metric fixed point theory. Operator Theory: Advances and Applications, **99**. Birkhäuser Verlag, Basel, 1997. viii+211 pp.

[6] J. Banaś and Z. Knap. Measures of weak noncompactness and nonlinear integral equations of convolution type, J. Math. Anal. Appl. **146** (1990), no. 2, 353–362.

[7] J. Banaś and K. Goebel. Measures of Noncompactness in Banach Spaces, Lecture Notes in Pure and Applied Mathematics, **60**. Marcel Dekker, Inc., New York, 1980.

[8] V. Barbu and T. Precupanu. Convexity and optimization in Banach spaces. Revised edition. Translated from the Romanian.

417

Editura Academiei, Bucharest; Sijthoff & Noordhoff International Publishers, Alphen aan den Rijn, 1978. xi+316 pp.

[9] R.G. Bartle. On compactness in functional analysis, Trans. Am. Math. Soc., **79** (1955), 35–57.

[10] S. Benslimane, S. Djebali and K. Mebarki. On The fixed point index for sums of operators, Fixed Point Theory **23** (2022), no. 1, 143–162.

[11] S. Benslimane, S.G. Georgiev and K. Mebarki. Expansion-Compression fixed point theorem of Leggett-Williams type for the sum of two operators and application in three-point BVPs, Studia Universitatis Babeş-Bolyai Mathematica, Accepted.

[12] L. Benzenati, K. Mebarki and S.G. Georgiev. Existence of positive solutions for some kinds of BVPs in Banach spaces, Studia Universitatis Babeş-Bolyai Mathematica, **66** (2021), no. 4, 723–738.

[13] L. Benzenati, K. Mebarki and R. Precup. A vector version of the fixed point theorem of cone compression and expansion for a sum of two operators, Nonlinear Studies **27** (2020), no. 3, 563–575.

[14] L. Benzenati and K. Mebarki. An extension of Krasnoselskii's cone fixed point theorem for a sum of two operators and applications to nonlinear boundary value problems, Studia Universitatis Babeş-Bolyai Mathematica, Accepted.

[15] F. Cianciaruso, V. Colao, G. Marino and H.-K. Xu. A compactness result for differentiable functions with an application to boundary value problem, Annali di Matematica, **192** (2013), 407–421.

[16] K. Deimling. Nonlinear Functional Analysis, Springer-Verlag, Berlin, Heidelberg, 1985.

[17] S. Djebali and K. Hammache. Fixed point theorems for nonexpansive maps in Banach spaces, Nonlin. Anal., **73** (2010) 3440–3449.

[18] S. Djebali and K. Mebarki. Fixed point index theory for perturbation of expansive mappings by k-set contractions, Topol. Methods Nonlinear Anal. **54** (2019), no. A2, 613–640.

[19] S. Djebali and K. Mebarki. Fixed point theory for sums of operators, J. Nonlinear Convex Anal. **19** (2018), no.6, 1029–1040.

[20] S. Djebali and K. Mebarki. Fixed point index on translates of cones and applications, Nonlinear Studies, **21** (2014), no. 4, 579–589.

[21] D. Duffy. Green's functions with applications, CRC, 2001.

[22] J. Garcia-Falset and O. Muñiz-Pérez. Fixed point theory for 1-set contractive and pseudocontractive mappings, Appl. Math. Comput., **219** (2013) 6843–6855.

[23] S.G. Georgiev and Z. Khaled. Multiple fixed-point theorems and applications in the theory of ODEs, FDEs and PDEs, CRC Press, 2020.

[24] S.G. Georgiev and K. Mebarki. Existence of positive solutions for a class ODEs, FDEs and PDEs via fixed point index theory for the sum of operators, Commun. on Appl. Nonlinear Anal., **26** (2019), no. 4, 16–40.

[25] S.G. Georgiev and K. Mebarki. Existence of positive solutions for a class of boundary value problems with p-Laplacian in Banach spaces, Journal of Contemporary Mathematical Analysis, **56** (2021), no. 4, 208–211.

[26] S.G. Georgiev and K. Mebarki. A New Multiple Fixed Point Theorem with Applications, Advances in the Theory of Nonlinear Analysis and its Applications, **5** (2021), no. 3, 393–411.

[27] S.G. Georgiev and K. Mebarki. On fixed point index theory for the sum of operators and applications to a class of ODEs and PDEs, Applied General Topology, **22** (2021), no. 2, 259–294.

[28] S.G. Georgiev and K. Mebarki. Existence of Solutions for a Class of IBVP for Nonlinear Parabolic Equations via the Fixed Point Index Theory for the Sum of Two Operators, New Trends in Nonlinear Analysis and Applications, Accepted.

[29] S.G. Georgiev and Z. Khaled. Classical solutions for a class of IVP for nonlinear two-dimensional wave equations via new fixed point approach, Partial Differential Equations in Applied mathematics, **48** (2021), no. 2, 257–272.

[30] D. Guo, Y. I. Cho and J. Zhu. Partial Ordering Methods in Nonlinear Problems, Shangdon Science and Technology Publishing Press, Shangdon, 1985.

[31] D. Guo and V. Lakshmikantham. Nonlinear Problems in Abstract Cones, vol. **5**, Academic Press, Boston, Mass, USA, 1988.

[32] L. Guozhen. The fixed point index and the fixed point theorems of 1-set contraction mappings, Proc. Amer. Math. Soc. **104** (1988), no. 4, 1163–1170.

[33] J.L. Kelley, *General Topology,* Springer, Graduate Texts in Mathematics, **27** (1955).

[34] R.D. Nussbaum. The fixed point index and asymptotic fixed point theorems for k-set contractions, Bull. A.M.S., **75** (1969) 490–495.

[35] R.D. Nussbaum. The fixed point index and fixed point theorems. Topological methods for ordinary differential equations, (Montecatini Terme, 1991), 143–205, Lecture Notes in Math., 1537, Springer, Berlin, 1993.

[36] B. G. Pachpatte. Inequalities for Differential and Integral Equations, Academuc Press, 1998.

[37] A. Polyanin and A. Manzhirov. Hoandbook of integral equations, CRC Press, 1998.

[38] R. Precup. Componentwise compression–expansion conditions for systems of nonlinear operator equations and applications, in Mathematical Models in Engineering, Biology, and Medicine, AIP Conference Proceedings 1124, Melville-New York, (2009), pp 284–293.

[39] B. Przeradzki. The existence of bounded solutions for differential equations in Hlbert spaces, Annal. Pol. Math., **LVI(2)** (1992), 103–121.

[40] L. Royden and P. Fitzpatrick. Real Analysis, Fourth Edition. Pearson Education, Asia Limited and China Machine Press, 2010. xiv+516 pp.

[41] L. Schwartz. *Analyse: Topologie Générale et Analyse Fonction-nelle,* Hermann, Paris (1970).

[42] D.R. Smart. Fixed Point Theorems, Cambridge University Press, Cambridge, (1980).

[43] H. Steinlein. Two results of J. Dugundji about extension of maps and retractions, Proc. AMS **77** (1979), 289–290.

[44] T. Xiang and R. Yuan. A class of expansive-type Krasnosel'skii fixed point theorems, Nonlinear Anal. **71** (2009), no. 7–8, 3229–3239.

[45] T. Xiang and S.G. Georgiev. Noncompact-type Krasnosel'skii fixed-point theorems and their applications, Math. Methods Appl. Sci. **39** (2016), no. 4, 833–863.

[46] F. Wang and F. Wang. Krasnosel'skii type fixed point theorem for nonlinear expansion, Fixed Point Theory (Cluj), **13** (2012), no.1, 285–291.

[47] Y.B. Xiao, J.K. Kim and N.J. Huang. A generalization of Ascoli-Arzelà theorem with application, Nonlin. Funct. Anal. Appl., **11** (2006), no.2, 305–317.

[48] E. Zeidler. Nonlinear Functional Analysis and its Applications. Vol. I: Fixed Point Theorems, Translated from the German by Peter R. Wadsack. Springer-Verlag, New York, 1986. xxi+897 pp.

[49] K. Zima. Sur l'existence des solutions d'une équation intégro-différentielle, Annal. Pol. Math., **XXVII** (1973), 181–187.

Index

.

Printed in the United States
by Baker & Taylor Publisher Services